Stentless Bioprostheses

Stentless Bioprostheses

Edited by

Armand Piwnica

Professor of Thoracic and Cardiovascular Surgery
Hôpital Lariboisière, Paris, France

Stephen Westaby BSc FRCS MS

Cardiac Surgeon, Oxford Heart Centre,
John Radcliffe Hospital, Oxford, England

ISIS
MEDICAL
MEDIA

Oxford

Saxon Beck, 58 St Aldates
Oxford OX1 1ST, UK

First published 1995

British Library Cataloguing in Publication Data.
A catalogue record for this title is available from
the British Library

ISBN 1 899066 08 X

Library of Congress
Cataloguing-in-Publication Data

Piwnica A. (Armand)
Stentless Bioprostheses/
Armand Piwnica and Stephen Westaby

Always refer to the manufacturer's Prescribing
Information before prescribing drugs cited in this book.

Typeset by
Creative Associates, Oxford, England

Printed and bound by
Dah Hua Printing Press Co. Ltd,
Hong Kong

Distributed by
Times Mirror International Publishers,
Invicta House, Sir Thomas Longley Road,
Medway City Estate, Rochester,
Kent ME2 4DU

Contents

List of Contributors viii
Preface xv

Section 1: From Homograft to Stentless Bioprostheses

Chapter 1: Why stentless valves?
S. Westaby and A. Piwnica 3

Chapter 2: From homograft to stentless bioprostheses
D. Ross 17

Chapter 3: Stented porcine bioprostheses durability and outcomes 1975–1993
W.R.E. Jamieson 24

Chapter 4: Pericardial bioprostheses: an overview
F. Sebening and A. Schroll 36

Section II: Stentless Valve Technology

Chapter 5: Toronto SPV™ valve: the engineer's standpoint
W. Mirsch 43

Chapter 6: Hydrodynamic performance of the Medtronic Freestyle™ Aortic Root Bioprosthesis
C.E. Eberhardt, A.P. Yoganathan, M. Capps and C. Toomes 48

Chapter 7: Edwards Prima™ stentless bioprosthesis: from concept to reality
L. Quintero 58

Chapter 8: Pericarbon™ stentless valve
S. Rinaldi, M. Curcio, E. Pasquino, C. Stacchino and F. Vallana 64

Chapter 9: Labcor stentless aortic valve: the engineer's standpoint
I.S.J. Casagrande, H.G. Menezes, M.L.G.N. Coelho, A.N.J. Casagrande and P. Cardoso Neto 69

Chapter 10: O'Brien-Angell Model 300 – stentless valve
O.J. May 77

Chapter 11: Wear-tested pulmonary porcine valve: mechanical properties and ultrastructural findings
G. Gerosa, H. David, M. Valente, D. Boughner, I. Vesely, G. Thiene, J. Davies, C. Muneretto and D. Casarotto 81

Chapter 12: Stentless heart valve concept – new anticalcificant tissue tanning – preclinical trial
M. Vrandecic, B.F. Gontijo, F.A. Fantini, S. Gabbay, E.A. Vrandecic and E. Vrandecic 93

Chapter 13: Stentless heart valves: biocompatibility issues associated with new antimineralization and fixation agents
D.J. Myers, J. Gross and G. Nakaya 100

Chapter 14: Acylazide treatment: an alternative to glutaraldehyde processing
H. Petite, P. Menasché, A. Piwnica and A. Huc 118

Section III: Clinical Experience

Chapter 15: Homograft valves in congenital heart disease: a 7-year experience
G. Bloch, A. Bel, B. Faris, A. Nguyen Van-Cao, T. du Roy de Chaumaray and P. Menasché 127

Chapter 16: Pulmonary autograft (Ross) operation
S. Westaby 130

Chapter 17: Aortic valve replacement with a stentless porcine aortic valve
C.M. Feindel, T.E. David, J. Bos, Z. Sun, H.E. Scully and H. Rakowski 139

Chapter 18: Early experience with the Toronto stentless porcine valve in aortic valve replacement
K. Wong, S. Brecker, A. Khaghani, M.H. Yacoub and J.R. Peppe, 146

Chapter 19: Stentless and stented valves in the aortic position: an early postoperative assessment
X.Y. Jin, D.G. Gibson, M.H. Yacoub and J.R. Pepper 149

Chapter 20: Medtronic Freestyle™ bioprostheses: clinical study
H.A. Huysmans 152

Chapter 21: 'Cylinder within a cylinder' method for porcine root implantation
S. Westaby 157

Chapter 22: Early results of the Edwards Prima™ stentless aortic bioprosthesis in 200 patients
H. Vanermen, W. Daenen, R. Pillai and W. Konertz 165

Chapter 23: Early valve haemodynamics following implantation of the Edwards Prima™ stentless aortic bioprosthesis
G.W. Parry, N.C. Batrick and F.P. Shabbo 172

Chapter 24: Indications, implantation techniques and results of the various types of stentless bioprostheses: the surgeon's standpoint and the Sorin Pericarbon™ stentless valve
R. De Paulis, G.F. Zattera, F. Parisi, E. Leotta, R. Casabona, M. Galloni, G. Pugliese, M. di Summa and M. Morea 176

Chapter 25: O'Brien–Angell stentless valve: early results of 120 implants
U. Hvass 182

Chapter 26: Early experience with the Labcor stentless porcine aortic bioprosthesis
J.M. De Souza and S.A. Oliveira 190

Chapter 27: Porcine stentless heart valve substitutes, technical and clinical data: short- and mid-term follow-up
M. Vrandecic, B.F. Gontijo, F.A. Fantini, O.C. Oliveira, I.C.J. Martins, M.H.C. Oliveira, S.O.S. Avelar, E.A. Vrandecic and E. Vrandecic 195

Chapter 28: Echocardiography and the aortic stentless valves
A. Nguyen Van-Cao 201

Index 207

List of Contributors

Sandra O. S. Avelar
Department of Cardiology, Biocor Institute, Rua Da Paisagem No. 310, Villa Da Serra, Nova Lima, MG CEP: 34000-100, Brazil

N. C. Batrick FRCS
Regional Cardiothoracic Surgical Unit, Brook Hospital, London, UK

Alain Bel MD
Assistant Professor, Department of Cardiovascular Surgery, Hôpital Lariboisière, 2 Rue Ambroise Paré, 75010, Paris, France

Gerard C. Bloch MD
Professor of Cardiac Surgery, Head of Department of Cardiac Surgery, Hôpital Lariboisière, 2 Rue Ambroise Paré, 75010, Paris, France

Joanne Bos
Registered Nurse, Toronto General Hospital, 200 Elizabeth Street, Toronto, Ontario, M5G 2C4, Canada.

Derek R. Boughner MD, PhD, FRCP(C)
Professor of Medicine, Professsor of Medical Biophysics, University of Western Ontario, University Hospital, Box 5339, 339 Windermere Road, London, Ontario, Canada, N6A 5A5

Stephen J. Brecker MD, MRCP
Senior Registrar in Cardiology, The London Chest Hospital, Bonner Road, London, E2 9JX, UK

Mark J. Capps
Reliability Engineer, Medtronic Heart Valve Division, 108011 So. Mitchell, Irvine, California, 92714, USA

Riccardo Casabona MD
Department of Cardiac Surgery, University of Torino, Torino, Italy

D. Casarotto MD
Professor of Cardiac Surgery and Chief, Division of Cardiovascular Surgery, Istituto di Chururgia Cardiovascolare, Universita di Padova, Via Giustiniani 2-35128, Padova, Italy

Alexandre N.J. Casagrande
Mechanical Engineer, MSc BioEngineering, Labcor LaboratÛrios Ltda, Av. Silva Lobo, 1670-Nova Granada, Belo Horizonte, 304060-00, MG, Brazil

Ivan S.J. Casagrande MD
Assistant Professor of Medicine School, Federal University of Minas Gerais, Avenida Silva Lobo, 1670-Nova Granada, Belo Horizonte-30460-00, MG, Brazil

Maria L.G.M. Coelho
Veterinary, Labcor Laboratorios Ltds, Av. Silva Lobo, 1670 Nova Granada, Belo Horizonte, Minas Gerais, 30460-000, Brazil

Maria Curcio
Sorin Biomedica Cardio S.p.A., 13040 Saluggia (VC), Italy

Wim Daenen
Professor of Surgery, Head of the Department of Cardiac Surgery, University Hospital Gasthuisberg, B-3000, Leuven, Belgium

Huw David BESc MSC
Robarts Research Institute and Departments of Medicine and Medical Biophysics, University of Western Ontario, London, Canada

Tirone E. David MD, FRCS(C), FACS
Professor of Surgery, University of Toronto; Head, Division of Cardiovascular Surgery, The Toronto Hospital, 200 Elizabeth Street, 13 EN-222, Toronto, Ontario, M5G 2C4, Canada

Jill Davies BSc Hons, PGCE
Oxford Heart Valve Bank Manager, Oxford Heart Centre, John Radcliffe Hospital, Headington, Oxford, OX3 9DU, UK

Thierry Du Roy De Chaumaray MD
Assistant Professor, Department of Cardiovascular Surgery, Hôpital Lariboisière, 2 Rue Ambroise Paré, 75010, Paris, France

Ruggero De Paulis MD
University of Rome Tor Vergata, European Hospital, Via Portuense 700, Rome, Italy

Januario M. De Souza
Cardiovascular Surgeon, Hospital Beneficencia Portugesa, Rua Parã, 49-CEP 01243-0320, São Paulo, Brazil

Michele di Summa MD
Department of Cardiac Surgery, University of Torino, Torino, Italy

Karl Dossche MD
Senior Registrar in Cardiothoracic Surgery, OLV Ziekenhuis, Moorselbaan 164, 9300 Aalst, Belgium

Carol E. Eberhardt
Product Development Manager, Medtronic Heart Valve Division, 109011 Sp. Mitchell, Irvine, California, 92714, USA

Fernando A. Fantini
Professor of Surgery, Federal University of Minas Gerais, Brazil. Senior Surgeon, Biocor Institute, Rua Da Paisagem No. 310, Villa Da Serra, Nova Lima, MG CEP: 34000-100, Brazil

B. Faris
Department of Cardiovascular Surgery, Hôpital Lariboisière, 2 Rue Ambroise Paré, 75010, Paris, France

Bayard G. Filho
Professor of Surgery, Federal University of Minas Gerais, Brazil. Senior Surgeon, Biocor Institute, Rua Da Paisagem No. 310, Villa Da Serra, Nova Lima, MG CEP: 34000-100, Brazil

Christopher M. Feindel MD, FRCS(C), FACS
Cardiovascular Surgeon, The Toronto Hospital; Assistant Professor of Surgery, University of Toronto, 200 Elizabeth Street, EN14-222, Toronto, Ontario M5G 2C4, Canada

Shlomo Gabbay MD
Cardiothoracic Surgeon, UMDNJ-New Jersey Medical School, Department of Surgery, Division of Cardiothoracic Surgery, 185 South Orange Avenue, Room G502, Newark, New Jersey, 07103, USA

Marco Galloni MD
Department of Cardiac Surgery, University of Torino, Torino, Italy

Geno Gerosa MD
Senior Registrar, Istituto di Chururgia Cardiovascolare, Universita di Padova, Via Giustiniani 2-35128, Padova, Italy

Derek G. Gibson MA MB FRCP
Consultant Cardiologist, Royal Brompton Hospital, Sydney Street, London, SW3 6NP, UK

Jeffrey M. Gross PhD
Principal Research Scientist, Medtronic Cardiac Surgery, Medtronic Heart Valves Inc., 18011 South Mitchell Avenue, Irvine, California, 92714, USA

Alain Huc
Scientific Manager, Coletica S.A., 32 Rue St. Jean de Dieu, 69007, Lyon, France

H. A. Huysmans
Professor of Cardio-thoracic Surgery, Head, Department of Cardio-thoracic Surgery, University Hospital, Rijusburgerweg 10, 2333 AA Leiden, The Netherlands

Ulrick Hvass
Professor of Heart Surgery, Head of Department, Hôpital Bichat, 46 Ruè Henri Huchard, Paris, 75018, France

W. R. E. Jamieson MD
Department of Surgery, Faculty of Medicine, The University of British Columbia, 910 West 10th Avenue, Vancouver,BC, V5Z 4B3, Canada

Xu Y. Jin MB MD
Research Fellow, Royal Brompton Hospital, Stanley Street, London SW3 6NP, UK

Asghar Khanghani MD, FRCS(Ed)
Senior Lecturer/Consultant Cardiac Surgeon, Harefield Hospital, Harefield, Middlesex, UB9 6JH, UK

W. Konertz MD, PhD
Wilhelmís University, Meunster, Germany

Eros Leotta MD
Department of Cardiac Surgery, University of Torino, Torino, Italy

Idail C. J. Martins
Department of Echocardiology, Biocor Institute, Rua Da Paisagem No. 310, Villa Da Serra, Nova Lima, MG CEP: 34000-100, Brazil

O. James May
Director, Cardiovascular Services, Cryolife Inc., 2211 New Market Parkway, Suite 142, Marietta, Georgia, 30067, USA

Philippe Menasché MD, PhD
Professor of Thoracic and Cardiovascular Surgery, Department of Cardiovascular Surgery, Hôpital Lariboisière, 2 Rue Ambroise Paré, 75010, Paris, France

Hamilton G. Menezes
Veterinary, Labcor Laboratorios Ltda, Av. Silva Lobo, 1670-Nova Granada, Belo Horizonte- MG, 30460-000, Brazil

William Mirsch
St. Jude Medical Inc., St. Paul, Minnesota, USA

Mario Morea MD
Department of Cardiac Surgery, University of Torino, Torino, Italy

C. Muneretto MD
Associate Professor of Cardiac Surgery, Istituto di Churugia Cardiovascolare, Universita di Padova, Via Giustiniani 2-35128, Padova, Italy

David J. Myers MA
Manager, Research and Technical Services, Medtronic Cardiac Surgery, Medtronic Heart Valves Inc. 18011 South Mitchell Avenue, Irvine, California, 92714, USA

Glenn Nakaya BS
Associate Engineer, Medtronic Cardiac Surgery, Medtronic Heart Valves Inc., 18011 South Mitchell Avenue, Irvine, California, 92714, USA

Pedro Cardoso Neto
Technic Perfusion, Labcor Laboratorios Ltda, Av. Silva Lobo 1670-Nova Granada, Belo Horizonte, MG, 30460-000, Brazil

Alix Nguyen Van-Cao
Consultant Cardiologist, Director, Laboratory of Echocardiography, Department of Cardio-Vascular Surgery, Hôpital Lariboisière, 2 Rue Ambroise Paré, 75010, Paris, France

Marcelo H.C. Oliveira
Department of Cardiology, Biocor Institute, Rua Da Paisagem No. 310, Villa Da Serra, Nova Lima, MG CEP: 34000-100, Brazil

Ozanam C. Oliveira
Department of Echocardiology, Biocor Institute, Rua Da Paisagem No. 310, Villa Da Serra, Nova Lima, MG CEP: 34000-100, Brazil

Sergio A. Oliveira
Professor of Surgery, Cardiovascular Surgeon (Head, Division of Cardiovascular Surgery), Hospital Beneficencia Portugesa, Rua Maestro Cardim, 794, São Paulo, Brazil

Francesco Parisi MD
Department of Cardiac Surgery, University of Torino, Torino, Italy

G. W. Parry BSc, FRCS
Regional Cardiothoracic Surgical Unit, Brook Hospital, London UK

Enrico Pasquino
Sorin Biomedica Cardio S.p.A., 13040 Saluggia (VC), Italy

John R. Pepper MA, MChir, FRCS
Consultant Cardiac Surgeon, Royal Brompton Hospital, Sydney Street, London, SW3 6NP, UK

Hervé Petite
Doctor of Science, Laboratoire di Recherche OrthopÈdique, FacultÈ di Medecine, Lariboisiere – St. Louis, 10 Rue de Verdun, 75010 Paris, France

Raui Pillai FRCS Ed, FRCS
Consultant Cardiothoracic Surgeon, Oxford Radcliffe and John Radcliffe Hospital, Headington, Oxford, OX3 9DU, UK

Armand Piwnica
Hôpital Lariboisière, 2 Rue Ambroise Paré, 75475, Paris Cedex 10, France

Gaetano Pugliese MD
Department of Cardiac Surgery, University of Torino, Torino, Italy

Lillian J. Quintero MS, Bioengineering
Senior Staff Engineer, Baxter Healthcare Corporation, 17221 Red Hill Avenue, M/S22, Irvine, California, 92714, USA

Stefano Rinaldi
Sorin Biomedica Cardio S.p.A., 13040 Saluggia (VC), Italy

Donald N. Ross
Consultant Cardiac Surgeon, 25 Upper Wimpole Street, London, WIM 7TA, UK

Harry Rakowski MD, FRCP (C)
Professor of Medicine; Director-Clinical Cardiology; Director-Echocardiography Laboratory, The Toronto Hospital, 12E-222, 200 Elizabeth Street, Toronto, Ontario, M5G 2C4, Canada

Fritz Sebening MD
Professor of Cardiac Surgery, German Heart Center, Department of Cardiovascular Surgery, Lothstrasse 11, D-80335, Munich, Germany

Armin Schroll MD
German Heart Center, Department of Cardiovascular Surgery, Lothstr. 11, D-80335, Munich, Germany

Hugh E. Scully MD, FRCS(C), FACS, FACC
Professor of Surgery, University of Toronto; Senior Surgeon, Toronto General Hospital, 14 EBN-224, 200 Elizabeth Street, Toronto, Ontario, M5G 2C4, Canada

F. P. Shabbo FRCS
Regional Cardiothoracic Surgical Unit, Brook Hospital, London, UK

Zhao Sun MA
The Toronto Hospital, 200 Elizabeth Street, Toronto, M5G 2C4, Canada

Carla Stacchino
Sorin Biomedica Cardio S.p.A., 13040 Saluggia (VC), Italy

Goetano Thiene MD
Professor, Cardiovascular Pathology; Professor of Pathology, Istituto di Anatotlia Patologica, Via Gabelli 61, 35121, Padova, Italy

Christopher G. Toomes
Product Development Engineer, Medtronic Heart Valve Division, 108011 So. Mitchell, Irvine, California, 92712, USA

Franco Vallana
Sorin Biomedica Cardio S.p.A., 13040 Saluggia (VC), Italy

Mazialuisa Valente MD
Assistant Professor-Ultrastructural Pathology; Consultant, Istituto Di Anatomia Patologica, Via Gabello 61, 35121, Padova, Italy

Hugo K. Vanermen MD
Head, Department of Cardiovascular and Thoracic Surgery, Onze-Lieve-Vrouw Kliniek, Moorselbaan 164, 9300, Aalst, Belgium

Ivan Vesely PhD
Associate Staff, Department of Biomedical Engineering, Cleveland Clinic Foundation, 9500 Euclid Avenue, Cleveland, Ohio, 44195, USA

Mario P. Vrandecic
Professor of Surgery, Federal University of Minas Gerais, Brazil. Chairman of Biocor Institute, Rua Da Paisagem No. 310, Villa Da Serra, Nova Lima, MG CEP: 34000-100, Brazil

Erika A. Vrandecic
Biocor Institute Research Center, Biocor Institute, Rua Da Paisagem No. 310, Villa Da Serra, Nova Lima, MG CEP: 34000-100, Brazil

Ektor Vrandecic
Biocor Institute Research Center, Biocor Institute, Rua Da Paisagem No. 310, Villa Da Serra, Nova Lima, MG CEP: 34000-100, Brazil

Stephen Westaby BSc FRCS MS
Consultant Cardiothoracic Surgeon, Oxford Heart Centre, Department of Cardiac Surgery, John Radcliffe Hospital, Oxford, OX3 9DU, UK

K. Wong FRCS
The Royal Brompton Hospital, Sydney Street, London, SW3, UK

Professor Sir Magdi Yacoub FRCS, FRCP, DSc, MCh
Professor of Surgery, Royal Brompton Hospital, Stanley Street, London SW3 6NP, UK

Ajit P. Yoganathan PhD
Regents Professor of Chemical and Mechanical Engineering, School of Chemical Engineering, Georgia Institute of Technology, Atlanta, Georgia, 30332-0100, USA

Giuseppe F. Zattera MD
Department of Cardiac Surgery, University of Torino, Torino, Italy

Preface

Valve prostheses continue to improve towards the goal of optimum haemodynamic function, indefinite durability and a minimal risk of thromboembolism. Most modern mechanical valves have indefinite structural integrity with excellent flow characteristics. Their propensity for thromboembolism remains however, and necessitates continuous anticoagulation. The current objective for bioprostheses is improved haemodynamics with prolonged durability and mitigation from early calcification in the young patients. They do not need anticoagulants but the stent of a bioprostheses is both obstructive and a focus for stress on the tissue component. Stents were adopted when the effect of prolonged cardiopulmonary bypass were prevalent and myocardial protection poorly developed. The stent and sewing ring facilitate implantation with a short ischaemic time but at a price. Stentless valves (epitomised by the autogenous pulmonary valve and aortic homograft) require a more complex surgical approach but leave the valve cusps within their natural environment and provide non-obstructive central flow. In addition to reduced stress on the tissue component, the use of zero pressure fixation, together with antimineralization agents should translate into improved durability. The potential benefit of this technology is to allow bioprosthetic valve replacement in the over 65 patients without the risk of early tissue failure and reoperation.

Recently there has been a resurgence of interest in the use of aortic homograft and pulmonary autograft aortic valve replacement. The use of stentless porcine valves was further stimulated by the work of Tirone David and his group. Others followed this lead with both aortic and mitral stentless bioprostheses, the majority of which appear to provide haemodynamic superiority over their stented counterparts. Only time will tell if improved flow characteristics will translate into less calcification and greater durability.

In January 1994 the proponenents and manufacturers of stentless bioprostheses came together in Paris to discuss the role and relative merits of these devices. A great deal of information and lively discussion are contained in the book from the Symposium and we are grateful to the contributors for providing their manuscripts. This information forms the baseline against which further developments in stentless valve technology may be compared.

Stephen Westaby　　Armand Piwnica

Dedication

To our wives and children.

I From Homograft to Stentless Bioprosthesis

Chapter 1

Why stentless valves?

S. Westaby and A. Piwnica

Some of the answers to the question 'Why stentless valves?' are provided in a paper published in *Guy's Hospital Reports* 25 years ago[1] (Figure 1.1). In this paper, Ross discusses his views on the preservation and storage of aortic homografts and his initial experience with 30 pulmonary autograft aortic replacements beginning in 1967. In this short extract Ross highlights the haemodynamic advantages of the freehand-sewn aortic homograft in comparison to mechanical and stent-mounted tissue prostheses (Figure 1.2). Ross, in tandem with Leonardo da Vinci and the Almighty, considered the aortic homograft to be of the perfect design to ensure central non-obstructive flow. Leonardo (1452–1519) characterized the flow characteristics through the aortic valve using a glass model of the aortic sinuses. With his prototype of the pulse duplicator and particles of carbon in the hydraulic fluid, Leonardo identified the mechanism of closure of the native aortic valve and would never have considered stent mounting in a patient (Figure 1.3). Clinical experience with aortic homografts over many years demonstrated minimal transvalvular gradients, absence of red cell damage and virtually zero thromboembolism. Only the patient's own pulmonary valve has produced superior results through greater durability and the property of growth in childhood.

OBSERVATIONS ON HOMOGRAFT AORTIC VALVES

By D. N. ROSS

Thoracic Unit, Guy's Hospital, S.E.1, and National Heart Hospital, W.1

THE sub-coronary replacement of the aortic valve with a homograft was first performed in June 1962, at Guy's Hospital (Ross 1962). Since that time the method has been used with varying technical modifications in over 350 patients.

Among the advantages of a homograft valve over a mechanical prosthesis are the perfect design characteristics of the valve (Figure 1) coupled with a total absence of embolic complications without the use of anticoagulants. It has certainly established itself as a practical alternative to the mechanical prosthesis and it can be inserted with a comparable operative mortality.

Among its disadvantages are the problems of collection and sterilization and the technical difficulties involved in achieving a competent valve. These are, however, all surmountable and with improvements in technique, notably the use of a double row of interrupted sutures, the early incidence of 70–80 per cent of regurgitation has fallen to 8 per cent, although diastolic murmurs are still common.

More recently there has been a tendency to use homograft and heterograft valves ready mounted on a rigid prosthesis (Geha, Titus and McGoon 1967), but although this has some advantages in relation to ease of insertion, it re-introduces the disadvantage

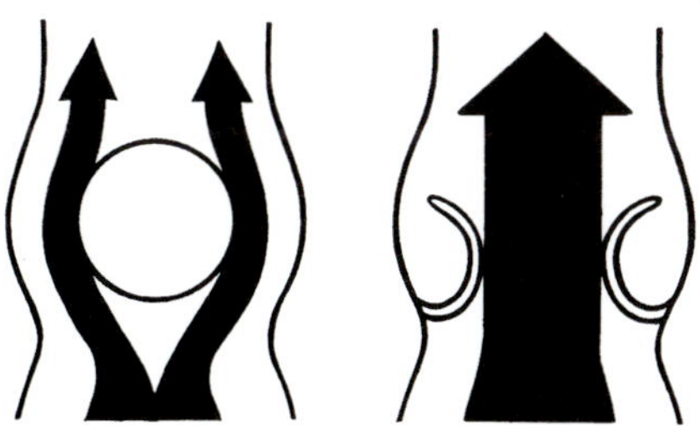

of a rigid ring in place of a flexible biological structure.

Figure 1.1. Excerpt from Ross's landmark report on early experience with homograft and autograft valves.

Figure 1.2. Homograft and stented valves.

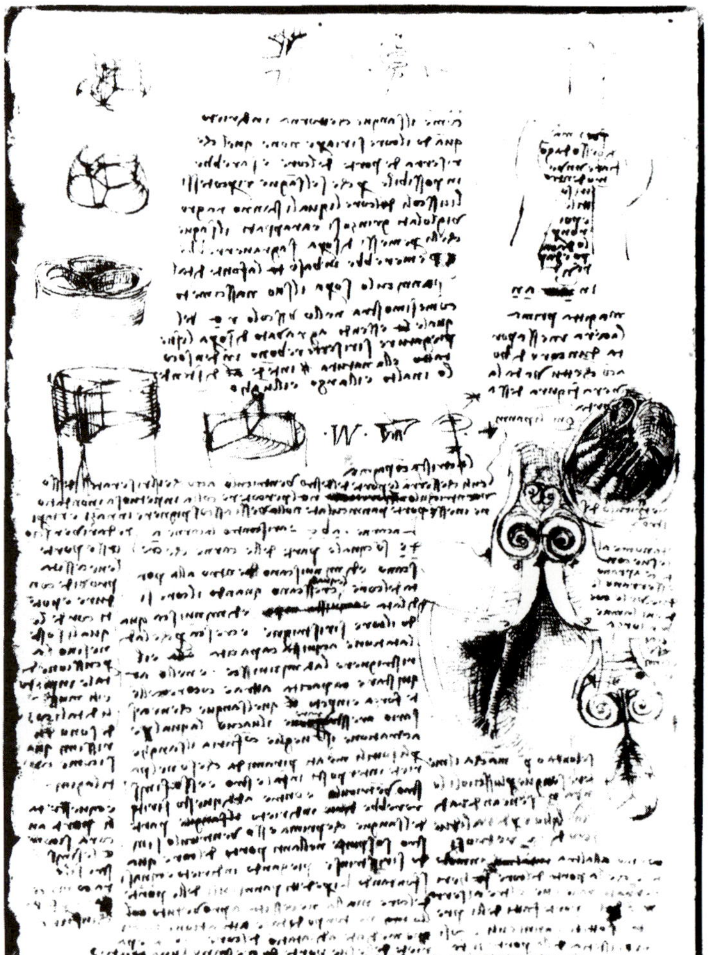

Figure 1.3. Early manuscript from Leonardo da Vinci describing the aortic valve and his models to test haemodynamics. (The Royal Collection © Her Majesty Queen Elizabeth II)

The first homograft operations were performed by Murray,[2] then Beale, Morris, Cooley and DeBakey,[3] who (as for the original Hufnagel ball valve) inserted the valve into the descending aorta. Ross's landmark subcoronary homograft implant (24 July 1962) followed the pioneering work of Gunning and Duran in Oxford, who established the method for valve harvesting and preparation in the animal laboratory.[4,5] Biggelow had previously implanted a homograft in the orthotopic position in 1961 but the patient died from coronary thrombosis in the early postoperative period. The same group attempted mitral replacement with an aortic homograft (23 March 1961): the surgeon Heimbecker succeeded initially but the patient died from chest infection 4 weeks after implantation, with the manuscript in preparation. Ross's successful case was followed (23 August 1962) by that of Barratt-Boyes at Green Lane Hospital.[6] Soon afterwards, Paneth, assisted by O'Brien at the Brompton Hospital, implanted a homograft that had been preserved in formalin for 6 months.[7] Others, such as Kirklin, Angell and Shumway, adopted homograft aortic replacement at a time when most preferred the simplicity of Starr valve insertion with a single row of sutures through a generous sewing ring.[8–10] Mechanical prostheses offered the prospect of indefinite durability whereas crude homograft preservation techniques caused early clinical failure.[11–14] Furthermore, the supply of suitable homografts was limited and a valve of appropriate size was not always available when needed. The homograft operation was more demanding, at a time when myocardial preservation was primitive, and there was less appreciation of the homograft's haemodynamic superiority. To simplify implantation, Weldon from Johns Hopkins published laboratory experience with aortic homografts mounted on a frame[15] (Figure 1.4). Angell adopted stent-mounted homografts for aortic, mitral and tricuspid replacement in the late 1960s.[9] In the short term these valves were easy to implant and proved acceptable in all but the smallest aortic annulus. Stent mounting significantly decreased the incidence of early postoperative aortic regurgitation in comparison with the 20–30% incidence with free aortic grafts.[16,17] However, medium-term follow-up showed limited durability of the stent-mounted homograft tissue (Figure 1.4).

Access to human cadaveric material remained limited and led to the search for other substitutes. In Zurich, Senning removed fascia lata from the thigh and

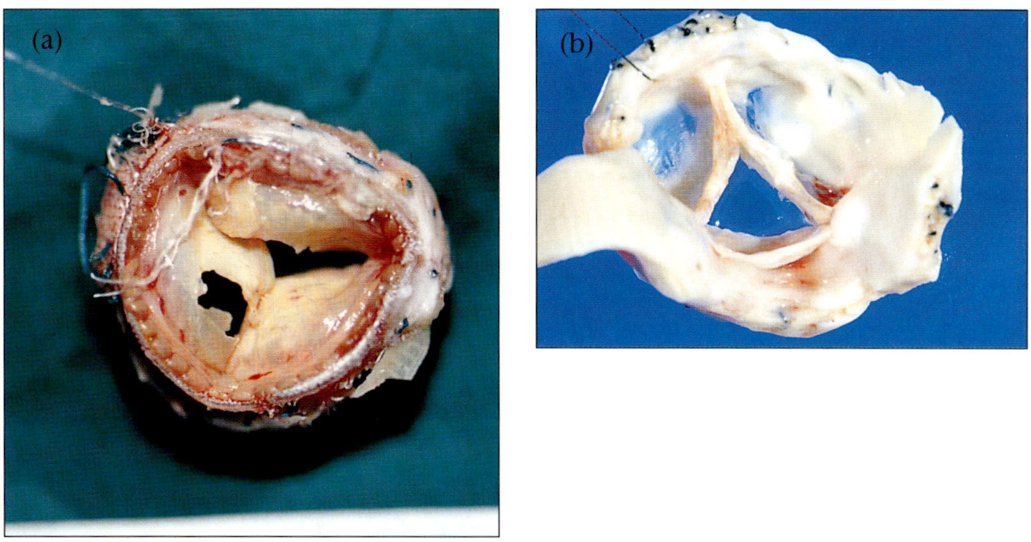

Figure 1.4. (a) Early failure of a frame-mounted aortic homograft (8 years); (b) freehand-inserted homograft explanted at 15 years.

Figure 1.5. Fascia lata cut by Senning for freehand construction of the aortic valve.

fashioned this into a trileaflet valve sutured freehand in the subcoronary position[18] (Figure 1.5). Ionescu in Leeds mounted fascia lata, then pericardium, on a frame with a sewing ring to facilitate implantation and suspension of the commissures[19–20](Figure 1.6). Lower, Stofer and Shumway began to investigate the potential for autogenous pulmonary valve aortic replacement in 1960 when they experimentally implanted pulmonary valve cusps in the descending thoracic aorta.[21] Ross (1967) used the pulmonary autograft in patients but the complexity of the operation precluded widespread use of the method before the era of cardioplegic arrest.[22] Soon, attention turned to the pig, who many consider to closely resemble the human. The diversity of early valve substitutes prompted Gerbode to comment 'Once a valve has been rendered inert through preservation, it matters little whether it was human or animal'.

Back in Oxford, Duran and Gunning transferred their attention to preservation of heterograft valves. On 23 September 1964 they performed the first human stent-mounted porcine valve implant. Their experimental work with insertion of freeze-dried pig aortic valves in the descending thoracic aorta of dogs was published the following year but did not mention their clinical cases.[23] In 1965, Binet, Duran, Carpentier and Langlois in Paris presented their early experience of mercurochrome- and formalin-preserved frame-mounted heterografts and acknowledged Gunning's contribution. Their report described five patients with heterografts preserved in mercurochrome, all of whom survived without anticoagulants.[24] In 1966 O'Brien, now in Melbourne, reported his experience

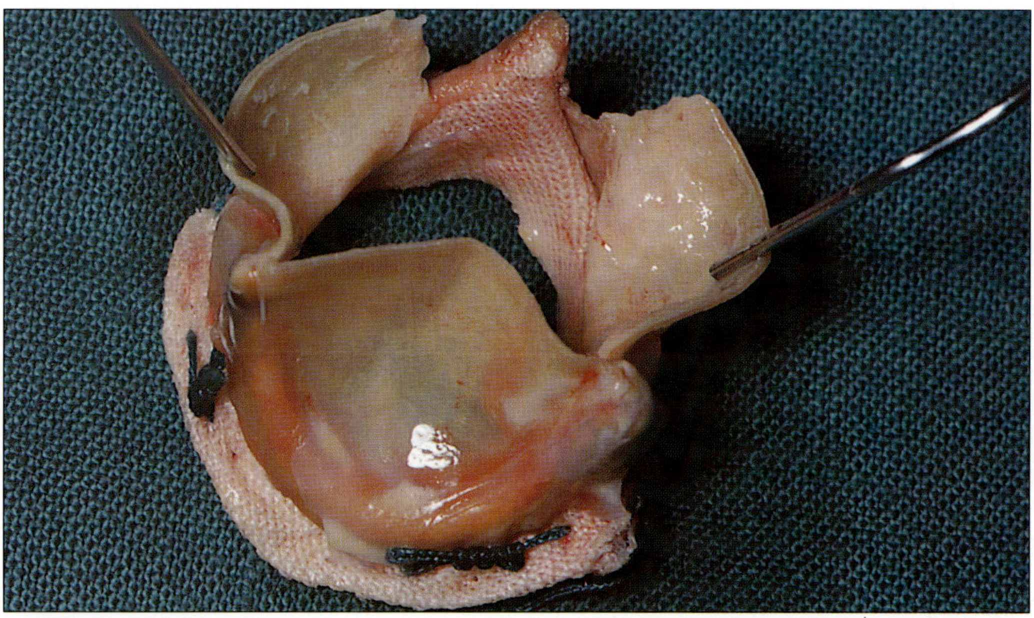

Figure 1.6. Pericardial prosthesis: early failure through tearing of the pericardium from the stent.

with formalin-preserved pig and calf aortic heterografts in nine patients and in 1968 Carpentier, DuBost and colleagues described stent-mounted heterograft replacements of the mitral and tricuspid valves.[25,26]

In Britain, Ionescu and Wooler used β–propriolactone-sterilized freeze-dried heterografts for aortic and mitral replacement followed by formalin-fixed stent-mounted valves. Although initial results with formalin fixation were satisfactory, the medium-term durability was poor. The preservation problem was soon resolved by Carpentier, who adopted glutaraldehyde fixation in 1968.[27] Following the efforts of Carpentier, Hancock and Angell, glutaraldehyde-fixed stented porcine valves became readily available commerically, did not require anticoagulants and were soon in great demand. They were easy to insert and had better flow characteristics than the first-generation mechanical valves (Figure 1.7). Although Binet had reported his experience with stentless aortic valve heterografts in 1965,[24] ease of implantation of stented xenografts rapidly superseded freehand stentless techniques. In discussion of Angell's paper on 'Five year study of viable aortic valve homografts' at the 52nd Annual Meeting of the American Association for Thoracic Surgery (1972), Carpentier made the following comments:

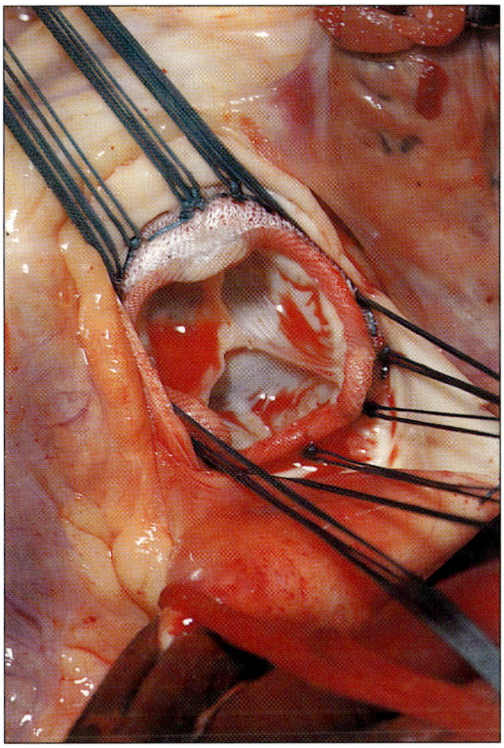

Figure 1.7. Carpentier–Edwards valve implanted in the aortic position.

> My only criticism of the homograft is the practical problem of getting enough specimens of different sizes. Because of this difficulty the technique cannot be routinely used. As a result we are still concentrating our efforts on heterografts. If the durability of the homograft depends upon the viability of the tissue, the durability of the heterograft depends on the stability of the tissue, with absence of collagen denaturation and immunologic reaction. This is achieved by an appropriate chemical treatment which decreases the antigenicity of the tissue by elimination of the antigenic components and prevents the denaturation of collagen by means of glutaraldehyde. The valve substitute so obtained is a 'bioprosthesis' rather than a graft. Its durability is based on the unfailing stability of the tissue and not upon regeneration by the host cells.[28]

In this same discussion, Cooley (not a pig valve enthusiast) recalled the case of a 12-year-old boy with rapid degeneration of a xenograft.

> About six months earlier this patient had had a porcine graft placed in the aortic annulus in a city in Louisiana. The poor, bewildered mother, knowing that her son had received a pork graft, asked me what type of valve we would now use. I told her we would use a Bjork valve. She said, 'What kind of animal is a Bjork?'

Influence of the stent

Stented porcine xenografts are hybrids of biological and mechanical structures, hence the term 'bioprosthesis'. Central unobstructed flow characteristics are retained but the stent is partially obstructive and produces turbulent flow. Much less judgement and skill are needed on the part of the surgeon to insert the valve and ensure competence. Their haemodynamic characteristics lie somewhere between the homograft and first-generation mechanical valves.[29] Long-term function of porcine xenografts proved to be less favourable than that of freehand homografts, with degeneration, calcification and cusp rupture tending to occur earlier, especially in children or young adults[30–32] (Figure 1.8). The difference in durability can be accounted for mainly by the nature of stent mounting, which leads to suboptimal valve geometry and maldistribution of stress on the leaflets. A similar rate of valve failure was found in preserved and mounted aortic homografts. Recent studies taking into consideration the concept of patient-orientated treatment failure and patient-related variables have suggested superior results in adult patients who received bioprostheses in contrast to mechanical valves in the aortic position.[33,34] The benign mode of bioprosthetic valve failure in conjunction with a low incidence of thromboembolic events of anticoagulation-related bleeding compensates for the risk of reoperation. Operative morbidity and mortality for the pulmonary autograft operation or aortic homograft insertion is related to the learning curve.[35] The same is to be expected of stentless xenograft insertion. The learning curve of the autograft and homograft operations is also reflected in the early phase of primary tissue failure, where cusp prolapse is not infrequently found at reoperation and probably evolves from imperfect commissural pillar alignment.[36] Such technical errors subject the replacement valve leaflets to increased stress, a contributory factor for primary failure in stent-mounted bioprosthetic valves. Although most observations on stentless valve behaviour are reported from the aortic position, the principles of stent mounting apply equally to the mitral and tricuspid positions. Structural failure requiring reoperation occurs earlier in the larger-sized prostheses of mitral patients.

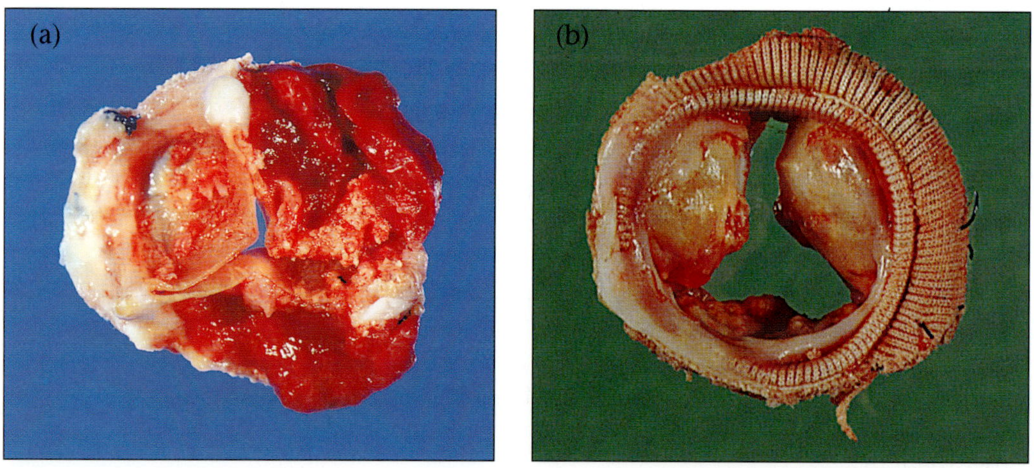

Figure 1.8. Modes of failure in stented xenografts: (a) severe calcification and stenosis — 14-year-old patient after 3 years; (b) attenuation of the leaflets and moderate calcification — 68-year-patient at 8 years.

In contrast to the patient with a mechanical prosthesis, the patient with a true biological valve substitute, such as the pulmonary autograft or aortic homograft inserted by the freehand technique, is able to live an unrestricted life without anticoagulants, threat of embolism or sudden death. Whereas the autogenous pulmonary valve remains vital and grows in children, with an indefinite life span,[37] structural failure of the aortic homograft occurs though calcification or cusp rupture, although considerably later than in the stented xenograft.[38] Although preservation methods in bioprostheses account for some differences in durability, accelerated fatigue tests show that the stent of a bioprosthesis is a major factor governing stress on the biological component.[39,40] Clearly, the best stent for a glutaraldehyde-fixed porcine xenograft is the native aortic root, where the structure of the sinuses is such that mechanical stress is dissipated during leaflet closure at high aortic pressure. This may account for the difference in durability between the freehand-sewn aortic valve homografts and stented aortic homografts used for valve replacement in humans. The average time for aortic homograft failure exceeds 12 years when hand sewn but is approximately 8 years when stent mounted.[41] The average time for failure of the stented porcine xenograft in the aortic position, although varying with age at implantation, is also 8 years.

The pathological features of explanted biological prostheses provide insight into mechanisms of degeneration. Calcification and stress-related tears and perforations are the predominant mode of failure and commissural stent dehiscence occurs predominantly in valve sizes greater than 27 mm. The last 10–15 years have seen considerable effort expended to elucidate the mechanisms of calcification in implanted biomaterials or polymer-based devices. Recent evidence suggests mechanisms for calcification that do not require cellular components but occur through interactions between individual molecules and ionic calcium and phosphate in the blood. Tissue crosslinking with glutaraldehyde provides a chemical environment for the initiation of calcification similar to that which occurs in polyurethanes. Glutaraldehyde in dilute aqueous solution is prone to formation of oligomers and polymers of glutaraldehyde, and computer modelling of glutaraldehyde interactions with calcium support the hypothesis for a calcium-initiated mechanism for calcification. Although the durability of stented porcine xenografts proves sufficient to prevent reoperation in the majority of patients over 70 years of age, haemodynamic function and freedom from structural valve failure remains less satisfactory than that for mechanical valves or homografts.

Resurrection of stentless valves

Given that the aortic wall and sinuses provide the best stent for valve leaflets, it is logical to expect that elimination of an artificial, albeit flexible, stent will reduce stress and improve durability. Absence of a stent and sewing ring provides reduced transvalvular pressure gradients, and the native aortic root and coronary sinus dissipate mechanical stress during leaflet closure at high pressure. This accounts for the difference in durability between the freehand-sewn aortic homograft and stent-mounted aortic homografts used for aortic replacement in humans. In 1988 David re-explored the use of stentless glutaraldehyde-fixed aortic xenografts, both experimentally and in humans.[42] Haemodynamic evaluation showed very small resting gradients and minimal aortic regurgitation in these early implants. David went on to develop a low-pressure-fixed valve with the aortic sinuses and subannular tissues trimmed down to 1.5–2 mm from the base of the leaflets. When sheep implants showed an important incidence of valve detachment at the inflow, the relatively fragile tissue was supported by Dacron cloth. This also provides a plane of separation between human tissue and the porcine implant (of potential value in reoperation). The cloth covering is said to prevent septal muscle bar reabsorption, and low-pressure fixation helps to retain preservation of collagen crimp and elasticity in the leaflets (although not to the degree afforded by zero-pressure fixation). The valves implanted into David's first 29 patients were manufactured by Johnson and Johnson Cardiovascular Products, later acquired by St Jude Medical. Their stentless bioprosthesis is now marketed as the Toronto SPV™ (stentless porcine valve) and has been used extensively by David and others with excellent medium-term results.[43] The SPV allows a larger valve to be inserted into an individual annulus through elimination of the stent and sewing ring. The flow characteristics are inherently superior to those of the stented xenograft and the larger valve orifice results in considerably better haemodynamic function. The human aortic sinuses are preserved, thereby retaining the shock-absorbing capacity of a distensible aortic root. The relationship between the porcine leaflet and the human versus porcine aortic sinus remains to be established.

David recently described the morphological findings in four reoperated patients with Toronto valves between 1 month and 5 years after implantation.[44] The suture lines were covered with pannus which extended onto the commissural areas adjoining the sinuses of Valsalva and onto the base of the cusps, similar to the findings in homografts. The cuspal tissue remained soft and pliable without important calcification or tears, although calcification had begun to affect the aortic wall. Already there is the suggestion that improved haemodynamics and reduced leaflet stress will translate into prolonged durability. Haemodynamic function and early improvement in left ventricular mechanics already parallel those seen for aortic homografts.

The stentless valves from Medtronic and Edwards are presented as porcine xenograft cylinders, directly equivalent to the homograft but with cloth covering of the inflow. These valves can be used for aortic root replacement. Glutaraldehyde-treated aortic roots are significantly less distensible than fresh human or porcine sinuses. The diameter of fresh aortic roots increases by 45% as the internal pressure is raised from zero to 120 mmHg; in contrast, the internal diameter of glutaraldehyde roots increases by less than 10%. Consequently, glutaraldehyde-treated aortic roots and valves have markedly higher forward-flow pressure drops compared with fresh aortic roots.

Fresh untreated porcine roots have a triangular open leaflet orifice with little or no bending deformation at the commissures, whereas higher open leaflet bending deformations can be measured in glutaraldehyde-treated aortic roots. These deformations may reduce the fatigue life of glutaraldehyde-treated valves and this applies to the Toronto SPV™ as well as the aortic cylinder Prima™ and Freestyle™ valves.[45] Glutaraldehyde-treated porcine roots do not reproduce the function of the fresh aortic root and valve and cannot be regarded as homograft equivalents (unless the homograft tissue is fixed). Differences in durability between glutaraldehyde-fixed aortic homografts and the so-called 'homovital homografts' may partially be explained by this phenomenon. The Edwards Prima™ valve has the pig coronaries excised and additional cloth applied to the internal aspect of the coronary interface.[46] This feature is potentially disadvantageous, since the porcine coronaries are closer together than in the human and it is safer to tailor the porcine cylinder to variable human coronary anatomy when the inflow suture line is complete. Attempts to make predetermined coronary orifices fit the human root can lead to distortion and malalignment of the commissures. This may account for the incidence of early aortic regurgitation with this valve in some series.[47] The haemodynamics of the porcine cylinder are similar to those of the Toronto SPV™. Transvalvular gradients are less than those for stented prostheses and decrease with time (Table 1.1 and Figure 1.9).

For the porcine cylinders, factors affecting valve gradients include implant technique, time of measurement, changes in left ventricular function due to resolving dynamic obstruction and remodelling of the valve cylinder itself

Table 1.1 Haemodynamic variables (at rest) for implanted porcine xenografts

Haemodynamic variable	Valve	Valve size (mm) 19	21	23	25
Gorlin area (cm^2)	Hancock MO	0.85	1.11	1.59	1.50
	Carpentier	0.77	0.84	1.14	1.37
	Freestyle	1.89	2.25	2.89	3.50
Mean gradient (mmHg)	Hancock MO	32.5	20.2	14.2	11.6
	Carpentier	31.7	22.4	13.8	15.1
	Freestyle	14.0	10.1	8.8	7.0

MO, modified orifice

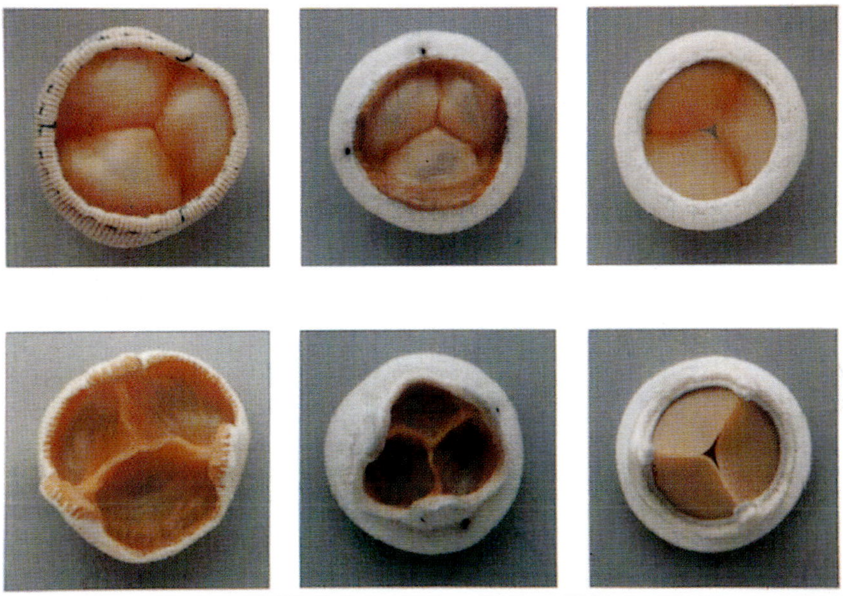

Figure 1.9. Inflow and outflow aspects of the Toronto SPV™ valve in comparison with stented porcine and pericardial valves of the same external diameter.

through resolution of the acute inflammatory response or perivalvar haematoma in the early postoperative period. A limited amount of distensibility persists after glutaraldehyde fixation of the porcine root and may account for remodelling of the porcine cylinder as perivalvar haematoma is absorbed. Involution of left ventricular hypertrophy, which occurs earlier in stentless valves, may contribute to a reduction in gradient at between 6 and 12 months postoperatively. Lastly, the implant technique is of considerable importance for the porcine cylinder xenografts. Differences in performance index may result if valves of a given mounting diameter are inserted into orifices of different sizes.

Aspects of surgical technique

The only conceivable argument against the use of stentless, as opposed to stented, xenografts is an element of difficulty in the surgical techniques. The stented xenograft or mechanical prosthesis with a sewing ring is easy to implant. In contrast, the stentless xenograft has been likened to an aortic homograft, which requires a certain amount of judgement during implantation. In practice the glutaraldehyde-stiffened porcine aortic root differs from the thin, flexible aortic homograft in behaviour. The porcine cylinder cannot be inverted for the inflow sutures but, in contrast, the cloth-covered xenograft inflow requires only the same number of sutures as a stented valve. It is important not to overcomplicate stentless valve insertion. With a carefully planned technique, ischaemic times for stentless valve implantation are only slightly longer than those for stented xenografts. A two-thirds transverse incision between 2 and 5 mm above the sinotubular junction allows the cylindrical porcine root to be inserted into the cylindrical aortic sinuses, guaranteeing spatial orientation of the valve commissures. This method also facilitates valve sizing for both the Toronto SPV™ and porcine root prostheses. In contrast a hockey-stick incision increases the degree of difficulty by demanding careful judgement of the size and orientation of the distorted sinotubular junction.

Whereas the Toronto SPV™ can be inserted only by subcoronary implantation, the porcine root prostheses (Freestyle™ and Prima™) can be used for aortic root replacement (Figure 1.10a–c). Some surgeons prefer this approach routinely with this device, whereas others only use it for the small aortic root (less than 21 mm). Full aortic root replacement clearly allows the largest possible device to be inserted with the lowest early transvalvular gradients. However, root replacement is undertaken with an increased operative risk over valve replacement and with a prospect of difficult reoperation. In contrast to root replacement for aortic root pathology, generous coronary buttons must be left with this method so that at reoperation uncalcified pericoronary tissue can be dissected for further reimplantation (as for surgery in homograft degeneration).

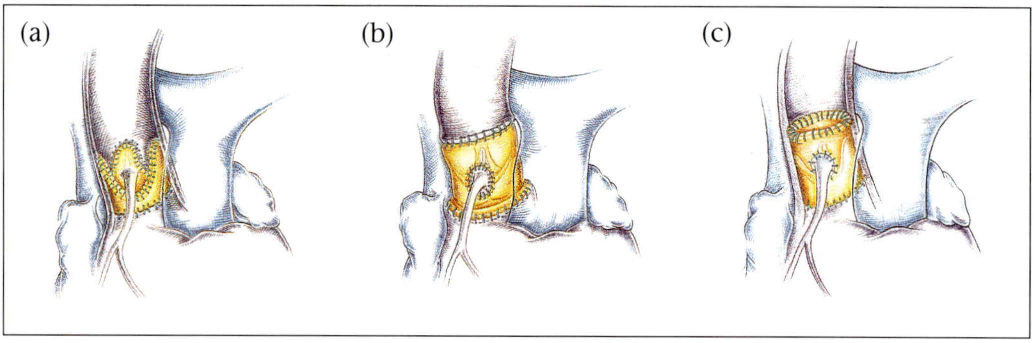

Figure 1.10. Porcine aortic cylinder implanted by (a) subcoronary method, (b) full root replacement, or (c) root inclusion technique.

Other methods have been employed for the porcine cylinders, including valve replacement by scalloping all three sinuses akin to homograft aortic replacement or the Toronto SPV™. Alternatively, some authors advocate leaving the outflow rim of the porcine cylinder intact in order to avoid commissural malalignment. In this case the porcine coronaries must be excised and dealt with by a pericoronary suture line. This method is associated with prolonged ischaemic times and haemodynamically is less satisfactory than full root replacement. The inappropriate and poorly descriptive term 'mini-root' technique has been applied to this method. The procedure is better described as the 'root inclusion' method. Perhaps the most simple and easily reproducible method is the so-called 'cylinder within a cylinder' technique whereby the aorta is two-thirds transected above the sinotubular junction and the intact porcine cylinder dropped into place in the human valve sinuses. The porcine coronaries can then be excised accurately in direct apposition to the human coronary ostia and the trimmed outflow of the porcine cylinder is sutured to the crest of the human aortic cylinder. This method is simple, provides a result similar to subcoronary homograft implantation and is achieved with ischaemic times of less than 45 min. The method ensures spatial orientation of the commissures, which guards against aortic regurgitation. If the right coronary artery is occluded in its proximal third and requires a bypass graft, there is no necessity to excise the porcine right coronary sinus, which greatly simplifies implantation by this method (Freestyle™ valve only). The technique also greatly facilitates implantation in a heavily calcified aortic root, which might otherwise be regarded as a contraindication to stentless valve insertion.

Whether the porcine valve leaflet is happiest within porcine sinuses or in either normal distensible or pathological human aortic sinuses will eventually be answered by this method. The cylinder within a cylinder operation leaves the porcine coronary leaflets within native human sinuses but the non-coronary leaflet in the porcine sinus, which is prone to late calcification. In practice, many aortic roots in patients with calcific aortic stenosis are already thickened or calcified and the theoretical advantage of siting the porcine leaflet within a distensible human aortic sinus is lost. The most important aspect for stentless valves is an easily reproducible and reliable implant method to prevent early valve failure through aortic regurgitation. For patients over 70 years old the theoretical advantage of siting the porcine leaflets in a human sinus is open to debate.

One of the advantages of the resurgence in interest in porcine xenografts resulting from stentless valve reintroduction has been a renewed concentration on the scientific aspects of bioprosthetic valves. Methods of antimineralization to prevent calcification and the introduction of zero-pressure leaflet fixation in particular must contribute to improved xenograft durability (Table 1.2).[48–51] Imperfect operative technique is the predominant cause of early aortic regurgitation in homograft aortic replacement and applies equally to the stentless porcine xenografts (Figure 1.11).

Table 1.2 Freedom from structural valve dysfunction (SVD) for different aortic bioprostheses at 8 years

Valve*	Source	*n*	Mean age (years)	Freedom from SVD (%)
Intact†	Green Lane	125	59	100
	Vancouver	153	66	97
Carpentier–Edwards Std	Vancouver	579	57	90
SAV		992	63	86
Hancock II	Toronto	376	63	93
Hancock MO	Boston	305	63	89

*Std, standard; SAV, supra-annular valve; MO, modified orifice; †the Intact valve has zero-pressure leaflet fixation and an anticalcification treatment. These advances appear to improve durability

Patient selection for stentless xenografts

Accelerated tissue failure occurs in the growing child and young adult through calcification.[52–54] The device chosen for aortic or atrioventricular valve replacement is age related and until it can be shown that the improved haemodynamics of stentless xenografts translates into improved durability, use of these valves is governed by the same principles as for stented xenografts. Stentless valves are ideal for patients over 70 years of age and for those for whom anticoagulant therapy is contraindicated (when a homograft is unavailable).

The haemodynamics of the stentless xenograft are equivalent to those of the homograft but durability and ease of reoperation are yet to be determined. Eradication of the stress of stent mounting,[54] together with the beneficial effects of zero-pressure fixation[47] and antimineralization[49] (Freestyle™ valve), may

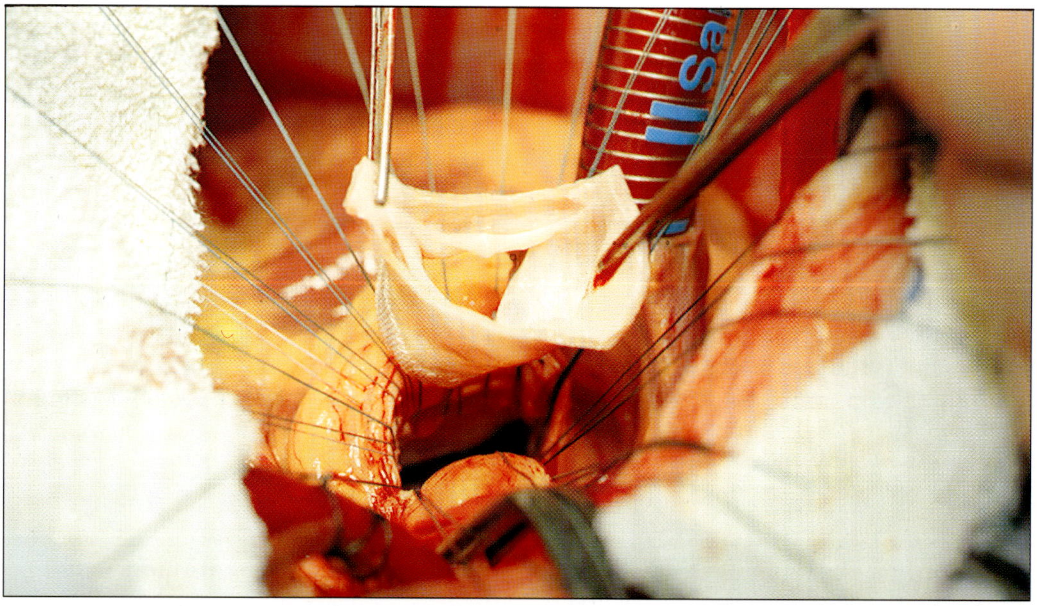

Figure 1.11. Freehand aortic homograft implantation with scalloping of the left and right coronary sinuses. The non-coronary sinus is left intact. Commissural pillar alignment must be accurate to prevent regurgitation.

greatly improve durability. In relatively early reoperations or at autopsy, a plane of cleavage remains between the porcine and human tissue and may facilitate separation. However, the porcine aortic wall will inevitably calcify with complete loss of compliance.

The authors' preference for aortic valve replacement by age group is to use the pulmonary autograft for infants, children and teenagers, the aortic homograft for young adults, bileaflet mechanical prostheses for the middle-aged patient and stentless xenografts for the elderly. In comparison with aortic homografts, the stentless xenograft is somewhat less versatile, with completely different handling characteristics but similar haemodynamics and thrombogenicity. Although the stentless porcine xenografts provide similar haemodynamic function, the authors do not regard them as substitutes for homografts in younger patients. In comparison to stented xenografts or mechanical prostheses, the stentless xenograft is technically more demanding but offers better haemodynamics and less thrombogenicity. The stentless xenograft will require a more difficult reoperation but may prove more resistant to infection through improved flow characteristics.

There are no absolute contraindications to stentless xenograft implantation. A large discrepancy between the diameter of the aortic annulus and sinotubular junction is a relative contraindication to the Toronto SPV™, whereas the dilated aorta can be tailored to the size of outflow for the porcine cylinder. For young patients with Marfan's syndrome a mechanical valved conduit is probably preferable, although aortic homografts have been used in this situation. Stentless xenografts may compensate for restricted availability of homografts but cannot be regarded as preferable if a homograft of appropriate size is available. The less pliable nature of the glutaraldehyde-preserved porcine cylinder also makes it less useful for right ventricular outflow surgery.

References

1. Ross D N 1969 Observations on homograft aortic valves. Guys Hosp Reports 118: 5–11
2. Murray G 1956 Homologous aortic valve segment transplants as surgical treatment for aortic and mitral insufficiency. Angiology 7: 466–471

3. Beall A C, Morris G C, Cooley D A, DeBakey M E 1961 Homotransplantation of the aortic valve. Thorac Cardiovasc Surg 42: 497–506
4. Duran C G, Gunning A J 1962 A method for placing a total homologous aortic valve in the subcoronary position. Lancet 2: 488–489
5. Ross D N 1962 Homograft replacement of the aortic valve. Lancet 2: 487–488
6. Barratt-Boyes B G 1964 Homograft aortic valve replacement and aortic incompetence and stenosis. Thorax 19: 131–150
7. Paneth M, O'Brien M E 1968 Transplantation of human homograft aortic valve. Thorax 21: 115–118
8. Hoeksema T D, Titus J L, Guilhari E R, Kirklin J W 1966 Use of homografts for replacement of the aortic valve in man. Circulation 34: 129–134
9. Angell W W, Ibsen A B, Stenson E B, Shumway M E 1968 Fresh aortic homografts for multiple valve replacement. J Thorac Surg 56: 323–328
10. Stenson E B, Angell W W, Ibsen A B, Shumway M E 1968 Aortic valve replacement with the fresh aortic homograft. Am J Surg 116: 204–210
11. Barratt-Boyes B G, Roche A H G, Brandt P W T, Smith J C, Lowe J B 1969 Aortic homograft valve replacement. A very long term follow up of an initial series of 101 patients. Circulation 40: 763–769
12. Bowman F O, Malm J R, Harris P D, Kaiser G A, Kovalik A T W 1969 Further evaluation of aortic homografts sterilised by electron beam energy. Circulation 39 (Suppl I): 40–46
13. Gonzalez-Lavin L, Ross D N 1970 Homograft aortic valve replacement. A five year experience at the National Heart Hospital, London. J Thorac Cardiovasc Surg 60: 1–8
14. Young W P, Kroncke G M, Dacumos G L, Rowe G G 1971 A follow up study of aortic valve homografts for two to four and one half year. Ann Thorac Surg 12: 155–160
15. Weldon C S, Amali M M, Morovati S S, Shaker I J 1966 A prosthetic stented aortic homograft for mitral valve replacement. J Surg Res 6: 548–552
16. Barratt-Boyes B G, Roche A H G, Subramaryan R, Pemberton J R, Whitlock R M L 1987 Long term follow up of patients with antibiotic sterilized homograft valves inserted freehand in the aortic position. Circulation 75: 768–775
17. Anderson E T, Hancock E W 1976 Long-term follow up of aortic valve replacement with fresh aortic homograft. J Thorac Cardiovasc Surg 72: 150–156
18. Senning A 1967 Fascia lata replacement of aortic valves. J Thorac Cardiovasc Surg 54: 465–470
19. Ionescu M I, Ross D N, Wooler G H, Deac R, Ray D 1970 Replacement of heart valves with autogenous fascia lata. Surgical technique. Br J Surg 57: 437–442
20. Ionescu M I, Pabraski B C, Holden M P, Mary D A, Wooler G H 1972 Results of aortic valve replacement with frame supported fascia lata and pericardial grafts. J Thorac Cardiovasc Surg 64: 340–353
21. Lower R R, Stofer R C, Shumway N E 1960 Autotransplantation of the pulmonic valve into the aorta. J Thorac Cardiovasc Surg 39: 680–687
22. Ross D N 1967 Replacement of aortic and mitral valve with a pulmonary autograft. Lancet 2: 959
23. Duran C G, Gunning A G 1965 Heterologous aortic valve transplantation in the dog. Lancet 2: 114–115
24. Binet J P, Duran C G, Carpentier A, Langlois J 1965 Heterologous aortic valve transplantation. Lancet 2: 1275
25. O'Brien M G, Clarebrough J K 1966 Heterograft aortic valve transplantation for human valve disease. Med J Aust 2: 228–230.
26. Carpentier A, Blondeau P, Laurens P, Mancel, P, Laurent D, Dubost C 1966 Replacement des valvules mitrales et tricuspides par des hétérogreffes. Am Chir Thorac Cardiovasc 7: 33–36
27. Carpentier A, Deloche A, Relland J, Fariani J N, Forman J, Camilleri J P, Soyer R, Dubost C 1974 Six year follow up of glutaraldehyde-preserved heterografts with particular reference to the treatment of congenital valve malformations. J Thorac Cardiovasc Surg 68: 771–782
28. Angell W W, Shumway N E, Kosek J C 1972 A five year study of viable aortic valve homografts. J Thorac Cardiovasc Surg 64: 329–353
29. Rashtian M Y, Stevenson D M, Allen D T, Yoganathan A P, Harrison E C, Edmiston W A, Rahimtoola S H 1990 Flow characteristics of bioprosthetic heart valves. Chest 98: 365–375
30. Magilligan D J, Lewis J W, Tilley B, Peterson E 1985 The porcine bioprosthetic valve. J Thorac Cardiovasc Surg 89: 499–507

31. Williams D B, Danielson G K, McGoon D C *et al* 1982 Porcine heterograft valve replacement in children. J Thorac Cardiovasc Surg 84: 446–450
32. Dunn J M 1981 Porcine valve durability in children. Ann Thorac Surg 32: 357–368
33. Hammermeister K E, Sethi G K 1991 Comparison of occurrence in bleeding, systemic embolism, endocarditis, valve thombosis and reoperation between patients randomised between a mechanical prosthesis and a bioprosthesis. J Am Coll Cardio 17: 362A
34. Hammond G L, Geha A S, Kopf G S *et al* 1987 Biological versus mechanical valves. Analysis of 1116 valves inserted in 1012 adult patients with a 4818 patient year and a 5326 valve year follow up. J Thorac Cardiovasc Surg 93: 182–198
35. Matsuki O, Okira Y, Almeida R S, McGoldrick J P, Hooper T L, Robles A, Ross D N 1988 Two decades experience with aortic valve replacement with pulmonary autograft. J Thorac Cardiovasc Surg 95: 705–711
36. Elkins R C, Santangelo K L, Stelzer P, Randolph J D, Knott-Craig C J 1992 Pulmonary autograft replacement of thc aortic valve: an evolution of technique. J Card Surg 7: 108–116
37. Elkins R C, Knott-Craig C J, Ward K E, McCue C, Lane M M 1994 Pulmonary autograft in children. Realised growth potential. Ann Thorac Surg 57. In press
38. Clarke D R, Cambell D N, Hayward A R *et al* 1993 Degeneration of aortic valve allografts in young patients. J Thorac Cardiovasc Surg 105: 934–941
39. Broom N D 1978 Fatigue induced damage in glutaraldehyde preserved heart valve tissue. J Thorac Cardiovasc Surg 76: 202
40. Reis F L, Hancock W D, Yarborough J W, Glancy D, Morrow A G 1971 The flexible stent. A new concept in the fabrication of tissue valve prostheses. J Thorac Cardiovasc Surg 62: 683–689
41. Angell W W, Grehl T M, Buch W, Wuerflein R D 1973 Mounted fresh homografts for aortic valve replacement. Med J Aust (Spec Suppl) 21: 74–76
42. Glower D D, White W D Hatton A C, Smith L R, Glenn Young W, Wolfe W G, Lowe J E 1994 Determinants of reoperation after 960 valve replacements with Carpentier–Edwards prostheses. J Thorac Cardiovasc Surg 107: 381–393
43. David T E, Ropchan G C, Butany J W 1988 Aortic valve replacement with stentless porcine bioprostheses. J Card Surg 3: 501–505
44. David T E, Pollick C, Bos J 1990 Aortic valve replacement with stentless aortic bioprosthesis. J Thorac Cardiovasc Surg 99: 113–118
45. Butany J, David T E 1994 Pathology of explanted stentless porcine valves (human experience). 3rd Scientific Meeting Int Assoc for Cardiac Biological Implants, New York
46. Thubrikar M J, Deck J D, Aouad J *et al* 1983 Role of mechanical stress in calcification of aortic bioprosthetic valves. J Thorac Cardiovasc Surg 86: 115–125
47. Pillai R, Sprigings D, Amarasena N, O'Regan D J, Parry A J, Westaby S 1993 Stentless aortic bioprosthesis. The way forward: early experience with the Edwards valve. Ann Thorac Surg 56: 88–91
48. Daenen W, Denef B, Herregeds C, Vanhoue M 1994 The Prima stentless aortic bioprosthesis. Preliminary results. 6th International Symposium on Cardiac Bioprostheses. Vancouver, p71
49. Vesely I 1991 Analysis of the Medtronic Intact bioprosthetic valve. Effects of zero pressure fixation. J Thorac Cardiovasc Surg 101: 90–99
50. Gott J P, Par-Chih, Dorsey L M A *et al* 1992 Calcification of porcine valves: a successful new method of antimineralisation. Ann Thorac Surg 53: 207–216
51. Giradot N, Giradot J–M 1991 Alpha-Aminoleic acid, a new compound prevents calcification of bioprosthetic heart valves. Trans Soc Biomater 14: 114
52. Schoen F J, Flomenbaum M A 1991 Morphologic characteristics of low and zero pressure fixed antimineralisation-treated porcine aortic bioprosthetic valves. Fifth International Symposium Cardiac Bioprostheses Avignon, France, p48
53. Geha A S, Laks H, Stansel H C *et al* 1979 Late failure of porcine heterografts in children. J Thorac Cardiovasc Surg 78: 351–364
54. Pupello D F, Bessone L N, Hiro S P, Lopez-Cuerca E, Glaterer M S, Ebra G 1988 The Carpentier Edwards bioprosthesis: a comparative study analyzing failure rates by age. J Card Surg 3(Suppl): 369–374
55. Ferrans V J, Boyce S W, Billingham M E, Jones M, Ishihara T, Roberts W C 1980 Calcific deposits in porcine bioprostheses. Structure and pathogenesis. Am J Cardiol 46: 721–734

Chapter 2

From homograft to stentless bioprosthesis

Donald Ross

The homograft was in fact the first stentless valve and this is undoubtedly one of its chief attributes (Figure 2.1). It was Dr David of Toronto who coined the term 'stentless' in relation to the porcine valve; however, in view of the various 'miniskirts' involved in the manufacture of such valves, the term stentless is somewhat flawed. Bioprosthesis was a term introduced by Carpentier to define the hybrid mixture of glutaraldehyde-treated leather cusps, a metal stent and Dacron cloth. The resultant device falls halfway between mechanical and true biological valves and combines the virtues and shortcomings of both. Perhaps the new variety of valve should be termed a 'skirted', rather than a stentless, valve.

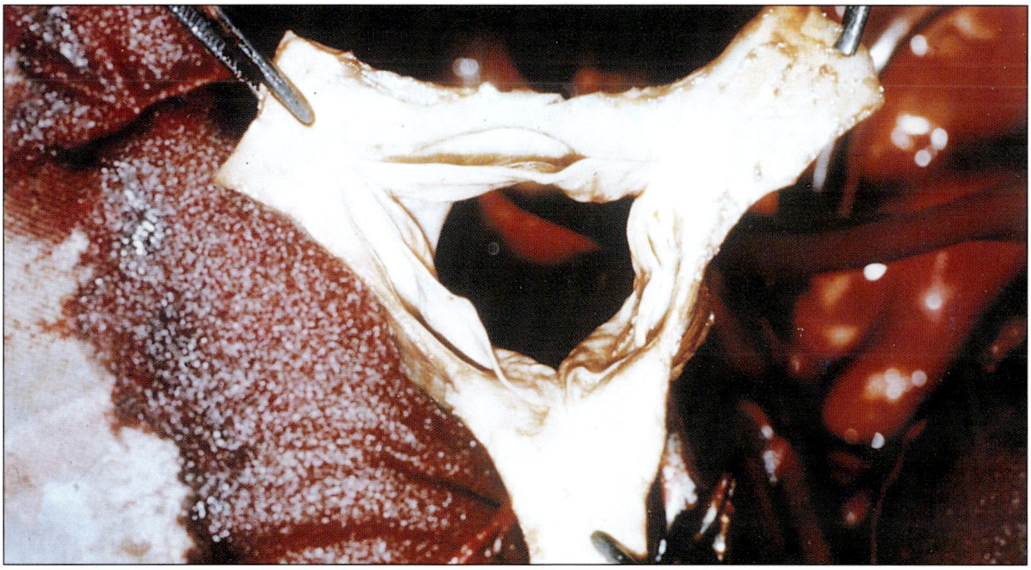

Figure 2.1. Homograft valve, showing its unique centrally flowing non-obstructive orifice.

New types of homograft

The homograft[1] remains the gold standard in the field of truly stentless valves, with well-established advantages; in introducing any new variety the unique features of the homograft should be safeguarded as jealously as possible (Table 2.1). Many workers in the sixties put homografts on frames for mitral and tricuspid replacement and some even tried this in the aortic position with a view to more rapid insertion. Without exception, the functional results were inferior to

Table 2.1 Advantages of homografts

- Free of emboli — no anticoagulants
- Low incidence of endocarditis
- Can be used in children
- Deteriorate slowly — no sudden deaths
- Quality of life is excellent
- Patient survival likely to be better than with bioprosthetic and mechanical valves

those with the uncluttered or stentless homograft, confirming the findings of Angell and Shumway in 1968.[2]

Realizing the limitations imposed by rigid stents, an attempt was made to avoid them altogether and a flexible 'lifebelt' was introduced particularly for mitral valve insertion. This was taken a step further by Yacoub, resulting in a 'top hat',[3] which certainly was an improvement but removal of the valve was extremely difficult. In fact, all three-cusp valve replacements are less successful in the mitral position than the aortic, which is hardly surprising in view of the different anatomical and physiological environmental factors relating to these two valve sites (Figure 2.2).

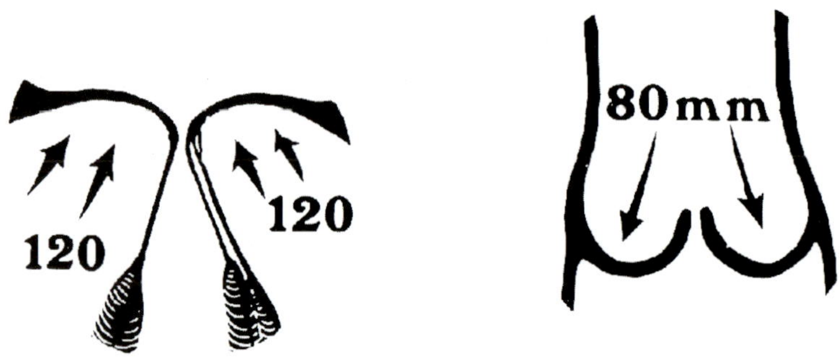

Figure 2.2. The systolic closing forces in the mitral area (left) are much higher than the aortic diastolic closing pressures (right) making the aortic configuration unsuitable for mitral valve replacement.

Similar experiments with unstented fascia, pericardium and dura mater aortic valves were abandoned as it slowly became recognized that only God can make a perfect valve — human or pig or calf. On the other hand, it is known that certain 'non-believers' in Europe are still producing man-made stentless pericardial valves; it is hoped that they will one day return to the 'true faith'.

Pulmonary autografts

A firm embrace of the God-made valve and adherence to its principles has been rewarded by increasingly improved homograft results; the same applies to that perfect natural stentless valve, the pulmonary autograft (Figure 2.3a).[4] This valve is at last receiving international acknowledgement, so much so that even British cardiac surgeons are starting to use it (Figure 2.3b) and their American cousins are

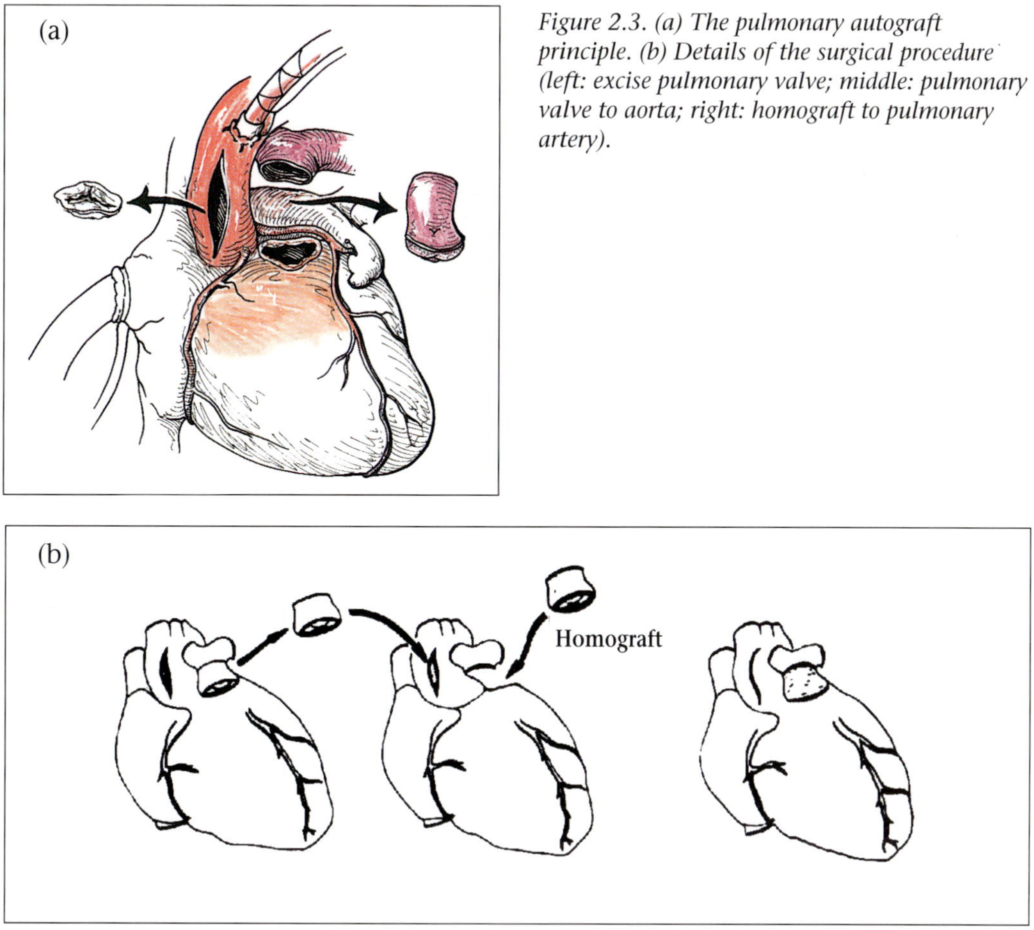

Figure 2.3. (a) The pulmonary autograft principle. (b) Details of the surgical procedure (left: excise pulmonary valve; middle: pulmonary valve to aorta; right: homograft to pulmonary artery).

almost inventing it![5] Although it has been used clinically since 1967 in over 400 patients (and the first patient is still alive), it is still used in only a few centres.

The chief indication for the pulmonary autograft is in young patients, to provide a permanent valve replacement and to avoid a lifetime of anticoagulants and haemorrhagic threats from the mechanical valve, or repeated dangerous operations with bioprostheses. Increasingly, it is being used in young children and even in infants, since the evidence is that it can grow and develop with the growth of the child.[6,7]

Aortic root replacement

In order to encourage growth and to ensure a competent valve the autograft is almost always inserted as a root replacement (Figure 2.4)[8,9] rather than in the conventional subcoronary position. In fact, the trend at present is to use root replacement with reimplantation of the coronaries also for homografts, and particularly in the presence of infection.[10]

It is particularly important that the transferred pulmonary autograft should be attached at the site of the excised aortic valve (Figure 2.5) so that it can enjoy the support of the surrounding left ventricular muscle and avoid progressive dilatation.

Advantages and disadvantages of stentless xenografts

Given the availability of almost perfect homografts and certainly perfect autografts, why should pig valves changed into leather with glutaraldehyde be

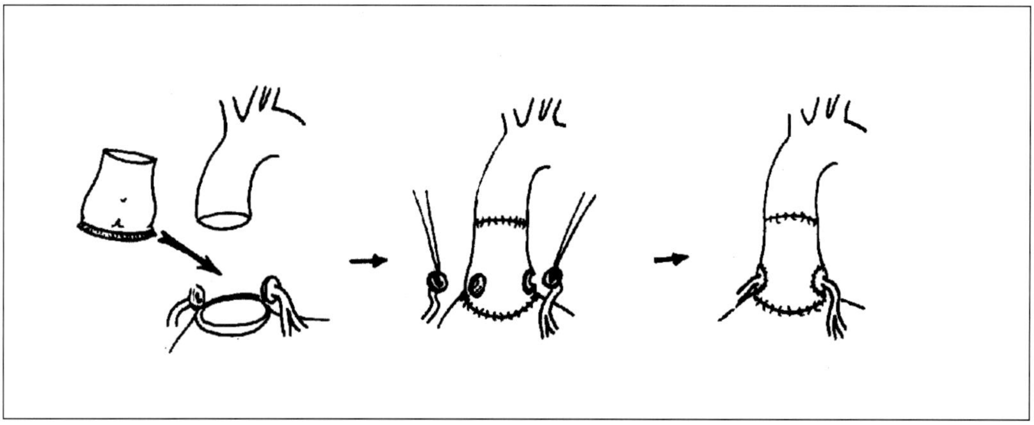

Figure 2.4. The freestanding aortic root replacement applicable to homograft, autograft and the stentless xenograft.

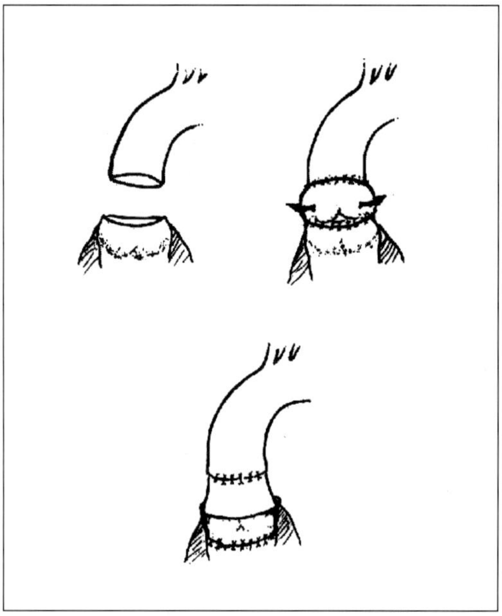

Figure 2.5. Unless the autograft is attached at the site of the original aortic valve ring there is a danger of dilatation of the unsupported autograft (top right: supra-annular; bottom: intra-annular).

used? One simple answer is that they also are made by God. Dr David was the first to see the need for them and many workers regret not using these stentless xenografts earlier, as they incorporate most of the advantages of the homograft. However, they do have the disadvantage of being stiffer than homografts; they are also denatured by glutaraldehyde and therefore are more likely to undergo tissue failure and calcification of the wall and the cusps, than the conventional stented bioprostheses (Table 2.2). On the other hand, given the increased shortage of homografts and the fact that homografts and bioprostheses deteriorate more slowly in elderly patients, they constitute an attractive substitute from the elderly patient's point of view. They are even more attractive to the commercial valve companies, whose stentless products and profits are proliferating rapidly.

As, in general, these values are less elastic and 'forgiving' than the homograft and autograft, their sizing is a little more critical; however, this author is not impressed by the interminable arguments about sinotubular junctions, for example, and prefers to 'eyeball' the aortic ring, sizing it by eye. The only logical valve sizers to use are the same as those that are used to size our homografts and xenografts; being cylindrical they average out the various aortic root constrictions and dilatations (Figure 2.6). Allowing for wall thickness, the inserted valve will always be slightly oversized which is thought to be an advantage (Figure 2.7).

Although some xenograft valves are inserted unashamedly naked, i.e. without a skirt or jacket, this author favours a Dacron or plastic cloth interposed between the xenograft and aortic wall (Figure 2.8a, b) to reduce tissue reaction or (more pessimistically) to facilitate later removal of the valve. The retained superior rim of

Table 2.2 Advantages of stentless xenografts, and disadvantages in comparison with homografts

Advantages	Disadvantages
Designed by God	Stiffer
Centrally flowing	Denatured by glutaraldehyde
Fully flexible	Undergo earlier tissue failure
Largely non-obstructive	
No need for anticoagulants	
No danger of haemorrhage	
Readily available supply	

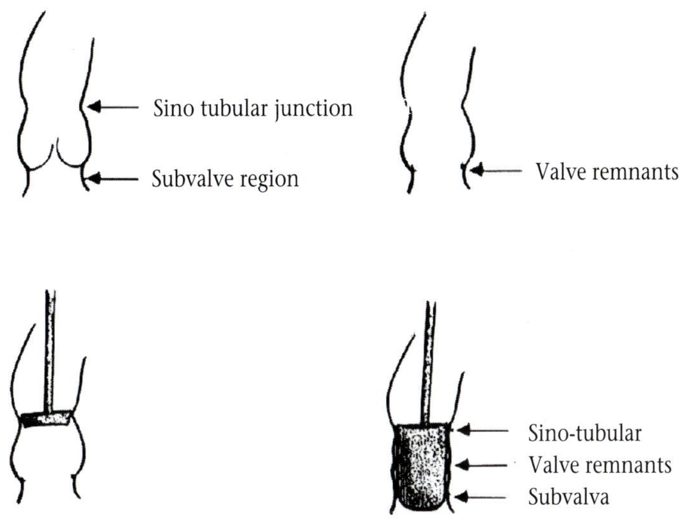

Figure 2.6. Conventional disc-like valve sizers are alternatively gripped or freed within the irregular left ventricular outflow tract. Thus, in the left panels, a short obturator sticks at or falls through, various potential measurement sites, whereas the special-design obturator (right panels) average out the various diameters.

xenograft tissue in one of the advertised valves is not recommended as it may form a calcified ring splinting the aorta, which can not have been the manufacturer's original intention.

Advantages of pulmonary xenografts

With the early model of stentless aortic valve, not only was the aortic wall tissue thick and somewhat rigid but also the postoperative transvalvular gradients were higher than expected. It is now known that a gradient should be expected and that, usually, this will be greater than after homograft replacement. However, this combination of thick semirigid wall and gradient encouraged the use of the porcine pulmonary valve as an alternative. The pulmonary xenograft (Figure 2.9b) has theoretical advantages in that the wall is thinner and is softer to handle, more like a homograft. More important, there is no muscle bar under the right coronary cusp (Figure 2.9a) to constitute an obstruction to outflow. Additionally, the pulmonary wall has a lower calcium content than the aortic, implying less probability of calcification.

Macroscopic and electron-microscopic investigations, coupled with stress/strain studies[11] of the porcine aortic and pulmonary valves, all implied that

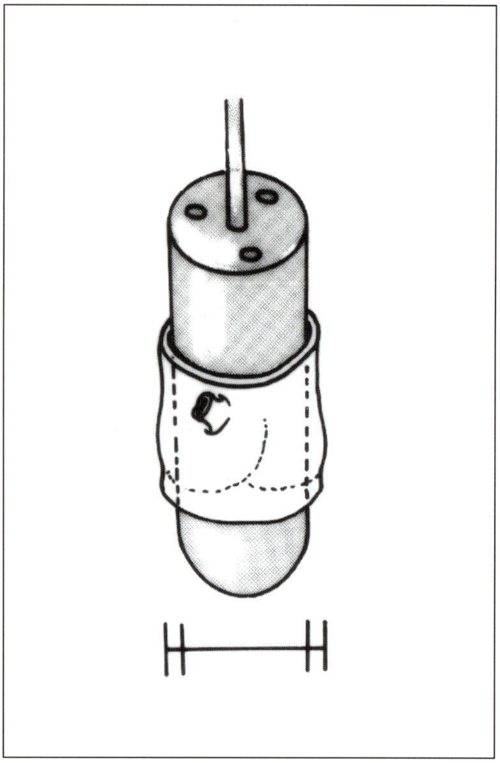

Figure 2.7. Cylindrical obturator used to size the homograft and also the valve orifice. Allowing for wall thickness means that the valve will be slightly 'oversized'.

there was no significant difference between the two valves. At the same time, fatigue testing *in vitro* showed no significant difference between aortic and pulmonary valves over equivalent 10-year cycles.

These stentless pulmonary valves have consequently been used clinically in parallel with stentless aortic xenografts: 341 have been distributed and 97 have been inserted, and there is now a user bias in favour of stentless pulmonary valves. Clinically, no significant differences have been reported so far.

Finally, the use of both aortic and pulmonary xenografts has been extended to root replacement (Figure 2.10a,b). Apart from ascending aortic replacement, there is plenty of potential for the pulmonary root for right ventricular outflow reconstruction and in the autograft operation when pulmonary homografts are not available.

Although some surgeons are opposed to the use of pulmonary valves in the aortic position, this author believes that if one remains scientifically objective and not too influenced by commercial pressures, the stentless xenograft principle has a promising future.

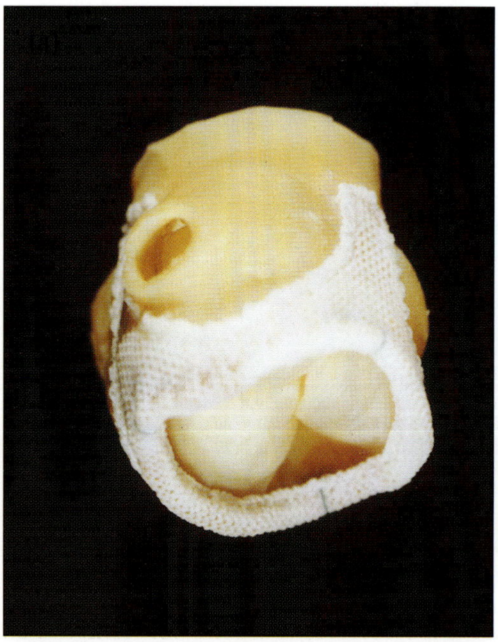

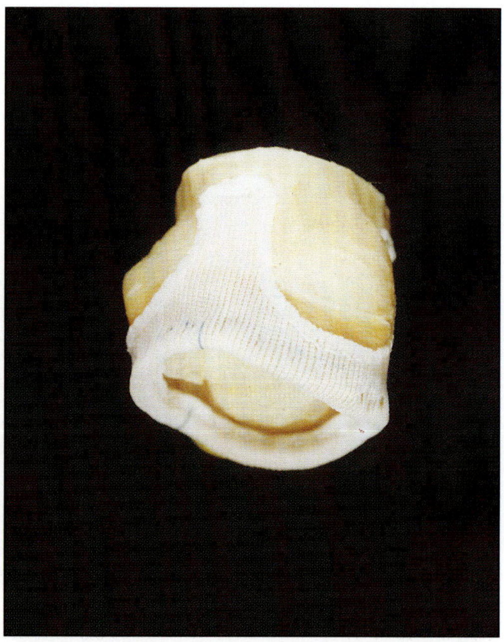

Figure 2.8. Aortic (a) and pulmonary (b) stentless xenograft, showing the brief Dacron covering. The uncovered wall is excised at surgery.

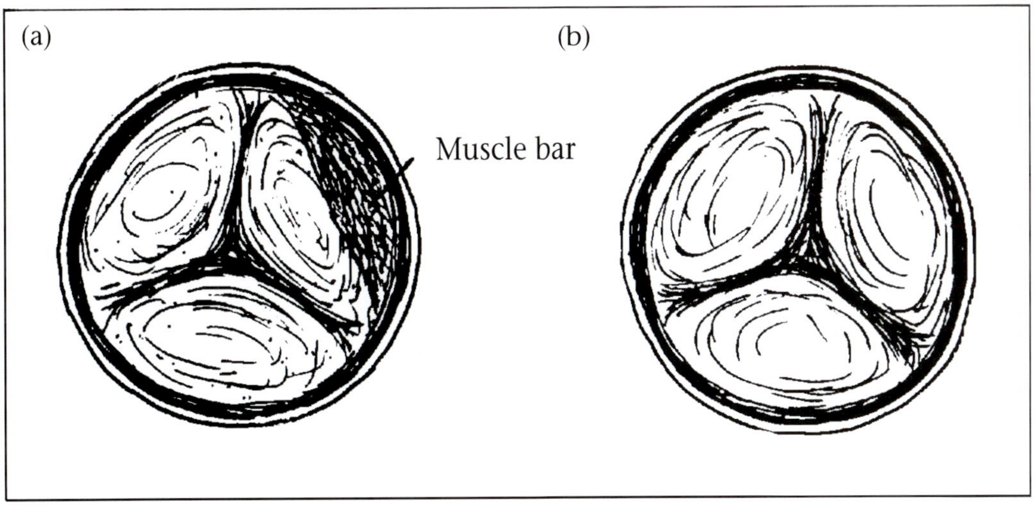

Figure 2.9. Aortic (a) and pulmonary (b) xenografts compared from the point of view of their inflow orifice.

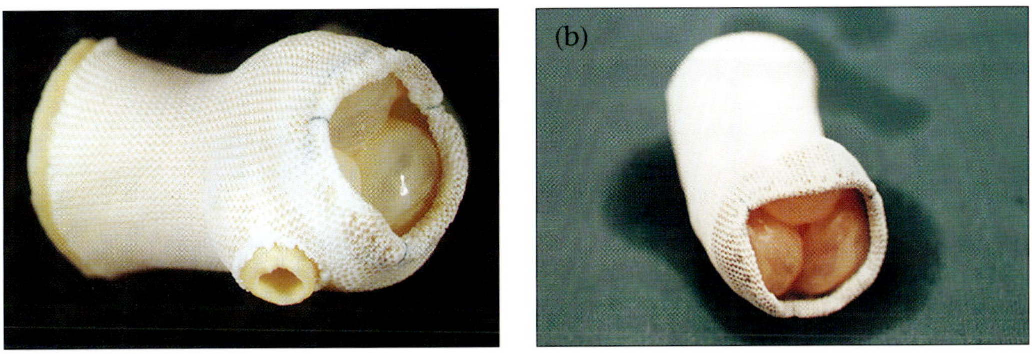

Figure 2.10. Aortic (a) and pulmonary (b) xenograft roots for left and right ventricular outflow reconstruction.

References

1. Ross D N 1962 Homograft replacement of the aortic valve. Lancet 2: 487
2. Angell W W, Iben A B, Shumway N E 1968 Fresh aortic homografts for multiple valve replacement. Arch Surg 97: 836
3. Yacoub M A, Kittle C F 1969 A new technique for replacement of the mitral valve by a semilunar valve homograft. J Thorac Cardiovasc Surg 58: 859
4. Ross D N 1967 Replacement of the aortic and mitral valves with a pulmonary autograft. Lancet 2: 956
5. Oury J H, Angell W W, Eddy C, Cleveland J C 1993 Pulmonary autograft — past, present and future. J Heart Valve Dis 2: 365
6. Gerosa G, McKay R, Davies J, Ross D N 1991 Comparison of the aortic homograft and pulmonary homograft for aortic valve or root replacement in children. J Thorac Cardiovasc Surg 102: 51–61
7. Elkins R C, Knott-Craig C I, Ward K E, McCue C, Lane M 1994 Pulmonary autograft in children: realized growth potential. In press
8. Somerville J, Ross D N 1982 Homograft replacement of aortic root with re-implantation of coronary arteries. Br Heart J 47: 473–482
9. Okita Y, Franciosi G, Matsuki O, Robles A, Ross D N 1988 Early and late results of aortic root replacement with antibiotic-sterilized aortic homograft. J Thorac Cardiovasc Surg 95: 696–704
10. Donaldon R M, Ross D N 1982 Homograft aortic root replacement for complicated prosthetic valve endocarditis. Circulation 70 (1): 178–182
11. David H, Broughner D R, Vesley I, Gerosa G 1994 The pulmonary valve — is it mechanically suitable for use as an ASAIO J. In press

Chapter 3

Stented porcine bioprostheses – durability and outcomes 1975–1993

W. R. E. Jamieson

Stented porcine bioprostheses have been utilized for valve replacement surgery since the introduction of the first-generation prostheses between 1972 and 1975. The long-term experience has been extensively documented, especially by Padova University, Stanford University and the University of British Columbia.[1–9] The major valve-related complication has been structural valve deterioration, occurring earlier and with increasing frequency in younger populations. Aortic porcine bioprostheses have outperformed mitral bioprostheses with regard to freedom from structural failure. Over the past decade mitral valve disease has been managed more frequently with mitral valve reconstruction or mechanical valvular replacement. Stentless porcine bioprostheses are currently under evaluation to determine the influence on durability. The aortic stentless bioprostheses currently being implanted are the Medtronic Freestyle™, St Jude Medical (Toronto SPV™), Bravo Stentless and the Edwards Prima™. These aortic stentless prostheses are investigational under regulatory control while a mitral stentless porcine bioprosthesis is experimental.

This paper reports the long-term experience with stented porcine bioprostheses that will be utilized in the future and with which the clinical performance of the stentless porcine bioprostheses will be compared. The teaching hospitals of the University of British Columbia, Vancouver Hospital and Health Sciences Centre and St Paul's Hospital have extended experience with stented porcine bioprostheses since 1975.

Patients and methods

Stented porcine bioprostheses were implanted in 4160 operations between 1975 and 1993. The prostheses implanted have been the first-generation Hancock standard and Carpentier–Edwards standard porcine bioprostheses and the second-generation Carpentier–Edwards supra-annular and Medtronic Intact™ porcine bioprostheses. The numbers of prostheses by prosthesis type and valve position are documented in Table 3.1. The prostheses have been evaluated by age groups and the number of prostheses by type and age groups are enumerated in Table 3.2.

The accompanying documentation in Figures 3.1–3.20 relates to the clinical performance of the prostheses for the various valve-related complications and composites of these complications. The *Guidelines for Reporting Morbidity and Mortality after Cardiac Valvular Operations,* approved by the Society of Thoracic Surgeons, the American Association for Thoracic Surgeons and the European Society of Cardiovascular and Thoracic Surgeons, were utilized in formulating this documentation.

Table 3.1 Long-term experience with stented porcine bioprostheses: patient population (prosthesis type and valve position)

	Valve position†					
Prosthesis type*	AVR	MVR	PVR	TVR	MR	Total
H-S	68	82	1	2	26	179
CE-S	578	512	1	8	115	1214
CE-SAV	1335	938	1	15	200	2489
MI	175	75	–	2	26	278
Total	2156	1607	3	27	367	4160

*H-S, Hancock Standard; CE-S, Carpentier–Edwards Standard; CE-SAV, Carpentier–Edwards supra-annular; MI, Medronic Intact
†AVR, aortic valve replacement; MVR, mitral valve replacement; PVR, pulmonary valve replacement; TVR, tricuspid valve replacement; MR, multiple valve replacement

Table 3.2 Long-term experience with stented porcine bioprostheses: patient population (prosthesis type and age group)

Prosthesis	Age (years)					
type*	≤35	36–50	51–64	65–69	≥70	Total
H-S	33	40	74	17	15	179
CE-S	101	204	506	195	208	1214
CE-SAV	83	245	728	458	975	2489
MI	5	25	68	51	129	278
Total	212	514	1376	721	1327	4160

*As in Table 3.1

The mean ages of the populations for the various prostheses differ [Hancock Standard (H-S), 57.2 years; Carpentier–Edwards Standard (CE-S), 57.3 years; Carpentier–Edwards supra-annular (CE-SAV), 64.1 years; Medtronic Intact™ (MI), 66.6 years]. The evaluation time was 18 years for the CE-S, 15 years for the H-S, 12 years for the CE-SAV and 7 years for the MI prostheses. The current documentation provides evaluation of each prosthesis type without statistical comparison.

Results

The freedom from thromboembolism (overall) is shown in Figures 3.1–3.3. the freedom from thromboembolism (TE) ranges between 81.6 (CE-SAV) and 85.0% (CE-S) at 10 years and 78.1 (CE-S) and 80.5% (H-S) at 15 years. There was no appreciable difference (not shown in the accompanying figures) in the freedom from TE by valve positions (aortic, mitral and multiple replacements). The freedom from fatal TE ranged between 99.2 (MI) and 97.1% (CE-SAV) at 7 years and 92.7 (CE-S) and 96.1% (H-S) at 15 years. The freedom from major TE and anti-thromboembolic-related haemorrhage was 87.1 (CE-SAV) and 91.0% (H-S) at 10 years and 86.8 (CE-S) and 87.9% (H-S) at 15 years.

The freedom from prosthetic valve endocarditis (overall) was similar for all the prosthesis types (Figure 3.4). At 15 years the freedom was 89.4% for H-S and 91.8% for CE-S. The freedom from non-structural dysfunction was also similar for the prostheses — 88.5 (H-S) to 91.9% (CE-S) (Figure 3.5).

Figures 3.6–3.10 show the freedom from structural valve deterioriation (SVD) for aortic valve replacement (AVR) by the five age groups. The freedom from SVD for AVR for the age groups ≤35 years was 84.7 (CE-S) and 87.5% (H-S) at 7 years compared with 95.0 (CE-SAV) and 100% (MI) for the second-generation prostheses (Figure 3.6). For the age group 36–50 years the freedom from SVD, at 10 years, was 84.9% for CE-SAV and 72.0% for CE-S (Figure 3.7). In the 51–64 years age group

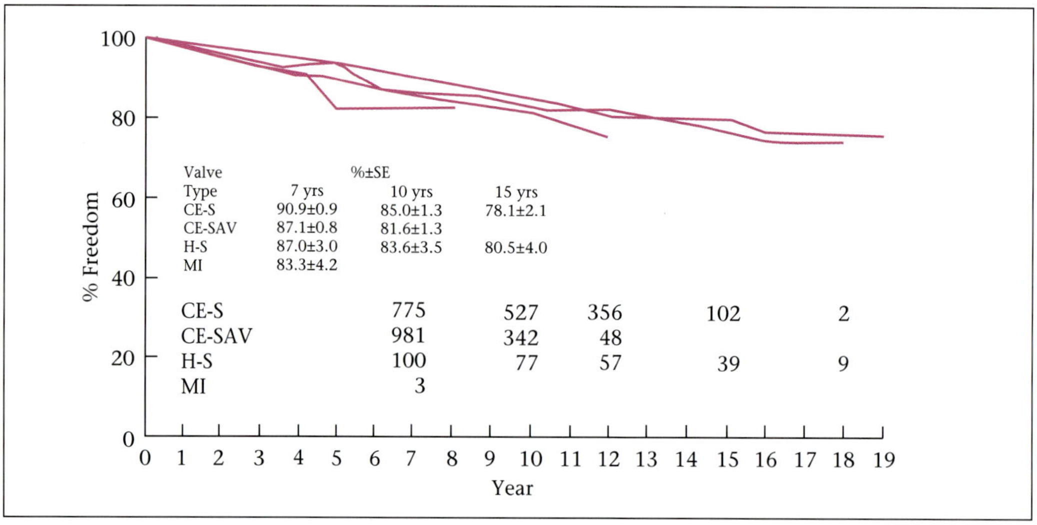

Figure 3.1. Freedom from thromboembolism: overall. Abbreviations of prosthesis type (in Figures 3.1–3.20): H–S, Hancock Standard; CE–S, Carpentier–Edwards Standard; CE–SAV, Carpentier–Edwards supra-annular; MI, Medronic Intact™.

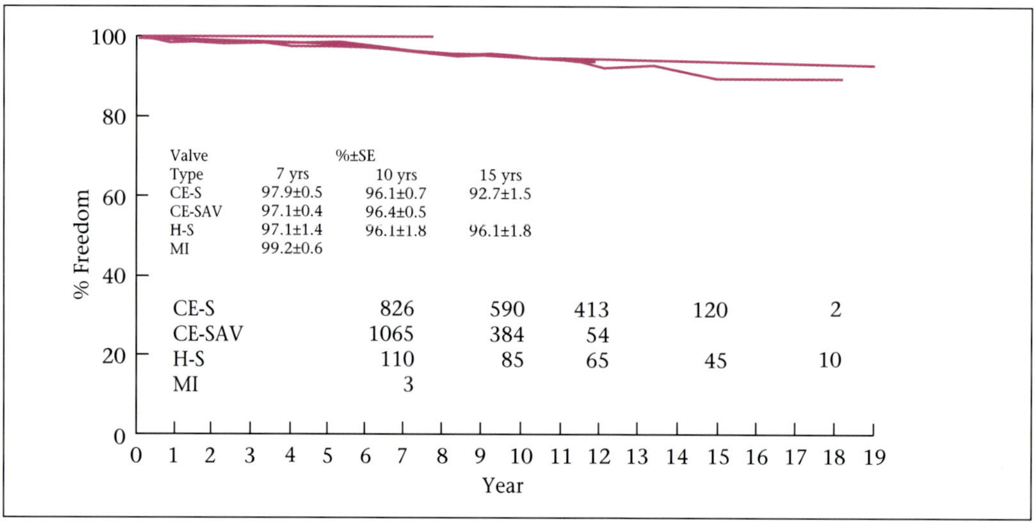

Figure 3.2. Freedom from fatal thromboembolism: overall.

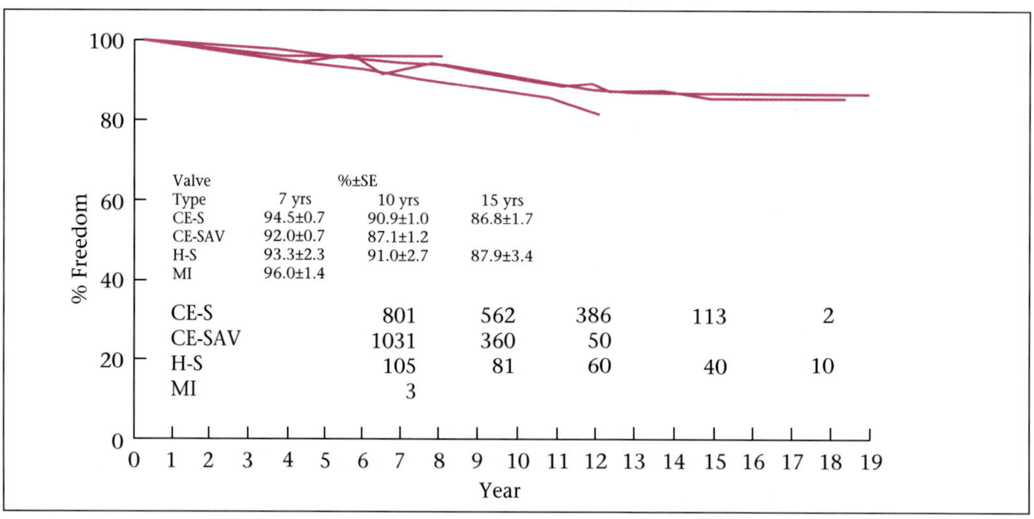

Figure 3.3. Freedom from major thromboembolism and antithromboembolic haemorrhage.

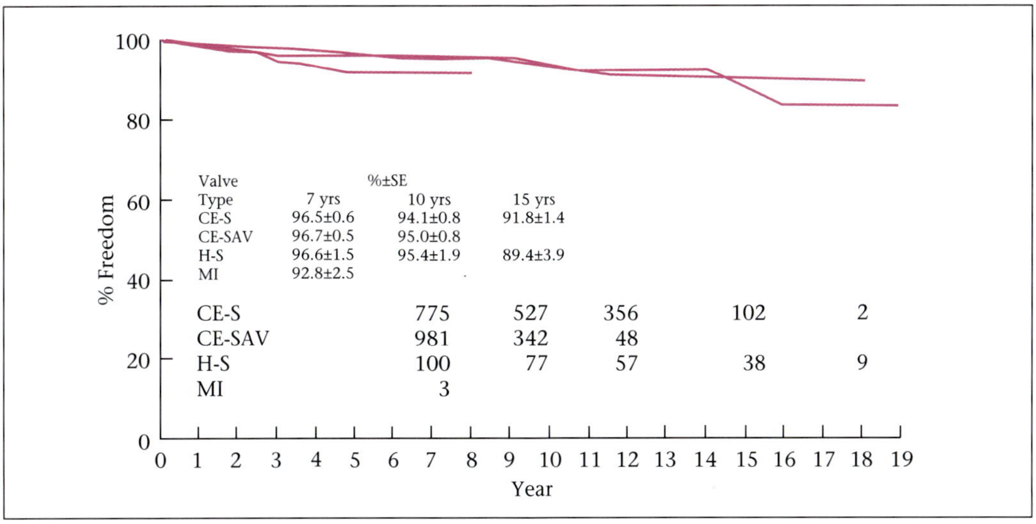

Figure 3.4. Freedom from prosthetic valve endocarditis.

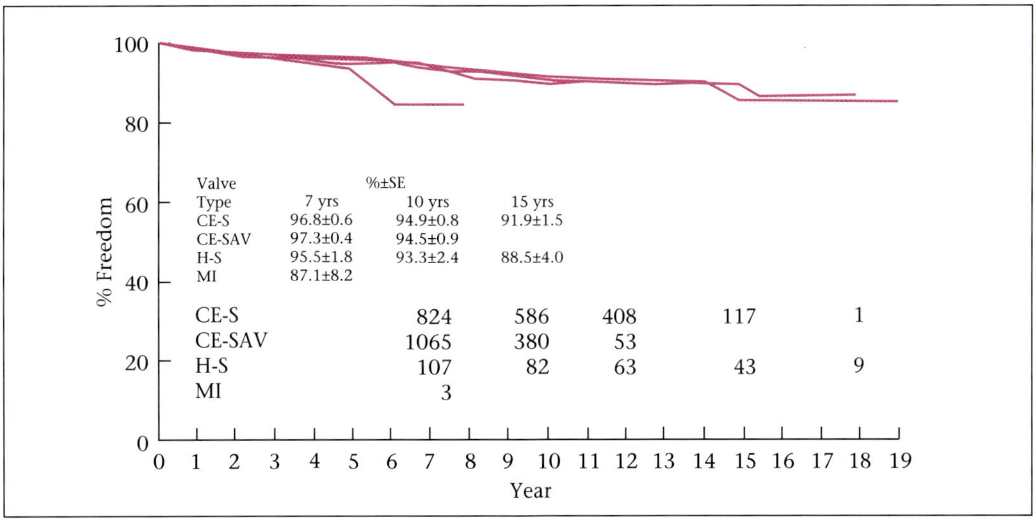

Figure 3.5. Freedom from non-structural dysfunction.

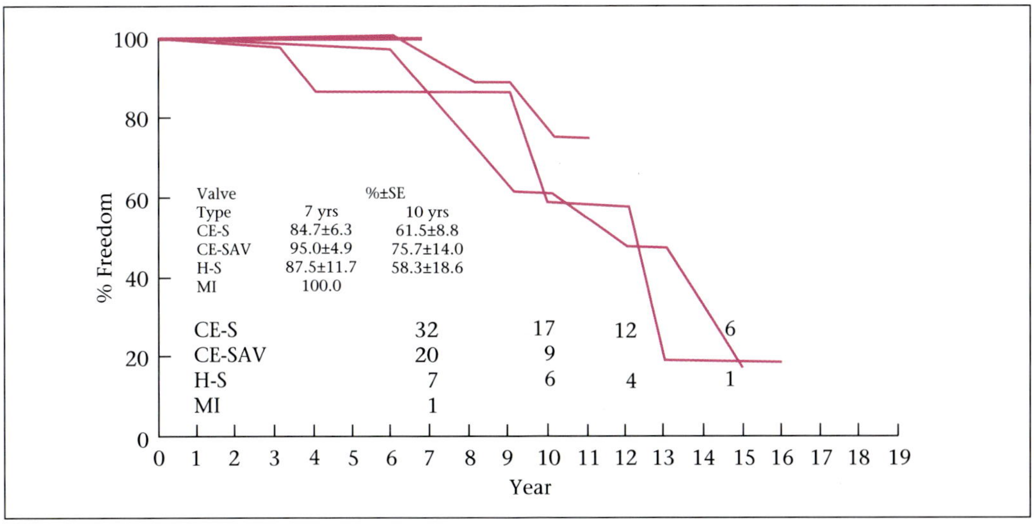

Figure 3.6. Freedom from structural valve deterioration: aortic valve replacement p (≤ 35 years).

the freedom from SVD, at 10 years, was the same for the CE prostheses, 84.8% for CE-S and 84.3% for CE-SAV, while for H-S it was 90.1% (Figure 3.8). For the CE groups for AVR at 10 years, for 65–69 years the freedom from SVD was also similar — 95.6 (CE-S) and 96.2% (CE-SAV) (Figure 3.9). For the age group ≥70 years there was essentially complete freedom from SVD for all prostheses — 100% for MI at 6 years, 100% for H-S and CE-S at 10 years and 99.4% for CE-SAV at 10 years (Figure 3.10).

The mitral prostheses' freedom from SVD is shown in Figures 3.11–3.15. The freedom for the ≤35 years group, at 10 years, was 51.4% for CE-S and 50.8% for CE-SAV (Figure 3.11). For the 36–50 years group the freedom, at 10 years, was 64.1% for CE-S and 70.3% for CE-SAV (Figure 3.12). The freedom from SVD was also similar in the 51–64 years group, at 10 years, 73.0 (CE-S) and 71.6% (CE-SAV) (Figure 3.13). In the 65–69 years group the freedom from SVD for MVR, at 10 years, was 66.6% for CE-S and 74.0% for CE-SAV (Figure 3.14).

The prostheses performed similarly in the ≥70 years group: the freedom from SVD for MVR at 10 years was 91.5% for CE-S and 90.4% for CE-SAV (Figure 3.15).

The freedom from residual morbidity or permanent impairment was, at 10 years, 95.0 (CE-S), 91.8 (CE-SAV) and 92.6% (H-S) and, at 7 years, 96.2% (MI) (Figure 3.16). At 15 years the freedom was 94.0% for CE-S and 89.4% for H-S.

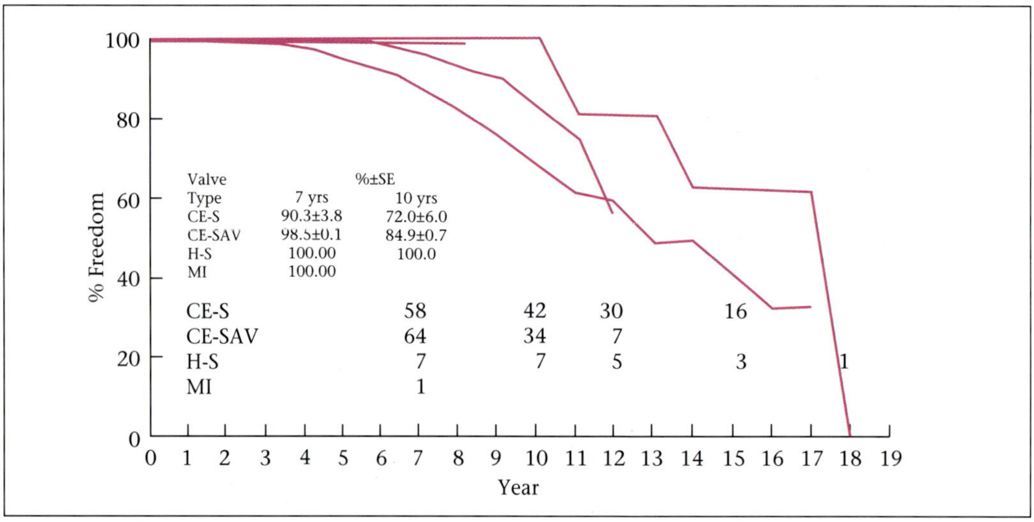

Figure 3.7. Freedom from structural valve deterioration: aortic valve replacement (36–50 years).

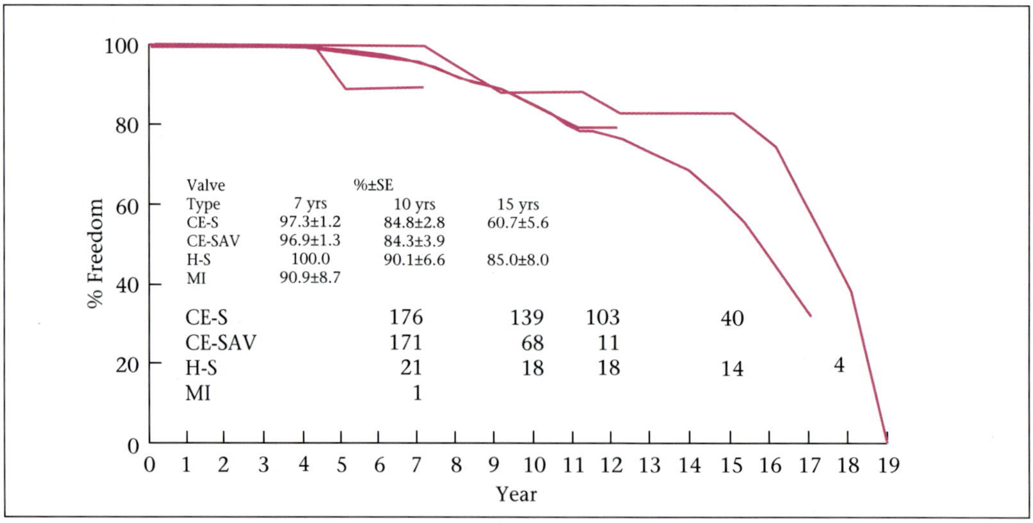

Figure 3.8. Freedom from structural valve deterioration: aortic valve replacement (51–64 years).

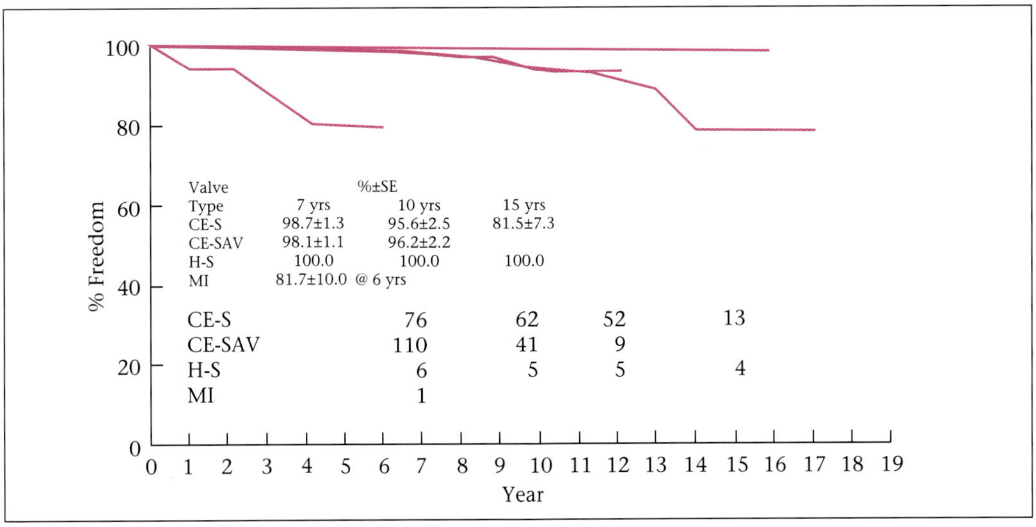

Figure 3.9. Freedom from structural valve deterioration: aortic valve replacement (65–69 years).

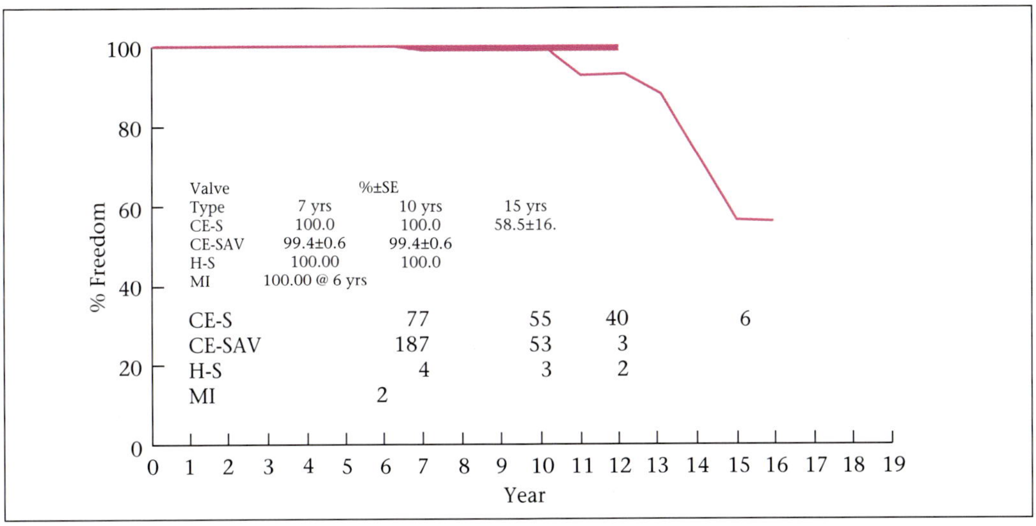

Figure 3.10. Freedom from structural valve deterioration: aortic valve replacement (≥ 70 years).

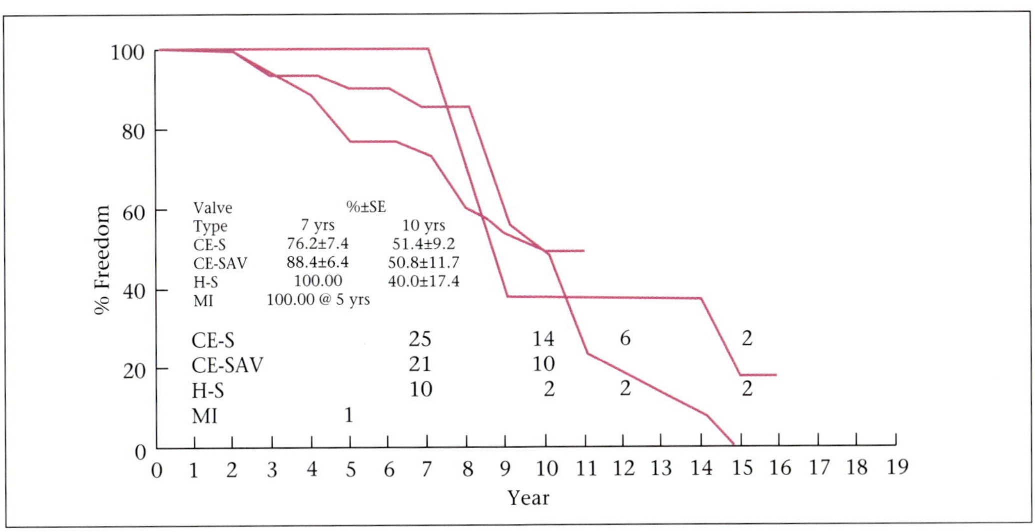

Figure 3.11. Freedom from structural valve deterioration: mitral valve replacement (≤ 35 years).

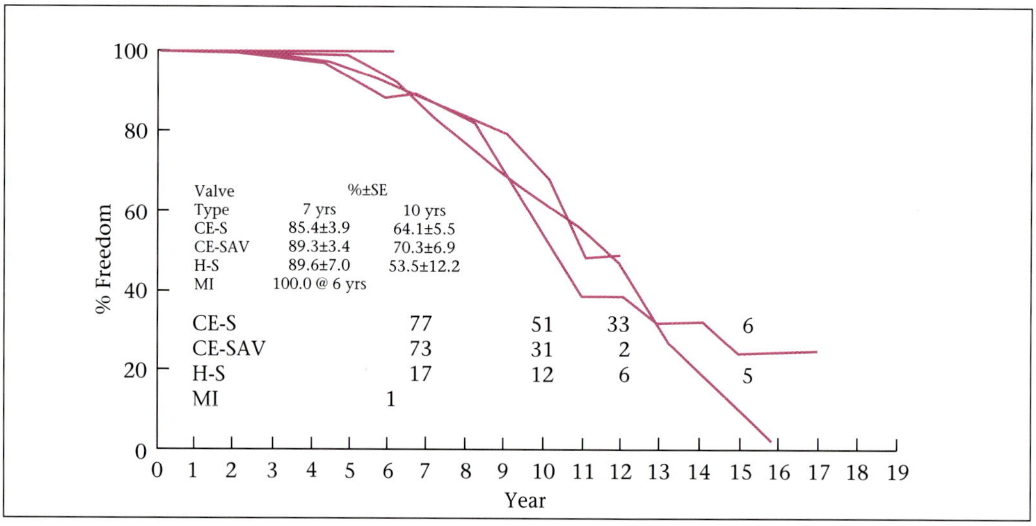

Figure 3.12. Freedom from structural valve deterioration: mitral valve replacement (36–50 years).

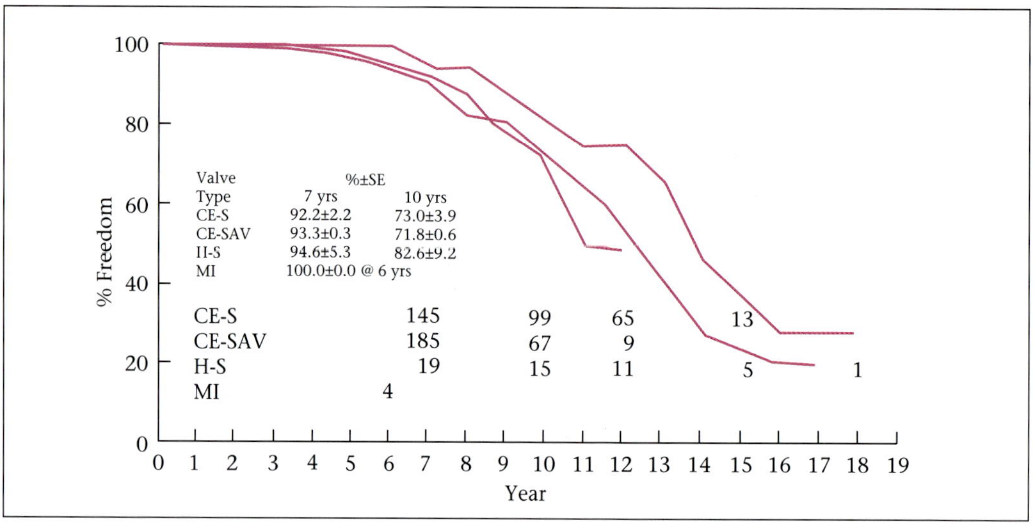

Figure 3.13. Freedom from structural valve deterioration: mitral valve replacement (51–64 years).

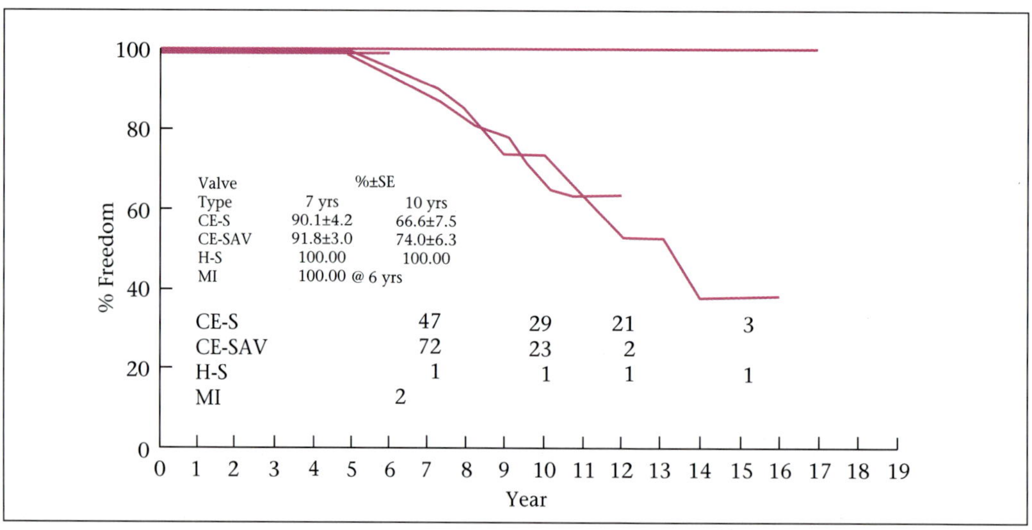

Figure 3.14. Freedom from structural valve deterioration: mitral valve replacement (65–69 years).

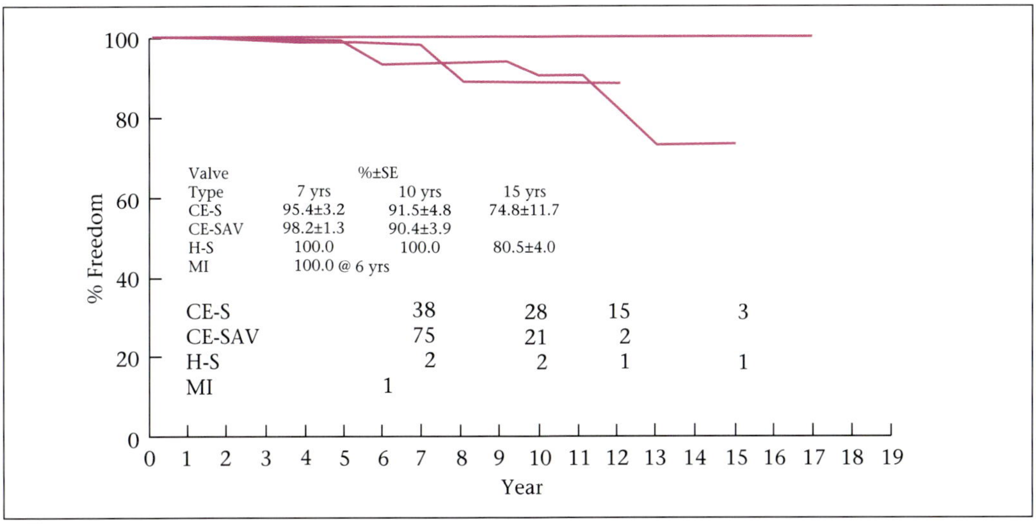

Figure 3.15. Freedom from structural valve deterioration: mitral valve replacement (≥ 70 years).

The freedom from valve-related reoperation (REOP) by valve position is shown in Figures 3.17–3.19. The freedom from REOP for AVR was 81.9 (CE-S), 80.5 (H-S) and 87.6% (CE-SAV) (Figure 3.17); that for the MI prostheses, at 7 years, was 89.6%, similar to the other prostheses. For MVR at 10 years the freedom from REOP was 65.4 (CE-S), 58.9 (H-S) and 68.5% (CE-SAV) and, at 6 years, 95.2% (MI) (Figure 3.18). For multiple replacements (MR), the freedom from REOP was, at 10 years, 80.9 (H-S), 67.5 (CE-S) and 49.3% (CE-SAV) and, at 6 years, 93.9% (MI) (Figure 3.19).

The freedom from all valve-related complications (including morbidity and mortality), at 10 years, for the overall population with each prosthesis is shown in Figure 3.20. The freedom for CE-S was 57.1%, for H-S it was 56.5% and, for CE-SAV, 57.6%.

Discussion

Structural valve deterioration of stented porcine bioprostheses has been manifested by calcification and tears and perforations of the glutaraldehyde-preserved tissue. Calcification demonstrated at the time of explanation for failure

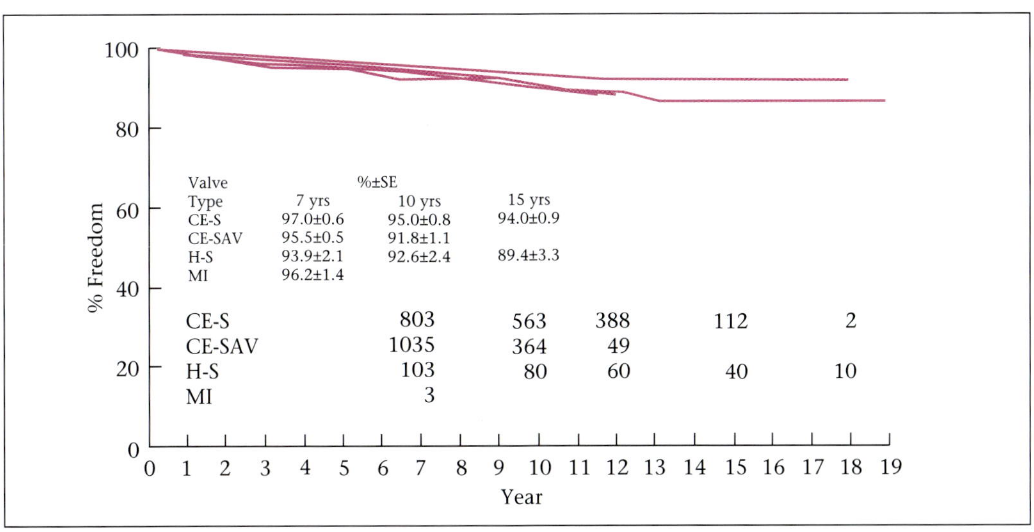

Figure 3.16. Freedom from residual morbidity: overall.

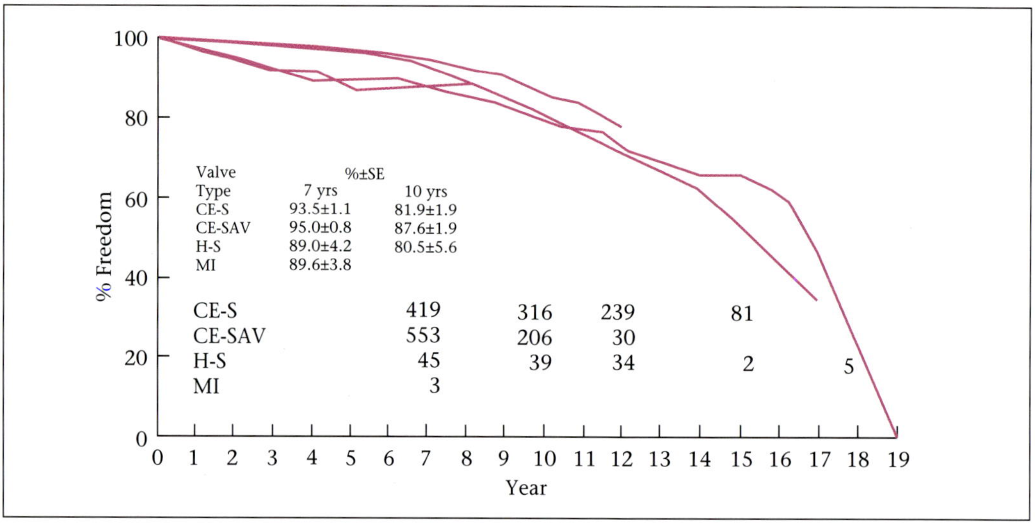

Figure 3.17. Freedom from valve-related reoperation: aortic valve replacement.

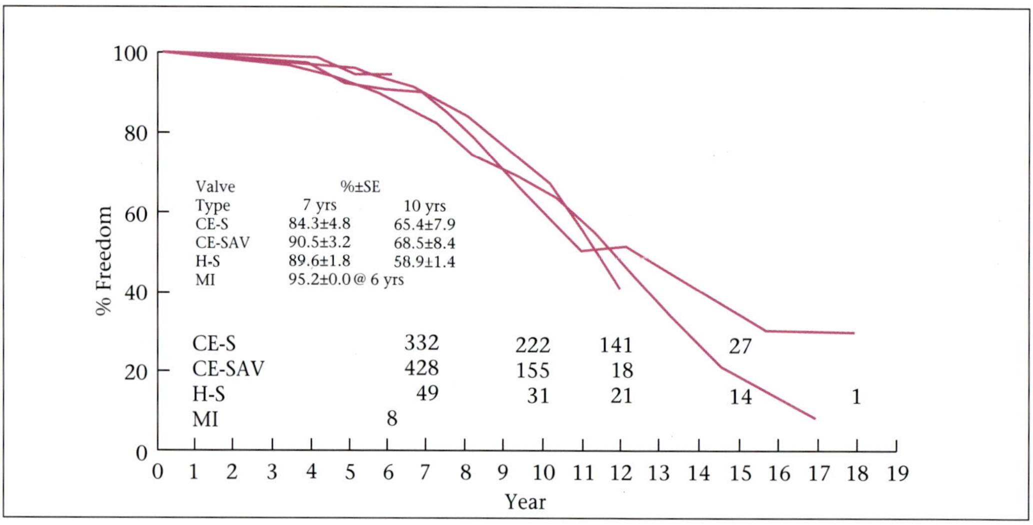

Figure 3.18. Freedom from valve-related reoperation: mitral valve replacement.

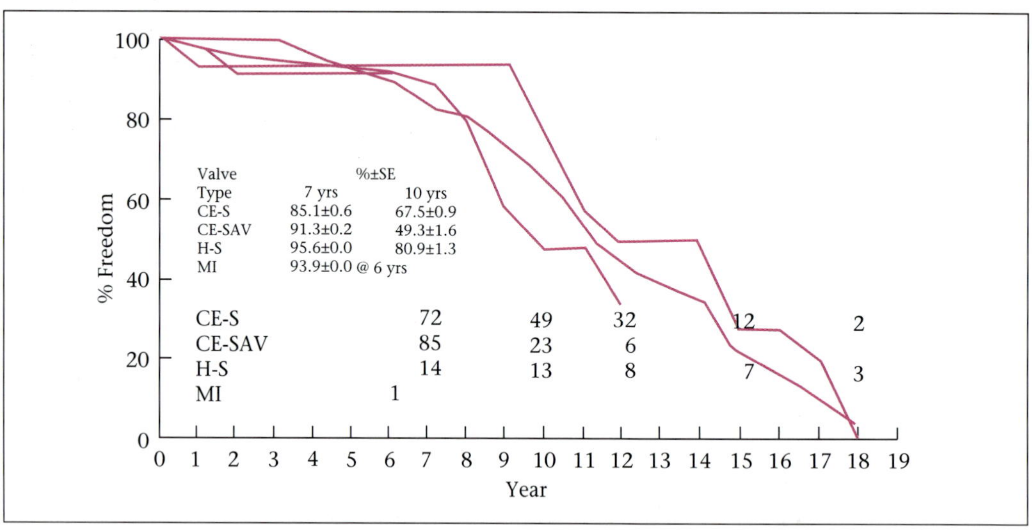

Figure 3.19. Freedom from valve-related reoperation: multiple valve replacement.

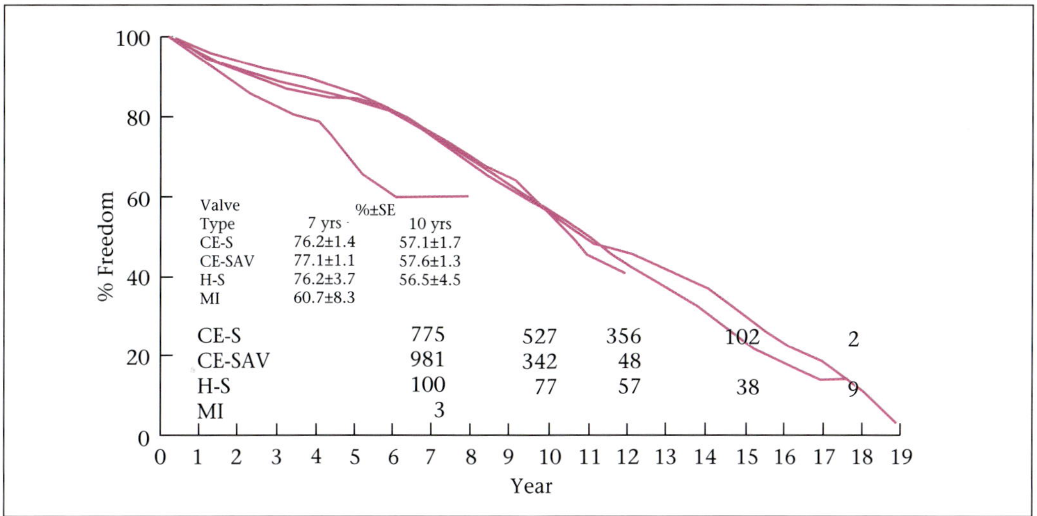

Figure 3.20. Freedom from valve-related complications: overall.

was either primary or accompanied by secondary tears. Tears and perforations occurring without the presence of calcification are considered to be related to stress-related fatigue injuries of the glutaraldehyde-preserved porcine tissue. The potential of the stentless bioprosthesis to reduce structural valve deterioration is currently under investigation.

Age has been demonstrated since 1988 to be the predominant risk factor for structural failure of porcine bioprostheses.[3] Elderly age groups have demonstrated a significantly greater freedom of structural valve deterioration to 15 years, much greater in the aortic than in the mitral position.[8] The role of the stent with biological tissue has been reported clinically by Angell and colleagues:[10] age correlated with structural valve deterioration with stented homografts but not with stentless homografts. The stentless porcine bioprostheses may also contribute to greater durability than stented prostheses for aortic valve replacement.

The extended experience with porcine bioprostheses has been documented from Padova University, Stanford University and the University of British Columbia.[1–9] Bortolotti and colleagues[1] reported on the long-term experience of the Hancock standard porcine bioprostheses in 665 patients with a mean age of 47 years, demonstrating a 4.3% per patient-year incidence of structural valve deterioration. The Stanford experience with the Hancock standard was last reported by Burdon and co-authors[2] in 1650 patients with a mean age of 57 years. The freedom from structural failure for aortic valve replacement was 85% at 10 years and 63% at 15 years, while for mitral valve replacement it was 78 and 45%, respectively. The Carpentier–Edwards standard experience from the University of British Columbia in 1195 patients, also with a mean age of 57 years, revealed a 15-year freedom from structural valve deterioration of 71% for aortic valve replacement and 41% for mitral valve replacement.[6] The 10-year experience was previously reported, with 10-year freedom of 78.6 and 71.6%, respectively, for aortic and mitral valve replacement.[4]

The first-generation porcine bioprostheses were formulated with high-pressure fixation (>60 mmHg) with glutaraldehyde; the second-generation Carpentier–Edwards supra-annular porcine bioprosthesis has tissue fixation at less than 2 mmHg. The Hancock II, also a second-generation prosthesis, is fixed at low pressure for 30 min followed by high-pressure preservation. The Medtronic Intact™ porcine bioprosthesis has the tissue pressure-free fixed with glutaraldehyde. There have been some trends to suggest that the new-generation bioprostheses may be experiencing a reduction in structural valve deterioration. David and colleagues,[11] reporting on the 8-year experience with the Hancock II in 614 patients, demonstrated an overall freedom from structural valve deterioration of 92%. Jamieson and colleagues[5] reported a similar performance with the Carpentier–Edwards bioprostheses at 7 years: the freedom from structural failure for aortic valve replacement was 96.1% for the standard prosthesis and 97.6% for the supra-annular prosthesis, whereas for mitral valve replacement it was 89.1 and 88.6% respectively. The current evaluation at 10 years revealed that the freedom for aortic valve replacement was 77.6% for the standard prosthesis and 80.5% for the supra-annular prosthesis, whereas for mitral valve replacement it was 70.1 and 72.6%, respectively.[9] These experiences demonstrate no appreciable difference between the first- and second-generation bioprostheses with regard to structural valve deterioration. The reports by Barratt-Boyes and colleagues[12–14] provide an early demonstration that the pressure-free glutaraldehyde-fixed Medtronic Intact™ prosthesis may have a reduced incidence of structural failure.

The current reports in this communication provide information permitting several conclusions to be drawn. The Carpentier–Edwards supra-annular prosthesis in the aortic position may have a lower incidence of structural failure than the standard prosthesis in patient groups below the age of 50 years. The freedom from structural valve deterioration over the age of 65 years was 96% and over the age of 70 years was 100%, at 10 years, for aortic valve replacement. Stentless porcine bioprostheses may not confer extended durability for those patients over 65 years of age where the freedom from prosthesis structural failure is already high. The effect may be most obvious in the 51–64 years group, where the current freedom from structural failure at 10 years is 84.8%. The clinical performance of stented bioprostheses with regard to structural valve deterioration provides the opportunity for new stented and stentless technologies to extend durability.

References

1. Bortolotti U, Milano A, Mazzucco A, Guerra A, Testolin L, Thiene G, Gallucci V 1991 Extended follow-up of the standard Hancock porcine bioprosthesis. J Card Surg 6(Suppl 4): 544–549
2. Burdon T A, Miller D C, Oyer P E, Mitchell R S, Stinson F B, Starnes V A, Shumway N E 1992 Durability of porcine bioprostheses at fifteen years in a representative North American patient population. J Thorac Cardiovasc Surg 103(2): 238–252
3. Jamieson W R E, Janusz M T, Miyagishima R T, Munro A I, Gerein A N, Burr L H, Tyers G F O 1988 Carpentier–Edwards standard porcine bioprostheses — primary tissue failure (structural valve deterioration) by age groups. Ann Thorac Surg 46: 155–162
4. Jamieson W R E, Allen P, Miyagishima R T, Gerein A N, Munro A I, Burr L H, Tyers G F O 1990 The Carpentier–Edwards standard porcine bioprosthesis — a first-generation tissue valve with excellent long-term clinical performance. J Thorac Cardiovasc Surg 99: 543–561
5. Jamieson W R E, Tyers G F O, Miyagishima R T, Germann E, Janusz M T, Ling H 1991 Carpentier–Edwards porcine bioprostheses — comparison of standard and supra-annular prostheses at 7 years. Circulation 84(5): (Suppl III) 145–151
6. Jamieson W R E, Hayden R I, Miyagishima R T, Tutassaura H, Munro A I, Gerein A N, Burr L H, MacNab J, Janusz M T, Chan F 1991 The Carpentier–Edwards standard porcine bioprosthesis: clinical performance to 15 years. J Card Surg (Suppl 4): 550–556
7. Jamieson W R E, Tyers G F O, Janusz M T, Miyagishima R T, Munro A I, Ling H, Burr L H, Tutassaura H 1991 Age as a determinant for selection of porcine bioprostheses for cardiac valve replacement: experience with Carpentier–Edwards standard bioprosthesis. Can J Cardiol 7(4): 181–188
8. Burr L H, Jamieson W R E, Munro A I, Miyagishima R T, Janusz M T, Ling H, Hayden R I, Tutassaura H, Fradet G, Gudas V M, Tyers G F O 1992 Structural valve deterioration in

elderly patient populations with the Carpentier–Edwards standard and supra-annular porcine bioprostheses. J Heart Valve Dis 1(1): 87–91
9. Jamieson W R E, Burr L H, Tyers G F O, Munro A I 1994 Carpentier–Edwards standard and supra-annular porcine bioprostheses: ten year comparison of influence of structural valve deterioration on valve performance. J Heart Valve Dis 3(1): 59–65
10. Angell W W, Pupelloo D F, Bessone L N, Hiro S P, Brock J C 1991 Effect of stent mounting on tissue valves for aortic replacement. J Cardiac Surg 6(Suppl 4): 595–599
11. David T E, Armstrong S, Sun Z 1992 Clinical and hemodynamic assessment of Hancock II bioprosthesis. Ann Thorac Surg 54: 661–668
12. Barratt-Boyes B G, Jaffe W M, Ko P H, Whitlock R M 1993 The zero pressure fixed Medtronic Intact porcine valve: an 8.5 year review. J Heart Valve Dis 2(5): 604–611
13. Barratt-Boyes B G 1992 Clinical experience with the zero-pressure fixed Medtronic Intact bioprosthetic valve. Eur J Cardiothorac Surg 6(Suppl 1): 579–581
14. Barratt-Boyes B G, Ko P H, Jaffe W M 1991 The zero pressure fixed Medtronic Intact porcine valve: clinical results over a 6-year period, including serial echocardiographic assessment. J Cardiac Surg 6(4 Suppl): 606–612

Chapter 4

Pericardial bioprostheses: an overview

F. Sebening and A. Schroll

Pericardial bioprostheses (PB) are valve substitutes that are usually constructed from bovine pericardium; according to the 1979 Munich Terminology meeting they are generally termed xenobioprostheses.

History

The first handmade specimens were implanted in March 1971 by Marian Ionescu at the General Infirmary, Leeds, UK, and from May 1976 they were available commercially, manufactured by Shiley and other companies. Together with the improved Ionescu–Shiley Low Profile valve (1981), the Hancock (1981–1985), the Mitroflow™ (March 1982) and the Carpentier–Edwards (since 1982), they belong to the first generation of PB, all being similar in design. They were often modified and some have been available until recently. Both Shiley PB were withdrawn in September 1988.

In 1982 the Meadox–Gabbay unicusp PB was presented. Its aim was to obtain a maximal orifice, but the unicusp principle has proved unsuitable for a cardiac valve substitute, owing to excessive stretching of the single leaflet with subsequent valve incompetence.

Another PB is the Bioflow™ valve, designed by D. Wheatley of Glasgow and produced by a Scottish manufacturer (1988). It was not intended for use in small sizes and showed a satisfactory maximal 5-year follow up; however, it has now been withdrawn as a result of bovine spongiform encephalopathy in English cattle.

The third and newest generation of PB is represented by the Pericarbon™ valve, introduced to the market by Sorin in 1989. This is constructed from two different pericardial sheets and shows some technical improvements. Apart from its short follow-up, the problem of calcification has not yet been overcome with this valve.[1]

Characteristics of PB

All PB, currently or formerly available, have only one characteristic in common — they are made from the same xenogeneic material, namely bovine pericardium from 6- to 18-month-old calves, which is fixed with glutaraldehyde and mounted on a stent by sutures.[2] Apart from this, the various types of PB have many differences and characteristic features, including, for example:

Design: one (Mitroflow™), two (Pericarbon™) or three (Hancock) separate pericardial sheets;
Stent: rigid, flexible, round or elliptical, with or without splay and pledgets;
Orifice: rigid or flexible;
Profile: standard or low profile;
Fixation: (a) with or without pressure, (b) before or after mounting on the stent, (c) with 0.2, 0.5 or 0.625% glutaraldehyde, (d) with or without a calcium-retarding process;
Ring: Dacron or a softer material;
Thickness: 0.25mm (Hancock), 0.5mm (Hancock and Edwards), 0.7mm (Mitroflow™);
Gap at the centre: small or large;
Storage: in glutaraldehyde or formaldehyde.

Thus, it is obvious that, compared with porcine bioprostheses, the variations between the numerous types of PB offered by the manufacturing companies are much greater.

New types of PB

Stentless bioprostheses

Stentless bioprostheses are a new development in the 1990s and there is also one type made from bovine pericardium — the Pericarbon™ Stentless Valve, which has been produced by Sorin since 1992. It is designed from two pericardial sheets: one is shaped to form the three leaflets, while the other lines the outflow side. All stentless bioprostheses are provided as an alternative for aortic valve replacement. They show a larger effective orifice area than stented bioprostheses: this has a particular advantage where there is a small aortic root, because the obstruction of the transvalvular forward flow and the velocity across the valve is less.[4] First results with a short follow-up of about 2 years have demonstrated a satisfactory haemodynamic performance.[3]

Autogenics™ system

Another innovative concept was presented by Dr Love with the Autogenics™ system (1990). A patch of autogenous pericardium is harvested immediately after the patient's chest has been opened. It is fixed by brief immersion in 0.6% glutaraldehyde and with a special kit appropriate to the patient's valve size, the pericardium is cut to a precise geometry, mounted on a prepared stent and tested for function. It remains dubious whether the resulting longer operation times will be acceptable or whether the advantages claimed for autogenous over xenogenic pericardium will be confirmed.

Advantages of PB

PB have been manufactured for more than 20 years for heart valve raplacement. Their advantages over porcine bioprostheses, relating to their superior haemodynamic properties, have been confirmed. Haemodynamic flow measurements *in vitro* have shown that the pressure gradients, even under high flow rates, are much lower than those of porcine valves. The opening areas are greater, resulting in better flow profile conditions and lower flow velocities across the valve orifice. The generally lower flow resistance is a major advantage for valves with small diameters between 15 and 19 mm, especially in small aortic roots.[4] PB are noted for their lower rate of thromboembolic events, with or without anticoagulants and they also present lower rates of endocarditis; both of these advantages are probably due to their superior haemodynamic properties.

Pericardium

In creating a valve device, the predominant aim should be to copy the human original. Apart from an allograft, a porcine trileaflet aortic valve is nearest to the ideal, even if structurally altered by processing. Any other valve, constructed from another collagenous material such as pericardium, obviously must be further away from the original. The manufacturers producing different types of PB make the implicit assumption that the material is isotropic. Because it is very flexible it seems to have the properties of a homogeneous membrane. When pericardium is stretched, however, it becomes very non-compliant and its stiffness increases over 100-fold, a sign of its partly anisotropic character. This also occurs when pericardium is stretched around a stent, thus reducing its compliance.

Anatomically, pericardium differs greatly from a natural valve leaflet, although both consist mainly of collagen. The pericardium presents dense collagen bundles parallel to the surface with occasional blood vessels, termed fibrosa. Only one surface is covered by mesothelial cells, termed serosa. This surface, analogous to porcine endothelium, which subsequently is destroyed during processing, then becomes the smooth outflow side of the PB. The collagen fibres vary in diameter and the variable thickness of the pericardium generally exceeds by far that of any natural valve leaflet. The free margins of a PB are cut surfaces and expose open collagen structures.[5]

Follow-up of PB

During the 1970s and early 1980s, PB, like porcine bioprostheses, were used worldwide for aortic, mitral and tricuspid valve replacement. The first short-term follow-ups of PB were very encouraging, because PB not only were equal to their porcine competitors but also possessed superior haemodynamic properties (especially in the small sizes) and gave rise to fewer thromboembolic problems and lower rates of endocarditis. However, subsequent long-term results with PB were clearly inferior: structural failure occurred earlier and with a greater incidence. Furthermore, deteriorated PB tend to break down acutely and unpredictably, often creating an emergency for the patient.[6,7]

For these reasons PB could no longer be recommended generally and their use declined rapidly. This applied mainly to the Hancock and Shiley PB, whereas the Edwards and Mitroflow™ valves, which had a superior performance in wear testing, were not withdrawn from the market. There were also differences in durability depending on the valve position: the mitral position gave worse results than the aortic or tricuspid positions. Freedom from valve-related degeneration improved with increasing age of the patients. Primary tissue failure is the dominant cause (about 60%) of structural valve deterioration in PB, depending on 'non-calcific' degenerative processes independent of any calcium uptake. Large leaflet tears in PB have often been reported as a result of primary tissue failure up to about 7 years after implantation. The tears manifest mainly at the flexion zone near the attachment to the stent, affecting one or more leaflets.[8,9]

Intrinsic calcification, mostly without but occasionally with secondary tearing, is another but less frequent failure mode of PB (about 36%). The mode of primary valve failure varies from one bioprosthesis to another, obviously owing to physiological differences in the pericardium used, with heterogeneous mechanical properties and durability.[10]

On the other hand, porcine valves mostly fail as a consequence of their calcification, with subsequent tears. It is also easy to understand that the thinner porcine leaflets are more resistant to mechanical stress and flexion than the thicker pericardial cusps.

The situation today

Analysis of the literature shows generally that the early enthusiasm for the PB has dwindled over the last decade. The intial teams to abandon the use of PB did so 10 years ago, judging them as unsatisfactory or unacceptable. Today, only a few teams worldwide continue to use and/or to recommend PB for particular indications, such as aortic (sometimes also tricuspid) valve replacement in the elderly patient of 70 years and over, where long-term durability is less important, and also where there is a small aortic root, from 15 to 19 or 21mm in diameter. In cases where the aortic root is of larger diameter, there is no haemodynamic or any other relevant advantage to justify the choice of PB.

Whereas the authors have implanted many thousands of porcine bioprostheses, their personal experience with PB is limited to one single patient. The three PB remaining on the market today — the Carpentier–Edwards, the Mitroflow™ and the Pericarbon™ — can therefore be regarded as dying species. All efforts to improve bioprostheses should now be directed towards the porcine bioprostheses. Their calcification is the most important problem, which must be solved.

References

1. Thiene G, Laborde F, Valente M, Gallix P, Talenti E, Calabrese F, Piwnica A 1989 Morphological survey of a new pericardial valve prosthesis (Pericarbon): long-term animal experimental model. Eur J Cardio-Thorac Surg 3: 65
2. Valente M, Minarini M, Ius P, Talenti E, Bortolotti U, Milano A, Thiene G 1992 Durability of glutaraldehyde-fixed pericardial valve prostheses: clinical and animal experimental studies. J Heart Valve Dis 1: 216
3. Casabona R, De Paulis R, Zattera G F, di Summa M, Botlone W, Stacchino C, Vrandecic MO, Morea M 1992 Stentless porcine and pericardial valve in aortic position. Ann Thorac Surg 54: 681
4. Reul H, Gersiepen M, Schindehütte H *et al* 1986 Comparative in vitro evaluation of porcine and pericardial bioprostheses. Z Kardiol (Suppl 2) 75: 223
5. Frater R W M, Bodnar E 1992 Towards understanding the pericardium as valve substitute. J Heart Valve Dis 1: 213
6. Christie G W, Stephenson R A 1990 Stress related failure modes of bovine pericardial heart valves. In: Bodnar E (ed) Proc. 1989 Symposium, London: Surgery for heart valve disease. ICR, London, p765
7. Schoen F J, Fernandez J, Gonzales-Lavin L, Cernaianu A 1987 Causes of failure and pathologic findings in surgically removed Ionescu–Shiley standard bovine pericardial heart valve bioprostheses: emphasis on progressive structural deterioration. Circulation 76: 618
8. Minami K, Schereika S, Körte H *et al* 1993 Long-term follow up of Mitroflow pericardial valve prostheses in the small aortic annulus. J Cardiovasc Surg 34: 189
9. Thowbridge E A, Lawford P V, Crofts C E *et al* 1988 Pericardial heterografts: why do these valves fail? J Thorac Cardovasc Surg 95: 577
10. Goffin Y A, Bartik M A, Hilbert S L 1986 Porcine aortic vs bovine pericardial valves: a morphologic study of the Xenomedica and Mitroflow bioprostheses. Z Kardiol (Suppl 2) 75: 213

II Stentless Valve Technology

Chapter 5

Toronto SPV™ valve: the engineer's standpoint

W. Mirsch

The Toronto SPV™ valve is an intact stentless porcine valve for use in subcoronary implantation. In 1987, Dr Tirone David developed the concept for the valve at the University of Toronto and the Toronto Hospital and implanted the first valves in 1988.[1] St Jude Medical, Inc. began collaborating with Dr David in 1990 and the first human implant of a St Jude Medical, Inc. manufactured valve took place in 1991. The FDA has recently approved an IDE submission and clinical studies in the USA are now starting.

Design

The stentless design of the Toronto SPV™ valve offers a significant advantage in sizing over traditional stented bioprostheses. By removing the sewing cuff and the stent, it is possible to implant a porcine aortic valve that is two to three sizes larger than correspondingly sized stented valves (Figure 5.1).

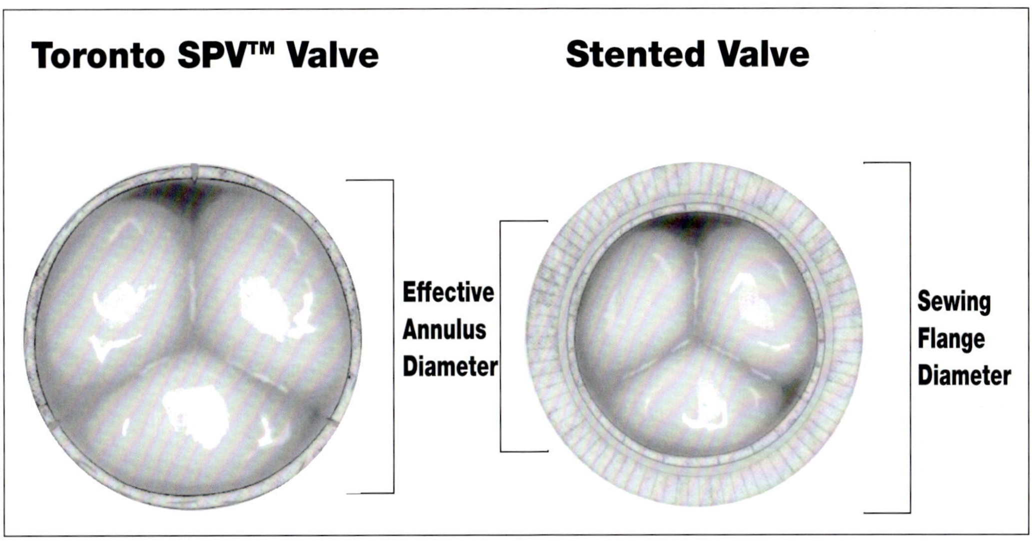

Figure 5.1. The Toronto SPV™ valve, showing difference in size from a stented valve.

The Toronto SPV™ valve has all three sinuses trimmed down to yield a bioprosthesis with similar implantation techniques to subcoronary freehand homograft valves (Figure 5.2). This design utilizes the native aorta's compliance to ensure proper valve function. During systole, the aorta expands and allows the valve to open more completely. This takes up slack in the leaflets, which decreases excess stress in leaflet tissue near the commissures. This stress causes collagen

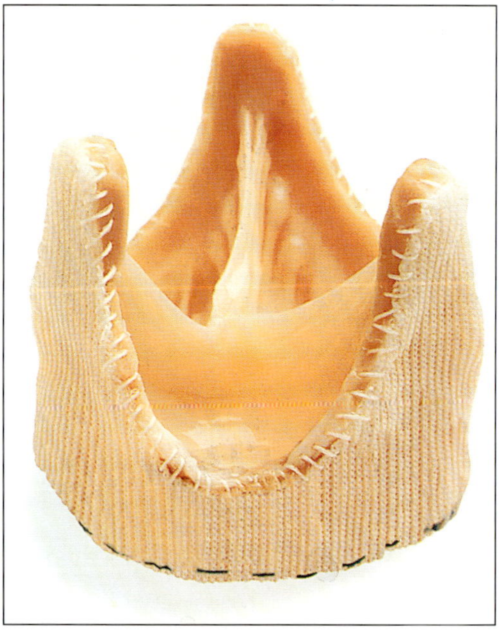

Figure 5.2. Toronto SPV™ valve, showing trimmed sinuses.

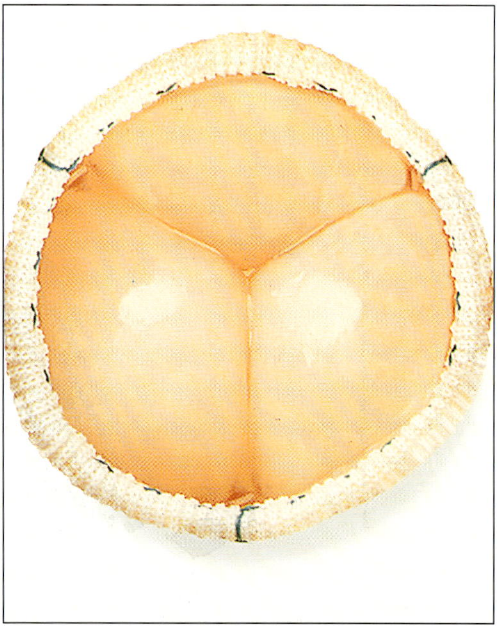

Figure 5.3. Toronto SPV™ valve, showing non-scalloped inflow edge.

damage and has been implicated in calcification and wear of stented bioprosthetic valves.[2]

The thin Dacron fabric that covers the outside of the prosthesis facilitates suturing and helps incorporate the valve into the aortic wall. Should the Toronto SPV™ valve ever need to be explanted, the fabric serves as a dissection plane and aids in complete removal of the valve with preservation of the tissue annulus. Unlike other valves, the Toronto SPV™ valve retains the surgeon's options for reimplantation with any other type of heart valve prosthesis.

A non-scalloped inflow edge accommodates normal or atypical anatomies (Figure 5.3). This eases the insertion and orientation of the valve. The small septal muscle shelf of the Toronto SPV™ valve provides improved haemodynamic performance.

The Toronto SPV™ valve has large leaflet coaptation areas. This reduces commissural stress levels and ensures complete valve closure, which minimizes the risk of aortic insufficiency (Figure 5.4). It can also reduce the incidence of central insufficiency due to technical error.

The Toronto SPV™ valve is preserved using low-pressure fixation with a glutaraldehyde solution. The crosslinking of the Toronto SPV™ valve maintains collagen crimp, enhances durability and provides for excellent haemodynamic properties.

Testing

The Toronto SPV™ valve has undergone extensive testing *in vitro* and *in vivo* to produce data for use in submissions to regulatory agencies. Tests *in vitro* include steady state and pulsatile flow testing for hydrodynamic performance. In addition, the Toronto SPV™ valve has been tested on accelerated wear testers for more than 200 million cycles to show durability.

The Toronto SPV™ valve been evaluated *in vivo* in both sheep and calf implant models to study haemodynamic performance, as well as durability. In the calf implants, a 25 mm Toronto SPV™ valve is implanted into an 8-week-old calf, weighing approximately 100 kg. Echographic and angiographic studies are

Figure 5.4. Toronto SPV™ valve, showing large leaflet coaptation areas.

performed at monthly intervals to examine valve performance. The calves grow to approximately double their weight at implant during the course of the study. Results to date have indicated that the Toronto SPV™ valve can withstand the added wear exerted by the increased cardiac output of the growing animal, without valve dysfunction.

Clinical trials

Clinical trials are being conducted on a worldwide basis. Early indications from follow-up data are that the Toronto SPV™ valve is performing extremely well.[3,4]

Measurements of transvalvular gradients at postoperative, 3–6 month and 11–14 month follow-up periods all indicate mean systolic gradients averaging below 10 mmHg (Figure 5.5). An interesting trend is that the gradients are falling over time.

Calculations of the effective orifice areas during the same follow-up periods show the improvement in overall haemodynamic performance of the Toronto SPV™ valve, during the first year of implantation (Figure 5.6).

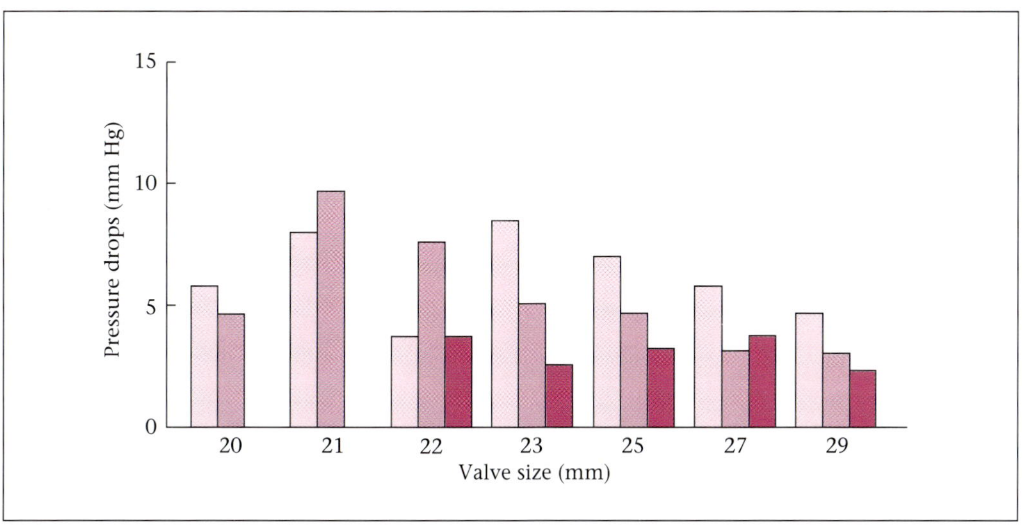

Figure 5.5. Mean systolic transvalvular gradients with the Toronto SPV™ valve at postoperative (□), 3–6-month (■) and 11–14-month (■) follow-up periods.

There has been considerable speculation as to the cause of these haemodynamic improvements in stentless valves after implantation. Dr David has postulated that a restructuring of the host aorta occurs when an inefficient valve is replaced with a stentless bioprosthesis. This may be reasonable, as the mechanics of the Toronto SPV™ valve closely approximate those of a healthy aortic valve.

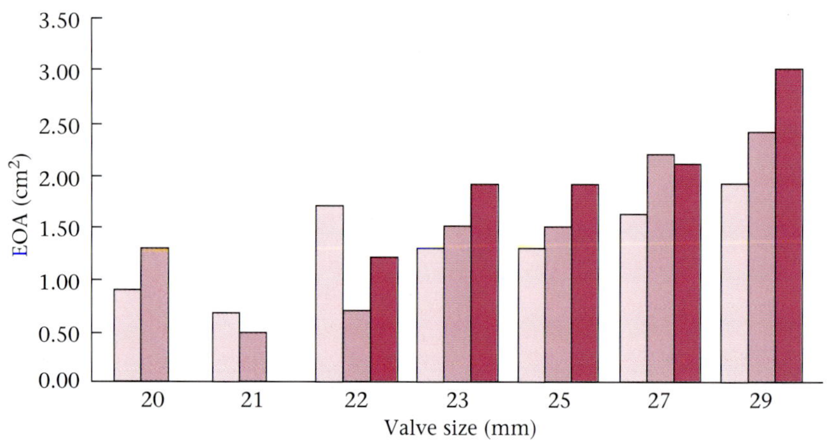

Figure 5.6. Effective orifice areas (EOA) of the Toronto SPV™ valve at postoperative (□), 3–6-month (■) and 11–14-month (■) follow-up periods.

The incidence of regurgitation of the Toronto SPV™ valve has been very low in the clinical implants to date (Figure 5.7). Approximately 90% of the valves have no regurgitation at any of the evaluation periods. Trivial, mild, and moderate regurgitation were all fairly insignificant in the valves studied so far.

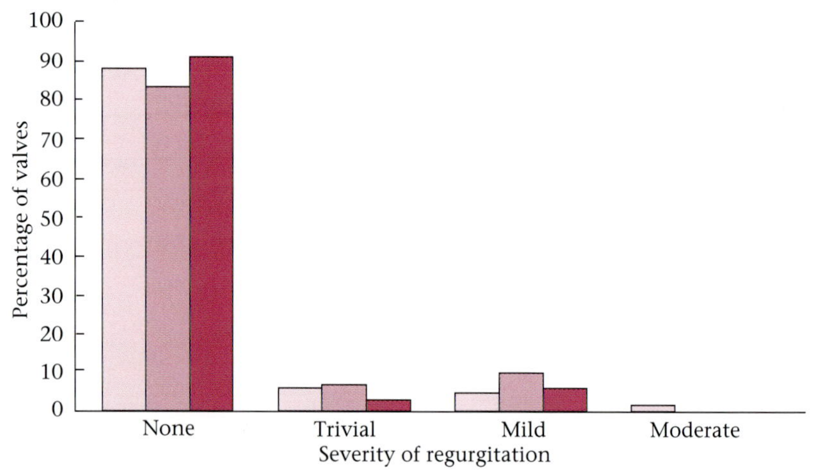

Figure 5.7. Percentage of Toronto SPV™ valves with no, trivial, mild, or moderate regurgitation at postoperative (□), 3–6-month (■) and 11–14-month (■) follow-up periods.

Another note of interest from the clinical trials is that the sizes of valves being implanted are larger than normally seen for aortic valve implants (Figure 5.8). This is probably due to the sizing technique for the Toronto SPV™ valve, which allows for repressurizing of the aorta after implant. This sizing technique guarantees proper function of the valve and results in improved performance.

The Toronto SPV™ valve offers many of the advantages of homograft valves. Its scalloped sinuses allow the commissures of the valve to follow the natural compliance of the host aorta, which improves haemodynamics and may lead to better durability than stented porcine valves. The availability of an extended range

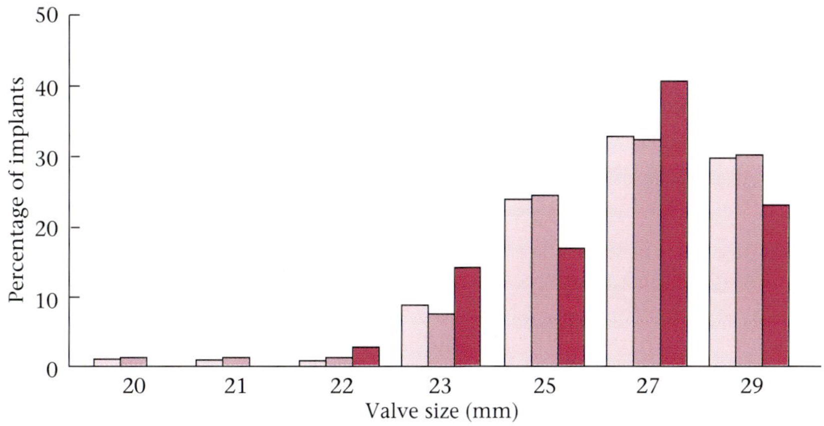

Figure 5.8. Sizes of Toronto SPV™ valves implanted, as percentage of implants assessed at postoperative (□), 3–6-month (▣) and 11–14-month (■) follow-up periods.

of sizes gives surgeons the ability to custom-fit the valve to the host aorta. In clinical use for 3 years, the Toronto SPV™ valve has demonstrated superior haemodynamics, with improvements in effective orifice areas over the first year after implant.

References

1. David T E, Ropchan G C, Butany J W 1988 Aortic valve replacement with stentless porcine bioprostheses. J Cardiac Surg 3(4): 501–505
2. Thubrikar M J, Deck J D, Aoad J, Nolan S P 1983 Role of mechanical stress in calcification of aortic bioprosthetic valves. J Thorac Cardiovasc Surg 86: 115–125
3. David T E, Bos J, Rakowski H 1992 Aortic valve replacement with the Toronto SPV bioprosthesis. J Heart Dis 1(2): 244–248
4. David T E, Pollick C, Bos J 1990 Aortic valve replacement with stentless porcine aortic bioprosthesis. J Thorac Cardiovasc Surg 99(1): 113–118

Chapter 6

Hydrodynamic performance of the Medtronic Freestyle™ aortic root bioprosthesis

C. E. Eberhardt, A. P. Yoganathan, M. Capps and C. Toomes

The Medtronic Freestyle™ aortic root bioprosthesis comprises a porcine aortic root crosslinked in buffered glutaraldehyde solution, with a 40 mmHg hydrostatic pressure applied to the root and a zero pressure differential across the valve leaflets (i.e. 'root pressure fixation'). The bioprostheses are treated with α-amino oleic acid (AOA), a derivative of the naturally occurring fatty acid (oleic acid), to reduce the potential for leaflet calcification. Animal studies *in vivo* have demonstrated that AOA processing of fixed porcine valves does mitigate leaflet calcification. Based on test data, AOA does not adversely affect the hydrodynamic performance, accelerated durability, collagen crosslink stability, mechanical properties, histological structure or biocompatibility of the porcine bioprosthesis. The preserved aortic root has a thin single layer of Dacron fabric covering the myocardium (muscle) and the inflow edge of the bioprosthesis. The Freestyle™ aortic root is supplied sterile in a buffered 0.2% glutaraldehyde solution.

The Freestyle™ aortic root incorporates the natural valve and approximately 2 cm of aorta with the right and left coronary arteries ligated with 5–6 mm remnants for aiding reimplantation of the patient's coronary arteries. A surgeon's suture guideline of green 5/0 suture is placed approximately 2 mm below the inflow of the valve. Green surgeon suture flags denote the base of each commissure and serve as identification reference points for the surgeon during implant.

The design is intended to provide improved haemodynamics and durability when compared with stented porcine bioprostheses. Stentless valves, in general, should provide superior haemodynamics compared with stented valves, owing to the elimination of the stent and suture ring, which are obstructive to flow. It is also believed that stentless valves offer the potential for increased durability, owing to elimination of the stent. Maintaining the structure of the natural coronary sinuses may decrease the mechanical stress on the leaflets by dissipation of energy to the aortic root during closure.

The development of the tissue fixation process used for the Freestyle™ aortic root is based on a process of pressurizing the native root in order to maintain natural root geometry, while at the same time maintaining a zero pressure differential across the leaflets. It is well documented that the pressure that valve leaflets are subjected to during fixation has a significant influence on the mechanical properties of the valve leaflets.

Zero pressure differential across the valve leaflets retains the natural functional collagen crimp. This maintains the tissue's 'shock absorbing' capacity, which aids in resisting cyclic fatigue, thus providing the potential for increased durability.

The work of Broom and Thompson[1] has defined mechanisms for achieving increased durability by retaining, as closely as possible, the structural characteristics of fresh porcine valve tissue.

Methods

Evaluation *in vitro* of stentless aortic bioprostheses requires that the valve be mounted in a test chamber (simulated aorta) that reasonably simulates the human aortic root. In order to develop a simulated aorta for use *in vitro*, several factors were considered, as follows.

The aortic valve in its natural state does not have a fixed shape and can only be described at a given time in the cardiac cycle, such as mid-systole or mid-diastole.

The aorta is anisotropic and expands quite easily as low internal pressure but stiffens at higher pressures to prevent ballooning, as reported by Thubrikar *et al.*,[2] in 1977, Brewer *et al.*[3] in 1976 and Ferguson *et al.*[4] in 1988. Since evaluation of a bioprosthesis *in vitro* requires extended testing, a material that provides bacterial stability was necessary. Materials such as rubber provide such bacterial stability and can easily be produced to exact geometric dimensions; however, these materials are isotropic and do not exhibit the same locking characteristics at high pressures that are seen with anisotropic materials. For these reasons, a simulated aorta of repeatable geometric design and the ability to vary the compliance chactertistics was developed that would provide reasonable model aortas *in vitro*.

The simulated aortas were patterned from dimensional data available through published clinical literature. The geometry of the aortas was based upon the normal human adult aorta, as reported by Reul *et al.*[5] in 1990, and Roman *et al.*[6] in 1989. A normal rather than diseased aorta was used since, as reported by Stefanadis *et al.*[7] in 1987 and also by Reul *et al.*,[5] the geometry of the diseased aorta varies as a result of the type and extent of the disease. With the use of the normal aorta as the model, a more normal distribution of the shape is reflected.[5] In order to produce simulated aortas of consistent, repeatable geometry and dimensions, steel compression moulds were made, and the material was formulated to produce the desired compliance for the tests *in vitro*.[8,9]

The inflow edge of the valve is attached to the aorta by a continuous line of blanket stitching around the valve diameter using a 5/0 suture. The outflow edge of the valve is attached to the aorta by the same method using a 4/0 suture. The line of stitching through the aorta is sealed on both inflow and outflow ends with a Silastic medical adhesive to prevent leakage.

The stented valve is placed inside the aorta with the sewing cushion sealed to prevent leakage.

Hydrodynamic protocols

Hydrodynamic testing was completed on three clinical-quality Freestyle™ aortic roots of each size and four Hancock standard (Model 242) aortic valves (3–31 mm and 1–19 mm) as control valves.

Tests were conducted at ambient temperature in normal saline in a modified version of the Georgia Tech. left heart pulse duplicator system interfaced to an on-line data collection system (Figure 6.1). The steady flow-rate pressure-drop testing was performed at a flow-rate range of 5–30 l/min. The pulsatile flow pressure drop testing was performed at a cardiac output range of 2.5–7.5 l/min, at a pulse rate of 70 bpm, with systole accounting for approximately 35% of the simulated cardiac cycle. Mean aortic pressure was 90–100 mmHg and mean left atrial pressure was 5–10 mmHg. An aqueous solution of saline with a viscosity of 1.0 cP (0.001Pa.s)

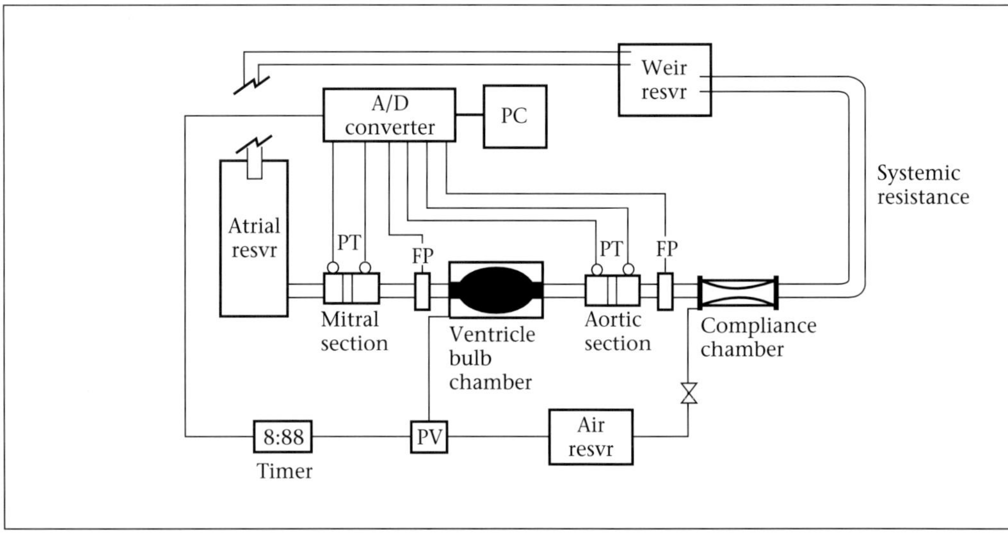

Figure 6.1. Georgia Tech. Left Heart Pulse Duplicator System, PT, Pressure transducer; FP, flow probe; PV, pneumatic valve.

was used as a blood analogue fluid at a temperature of 22°C. Pulsatile flow regurgitant volume versus back-pressure (closing pressures applied to the valve) data were obtained at back pressures (mean arterial pressure) of 90, 110, 130 and 150 mmHg. Details of the pulse duplicator and on-line data collection systems have been described previously.[10–12]

Additional studies were performed to determine the effects of accelerated wear testing on the hydrodynamic performance of the Freestyle™ aortic root using both AOA-treated, non-AOA treated, and the Hancock control valves (i.e. at zero, 60×10^6, 120×10^6 and 200×10^6 cycles) as part of the accelerated wear study. Comparisons were made between the effective orifice areas of the AOA treated vs non-AOA treated using an unpaired *t*-test and the zero cycle vs 200×10^6 cycles was made using a paired *t*-test (Execustal software package). Significance was set at $p = <0.05$.

Doppler ultrasound studies *in vitro* were performed on one of each size (19, 23 and 27 mm) of Freestyle™ aortic root bioprosthesis to determine if the Bernoulli relationship, $\Delta P = 4(V_2^2 - V_1^2)$, may be used to assess peak and mean pressure gradients non-invasively in patients implanted with a Freestyle™ aortic root bioprosthesis. Studies were conducted at cardiac outputs in the range of 2.5–8.0 l/min, at a pulse rate of 70 bpm, with a systolic duration of 280–320 ms. Cornstarch particles (1–2%) were added to the blood analogue fluid to act as acoustic scatterers. Peak and root mean square (RMS) systolic flow rates, peak and mean systolic pressure drops, and peak and RMS systolic centreline pulsed and continuous-wave Doppler velocities, were averaged over 10 cardiac cycles. Maximum pulsed and continuous-wave systolic Doppler velocities were measured proximal and distal to the aortic valves, respectively, from an ultrasound port on the flow loop, using a Vingmed SD-100 ultrasound Doppler system. The Vingmed Doppler system was interfaced, via an eight-channel analog-to-digital convertor system (Metrabyte DASH 16-F), to a 286 personal computer for on-line data collection.

Colour Doppler flow mapping (CDFM) was conducted in order to visualize qualitatively the flow fields downstream of a 27 mm Freestyle™ aortic root. It was not possible to conduct optical flow visualization or laser Doppler studies owing to the opaque nature of the porcine aortic root. The CDFM studies were conducted at a heart rate of 70 bpm, systolic duration of about 280 ms, mean aortic pressure of 100 mmHg and cardiac outputs of 3.0 and 5.0 l/min. The CDFM measurements were performed using a Toshiba SSH 270 colour Doppler ultrasound system with a 3.75 MHz phased-array transducer. The valve was imaged from the ultrasound viewing port on the flow loop fitted especially for this study. The imaging depth was 15–20 mm.

Results and discussion

Steady flow pressure drop

The steady flow gradients across the Freestyle™ aortic root are significantly lower than the gradients across the Hancock standard control valves.

Pulsatile flow pressure drop

The effective orifice areas as calculated from the pulsatile flow gradients across the Freestyle™ aortic root are significantly larger than the effective orifice areas of the Hancock standard control valves (p=0.0002). The effective orifice area of the Freestyle™ size 23 mm was found to be equivalent to the Hancock control valve size 31 mm (Figure 6.2). In the two larger sizes (25, 27mm) Freestyle™ valves had effective orifice areas equivalent to the St Jude bileaflet valves. However, in the three smaller sizes (19, 21, 23 mm), the effective orifice areas of the Freestyle™ design were significantly larger than the St Jude (p=0.0130) and Hancock Modified Orifice (p=0.0001) valve designs. In the smallest size (19 mm), the Freestyle™ valve is one size superior to the St Jude design as evidenced by the calculated EOAs (see Table 6.1).

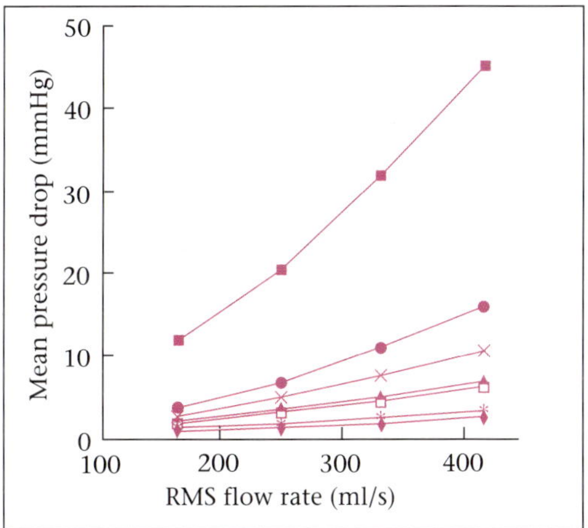

*Figure 6.2. Freestyle™ aortic root bioprosthesis hydrodynamics: pulsatile flow gradients across various sizes of valve, (•, 19 mm; ×, 21 mm; □, 23mm; *, 25 mm; ◆, 27mm) compared with two sizes of Hancock valves (▲, 31 mm; ■, 19 mm). The pulsatile flow gradients across the Freestyle™ aortic root were statistically significantly lower than those across the Hancock valves.*

Valvular regurgitation

The total regurgitant volume was equivalent to that for the Hancock control when compared with the Freestyle™ aortic root with *equivalent flow area*. For example, the Hancock 31 mm control valve has an equivalent flow area to the Freestyle™ aortic root size 23 mm. The 23 mm Freestyle™ aortic root has a slightly lower regurgitant volume than the Hancock 31 mm control valve. The valves had good coaptation, with little or no leakage (Table 6.2).

Doppler ultrasound *in vitro*

The peak and RMS velocities and the peak and mean systolic pressure drops indicate that the Bernoulli equation works very well within experimental error with the Freestyle™ aortic root design (i.e. with 5%) as shown by the results in Figures 6.3a and 6.3b. Based on these studies *in vitro* it can be concluded that the Bernoulli relationship may be used to assess peak and mean pressure gradients non-invasively in patients who have the Freestyle™ aortic root implanted. It should, however, be noted that if the velocity, V_1, proximal to the valve is not measured and used in the Bernoulli equation calculation, the Doppler-derived

Table 6.1 Comparison of effective orifice area (EOA) of aortic valves

	Valve EOA (cm^2)			
Valve (mm)	Freestyle	Modified Orifice	St Jude Medical	Hancock 242
19	1.84	1.22	1.21	1.14
20	—	—	—	—
21	2.17	1.43	1.81	1.31
23	2.69	1.94	2.24	1.73

Table 6.2 Valvular regurgitation in the Freestyle™ aortic root compared with the Hancock control valve

	Back-pressure leakage (ml/beat) of valves*						
Back pressure	Freestyle size					Hancock (control) size	
(mmHg)	19 mm	21 mm	23 mm	25 mm	27 mm	19 mm	31 mm
90	1.4 ± 0.4	1.8 ± 0.5	1.2 ± 0.2	2.4 ± 0.5	2.3 ± 0.9	0	0.4 ± 0.1
110	1.7 ± 0.4	2.1 ± 0.5	1.5 ± 0.3	2.7 ± 0.5	2.6 ± 0.9	0.5	0.2 ± 0.2
130	1.8 ± 0.3	2.3 ± 0.5	1.8 ± 0.3	2.9 ± 0.4	2.8 ± 0.9	0.3	0.2 ± 0.2
150	2.3 ± 0.4	2.5 ± 0.4	1.9 ± 0.3	3.3 ± 0.3	3.1 ± 0.9	0.1	0.3 ± 0.2

*Values are means ± SD

(a)

	RMS velocity		Peak velocity	
Valve size (mm)	M [mmHg/(m^2/s^2)]	R^2 (%)	M [mmHg/(m^2/s^2)]	R^2 (%)
19	3.92	99.99	3.91	99.99
23	3.91	99.97	3.87	99.98
27	3.97	99.93	3.95	99.97

(b)

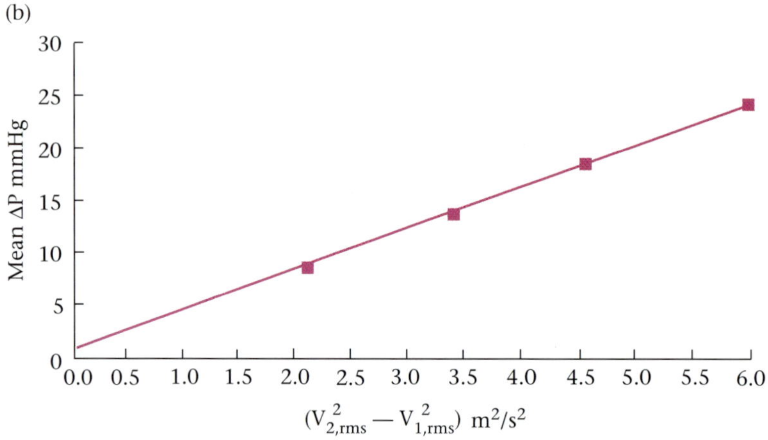

Figure 6.3. Linear regression analysis through the origin for Freestyle™ ultrasound Doppler velocity values: (a) Peak and (root mean square) velocities (liquid properties: viscosity 1.0 Cp, temperature 22°C, heart rate 70 bpm; systolic duration 280–320 ms). (b) Ultrasound Doppler measurements in vitro *of 19 mm Freestyle™ aortic root bioprosthesis, based on RMS velocity values;* $y = 3.92191x$, $R^2 = 99.99\%$ *[liquid properties and systolic duration as in (a)].*

peak and mean systolic gradients across these valves will be 'overestimated' by as much as 30–85%. The reason for this overestimation is that the pressure gradients across this valve design are relatively small, as shown in Table 6.1. In clinical practice, therefore, when evaluating the Freestyle™ valve design by echo-Doppler, it is imperative that the entire Bernoulli equation (i.e. with the proximal velocity term) be utilized.

Flow visualization

The studies showed good opening of the valve leaflets by M-mode, 2D echo and CDFM. The flow accelerates uniformly through the valve and appears to have a uniform flat profile downstream of the prosthesis. At the leaflet tips the maximum velocity was of the order of 75 and 125 cm/s, at the cardiac outputs of 3.0 and 5.0 l/min, respectively. There is no appearance of a jet with steep velocity gradients, as observed with stented porcine valves such as the Hancock and Carpentier–Edwards bioprostheses.[10] Approximately 5–10 mm from the leaflet tips the peak systolic velocity is of the order of 60–100 cm/s, at the cardiac outputs of 3.0 and 5.0 l/min, respectively.

Accelerated wear

The results show no difference in pulsatile flow gradients between treated and non-AOA treated Freestyle™ aortic roots (Figure 6.4). The effective orifice areas between zero cycles and 200×10^6 cycles were not significantly different (p=0.8802) as a result of the accelerated wear testing. The results show no significant difference in effective orifice areas between AOA treated and non-AOA treated Freestyle™ aortic roots (p=0.1024). Through visual evaluation, no structural differences were observed between the AOA treated Freestyle™ aortic roots and non-AOA treated Freestyle™ aortic roots, at the end of 200×10^6 cycles of accelerated wear.

Regurgitant volume

In most cases the total regurgitant volume was reduced at each test point (i.e. 60×10^6, 120×10^6, 200×10^6 cycles). In some valves where damage occurred

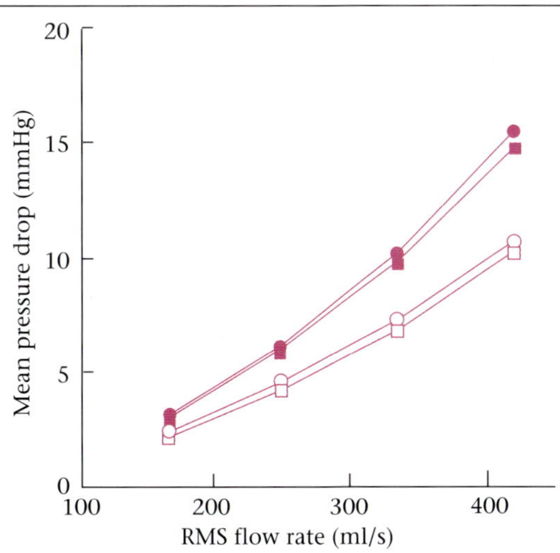

Figure 6.4. Freestyle™ aortic root bioprosthesis hydrodynamics: pulsatile flow gradients across AOA-treated (closed symbols) or untreated (open symbols) 19 mm (circles) or 21 mm (squares) Freestyle™ aortic roots, showing lack of effect of the antimineralization treatment on hydrodynamic performance.

during the accelerated wear testing, the regurgitant volume increased slightly from the previous test point. In most cases, the damage that occurred during the wear testing was due to contact of the wall of the root with the cut edge of the aorta test chamber. The resulting leakage originated at the wall and not at the location of leaflet coaptation.

Closing volume

In most cases the closing volume was reduced as a result of accelerated wear testing. The most significant reduction in closing volume was with the Hancock standard control valves, with a reduction of 6.8 ml/beat. This change in closing volume is attributed to the noticeable softening of the high-pressure-fixed leaflets of the Hancock standard control valves.

Leakage volume

Leakage volume was reduced in most cases at each test point (i.e. 60×10^6, 120×10^6, 200×10^6 cycles) as a result of accelerated wear testing.

Effects of sizing and implant technique on the hydrodynamic performance of the Freestyle™ aortic root bioprosthesis

Realizing that exact sizing may not be possible in the clinical situation, further studies were performed *in vitro* to assess the effects of sizing on the hydrodynamic performance of the Freestyle™ aortic root bioprosthesis. This study also includes the evaluation of four implant techniques – total root replacement, inclusion, scalloped (two sinuses removed) and valve (all three sinuses removed).

Performance assessments included pulsatile flow pressure drop and back-pressure leakage.

Methods

The test samples were attached to simulated aorta test chambers and sealed at the suture lines to reduce leakage at the attachments.

Transvalvular pressure drop (gradient) versus pulsatile flow rate and regurgitant volume data were obtained from three of each size (21, 23, 25 and 27 mm) Freestyle™ aortic roots. All valve sizes were mounted into the size 21 mm simulated aorta. Each valve was first configured and tested as a total root replacement, then configured and tested as an inclusion technique, then scalloped at two sinuses and tested, and finally fully scalloped as a valve configuration and tested again. The results were analysed using an ANOVA test (Executive software package). Significance was set at $p = <0.05$.

Total root replacement

The entire sinus section of the aorta is excised so that only the end pieces of the aorta remain attached to the valve.

Inclusion

A circular section of the aorta is excised to accommodate the extension of the coronaries.

Scalloped (two sinuses)

The valve is scalloped at the right and left coronary leaving the non-coronary sinus intact. The valve is then placed into the aorta test chamber aligning the commissure to the aorta.

Valve (scalloped three sinuses)

The valve is scalloped at all three sinuses then placed into the aorta test chamber aligning the commissure to the aorta.

Results and discussion

Pulsatile flow pressure drop

Sizing

Based on these analyses the following observations can be made:

1. Of the four valve types (21, 23, 25 and 27 mm), the mean EOA for the size 23 mm is the largest.
2. The data suggest that the EOA is optimized by the use of a valve one size larger, but generally tends to decrease as larger sizes are used (25 and 27 mm).

Configuration

The analysis of configurations reveals that the total root and inclusion techniques resulted in the highest EOA of the four implant configurations (Table 6.3). In the size 21 mm group that was properly sized, the performance by EOA ranked as follows: root, inclusion, scalloped, valve (Figure 6.5). High-speed photography (filmed at 500 frames/s) revealed a slight constructing phenomenon occurring at the inflow of the bioprosthesis with both the scalloped and valve configurations, due to the effect of the aorta compliance.

For the inclusion configuration it was noted that one valve of size 25 mm demonstrated a significantly higher gradient than the rest of the 25 mm group. Visual evaluation revealed a distinct mismatch between aorta and root shape, causing a notable buckle in the aortic wall.

There are statistically significant differences between the mean EOA due to valve size ($p < 0.002$) and configuration ($p < 0.012$). Size contributed to about 25% of the variation and configuration contributed to about 16%. There was not a significant contribution from the interaction between size and configuration in this experiment ($p < 0.94$).

Table 6.3 Performance ranking by effective orifice area (EOA) for each configuration group

Valve size (mm)	Configuration (in ranking order)	EOA (cm^2)*
21	Root	2.19 ± 0.05
	Inclusion	2.12 ± 0.02
	Scalloped	2.11 ± 0.04
	Valve	1.94 ± 0.03
23	Root	2.41 ± 0.07
	Inclusion	2.34 ± 0.09
	Scalloped	2.17 ± 0.09
	Valve	2.06 ± 0.11
25	Root	2.29 ± 0.11
	Inclusion	1.85 ± 0.04
	Valve	1.89 ± 0.15
	Scalloped	1.74 ± 0.14
27	Root	2.00 ± 0.19
	Inclusion	1.91 ± 0.26
	Valve	1.59 ± 0.18
	Scalloped	1.39 ± 0.16

*Valves are means ± SD

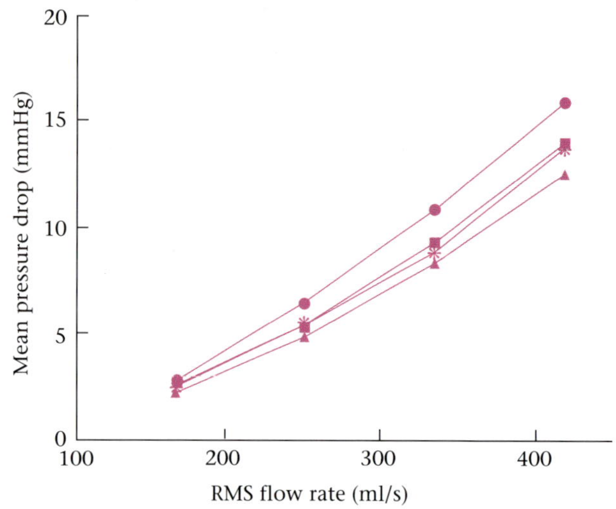

*Figure 6.5. Comparison of pulsatile flow gradients across 21 mm Freestyle™ aortic root bioprostheses implanted using various configurations of valve (▲ root, ■ inlcusion, * scalloped, ● valve). The total aortic root produced a slightly lower gradient than the four valve configurations tested.*

Valvular regurgitation

Regurgitant volume is typically affected by flow area, where the larger flow areas will result in a higher regurgitant volume (i.e. as valve area increases generally the regurgitant volume also increases). These studies demonstrated a similar situation in that the scalloped and valve techniques resulted in both lower EOA and lower regurgitant volume; the root and inclusion techniques, with higher EOAs, also resulted in higher regurgitant volumes (Table 6.4).

There are statistically significant differences between the mean regurgitant volumes due to valve sizing ($p < 0.05$), implant configuration ($p < 0.001$) and interaction between sizing and configuration ($p < 0.001$).

Conclusions

The pulsatile flow hydrodynamic studies *in vitro* described in this paper indicate that the Freestyle™ valve design is superior to the stented porcine valve designs currently in clinical use (Hancock and Carpentier–Edwards). Furthermore, in the smaller sizes (19, 21, 23 mm) it has better overall flow performance characteristics than the St Jude bileaflet valve design. In all sizes, the Freestyle™ valve design showed good coaptation with small leakage volumes.

Based on statistical analysis, both gradient and regurgitation are significantly affected by both valve sizing and implant configuration.

Sizing of the 23 mm valve into the 21 mm aorta resulted in the highest EOA of the four groups. Oversizing by more than two sizes compromised the hydrodynamic performance (the 25 and 27 mm valves inserted into the 21 mm aorta had a lower EOA than the 21 mm valves).

For the configuration study, the total root and inclusion implant techniques performed best, with the size 23 mm root having the highest EOA. The scalloped and valve techniques resulted in both the lowest EOA and lower regurgitant volume of the four configurations tested.

Optimal flow performance of the Freestyle™ and other stentless aortic valve designs *in vitro* will therefore be dependent upon their configuration of use and optimal sizing criteria.

Table 6.4 Performance ranking by total regurgitant volume for each configuration group

Valve (mm)	Configuration (in ranking order)	Total regurgitant vol. (ml/beat)*
21	Valve	4.6 ± 0.1
	Scalloped	5.0 ± 0.2
	Root	5.7 ± 0.1
	Inclusion	6.2 ± 0.1
23	Valve	4.7 ± 0.1
	Scalloped	5.0 ± 0.2
	Inclusion	5.8 ± 0.2
	Root	6.5 ± 0.2
25	Scalloped	4.9 ± 0.1
	Valve	4.3 ± 0.1
	Inclusion	6.0 ± 0.2
	Root	7.1 ± 0.2
27	Valve	4.5 ± 0.1
	Inclusion	5.4 ± 0.1
	Root	6.3 ± 0.1
	Scalloped	6.4 ± 0.2

*Values are means ± SD

References

1. Broom N D, Thompson, F J 1979 Influence of fixation conditions on the performance of glutaraldehyde treated porcine aortic valves; towards a more scientific basis. Thorax 34: 166–176
2. Thubrikar M, Harry R, Stanton PN 1977 Normal aortic valve function in dogs. Am Cardiol 40: 563–568
3. Brewer J, Deck DJ, Capati B, Stanton PN 1976 The dynamic aortic root. J Thorac Cardiovasc Surg 72(2): 413–417
4. Ferguson J, Miller MJ, Sahaglan P, Aroesty JM, McKay RG 1988 Assessment of aortic pressure volume relationships with an impedance catheter. Cathet Cardiovasc Diagn 15: 27–36
5. Reul H, Vahlbruch A, Gierstepen M, Schmitz-Rode T, Hirtz V, Effert S 1989 The geometry of the aortic root in health, at valve disease and after valve replacement. J Biomech 23: 181–191
6. Roman J, Devereux RB, Kramer-Fox R, O'Loughlin J 1989 Two dimensional echocardiographic aortic root dimensions in normal children and adults. Am J Cardiol 64: 507–512
7. Stefanadis C, Wooley CF, Bush CA, Kolibash AJ, Boudoulas H 1987 Aortic distensibility abnormalities in coronary artery disease. Am J Cardiol 59: 1300–1304
8. Collins R, Hu L 1972 Dynamic deformation experiments on aortic tissue. J Biomech 5: 333–337
9. Arndt J O, Stegall HF, Wicke HJ 1971 Mechanics of the aorta *in vivo*. Circ Res 37: 693–704
10. Woo Y R 1984 *In vitro* velocity and shear stress measurements in the vicinity of prosthetic heart valves. PhD thesis, Georgia Institute of Technology
11. Yoganathan A P, Strand D M, Woo Y R, Stevenson D M 1984 An on-line method for evaluation of the *in vitro* pulsatile pressure drop and regurgitation characteristics of prosthetic heart valves. Med Instrum 18: 109–113
12. Yoganathan A, Woo Y R, Sung H W, Williams F P, French R H, Jones M 1986 *In vitro* haemodynamic characteristics of tissue bioprostheses in the aortic position. Thorac Cardiovasc Surg 92(2): 198–209

Chapter 7

Edwards Prima™ stentless bioprosthesis: from concept to reality

L. Quintero

Current tissue valves have limitations that need to be overcome in order to provide patients with the best quality of life. Stented bioprosthetic valves have limited durability, which most commonly is due to calcification[1–4] or primary tissue failure.[5–7] In addition, the presence of the stent and sewing ring reduces the effective orifice area of bioprosthetic valves, resulting in suboptimal haemodynamics.[8–10] Homografts have shown a slower calcification rate,[11] especially in children, resulting in improved durability over that of bioprostheses. Most homografts are implanted without a stent or sewing ring, and the transvalvular pressure gradients across them are extremely low — less than 10 mmHg for most sizes. However, the availability of homografts is extremely limited, partially owing to storage issues. In addition, the double suture line technique most commonly used with these valves results in a more difficult and time-consuming surgical procedure.

Improvement of bioprosthetic valves

If bioprosthetic valves are considered first, it can be seen that there are several potential solutions for the limitations discussed above. If a treatment could be applied to the tissue that would effectively mitigate calcification, average valve durability would increase. An improved stent design could both improve the haemodynamic performance of the valve and increase its durability. Pericardial valves, such as the Carpentier–Edwards pericardial bioprosthesis can be engineered for improved opening characteristics and increased durability. Furthermore, the stent, the focus of this chapter, can be removed from the bioprosthesis to provide less obstruction to blood flow and more flexible motion of the device.

Improvement of homografts

The availability of homografts can be increased by the use of tissue preservatives such as glutaraldehyde, or improved antibiotics. Implantation could be facilitated by the addition of a stent/sewing ring combination. However, stenting and fixation with glutaraldehyde would also reduce the benefits of using a homograft, which are flexibility and good orifice area.

Stentless xenografts

The Baxter Company (Irvine, California, USA) started to examine the concept of a stentless porcine aortic valve in the mid-1980s. Removal of the stent from a bioprosthetic valve would create a more 'normal' function, with less restriction of the aortic wall and leaflets. The lack of a sewing ring would result in a larger effective orifice area and correspondingly lower transvalvular gradients. With less material in the path of the blood flow, turbulence would be reduced. If foreign material (i.e. cloth and suture) were minimized the thromboembolism rate might be lower, which could potentially make this type of valve useful in cases of bacterial endocarditis. Allowing the valve to expand with the patient's aorta could lower transleaflet stress differentials; this might translate to an increased durability and lowered rate of stress-mediated calcification.

As with a homograft, implantation of a stentless xenograft would require more technical judgment and expertise than that of a stented xenograft. The learning curve could be minimized by utilizing homograft surgeons for initial implants or initial time on the market, but the risk of early failures due to technique could not be eliminated. Limited historical experience with stentless xenografts means that the durability of these devices is largely unknown. Theoretically, the greatest haemodynamic benefits would be seen in the smaller sizes, but this could depend greatly on the implant technique. The fate of the porcine aortic wall, directly in contact with the patient's aortic wall, is another question that will be answered only with time. It will also be necessary to monitor whether this type of device is as useful in cases of endocarditis as theory suggests.

Valve design

Baxter investigated several potential valve designs. It is possible to retain all three coronary sinuses, allowing the surgeon complete freedom to customize, but increasing the amount of work required to prepare the valve for implantation and increasing the risk of damaging the leaflets during trimming. A second design possibility is removal of all three coronary sinuses, which essentially eliminates the work needed by the surgeon before implanting the valve. However, this also eliminates the possibility of using the device as a root or mini-root replacement and may increase the percentage of patients with central regurgitation, owing to the ease with which the commissures can be distorted. A third choice, and the one Baxter ultimately selected, is the retention of the non-coronary sinus and a band of aortic tissue, with the coronary arteries removed. The presence of the aortic tissue reduces the chance of distortion during implant, and may be removed after placement of the first suture line to allow the valve to move freely. Some customization is possible, and the device can be used as a mini-root replacement if desired.

Figure 7.1. Original prototype modified with cloth covering.

Baxter worked with Dr Delos Cosgrove of the Cleveland Clinic to create the Edwards Prima™ stentless bioprosthesis.[12] Dr Cosgrove's original design concept was a valve that retained the advantages of a homograft while providing availability in a wide size range. The original prototype did not incorporate any cloth covering, in order to maximize flexibility. The fragility of the muscle shelf after fixation resulted in the decision to cover the myocardial tissue with polyester cloth, as seen in Figure 7.1. The outflow end of the valve has been scalloped around all three

Figure 7.2. Modification of device shown in Figure 7.1, with cloth-reinforced inflow rim in area of muscle shelf.

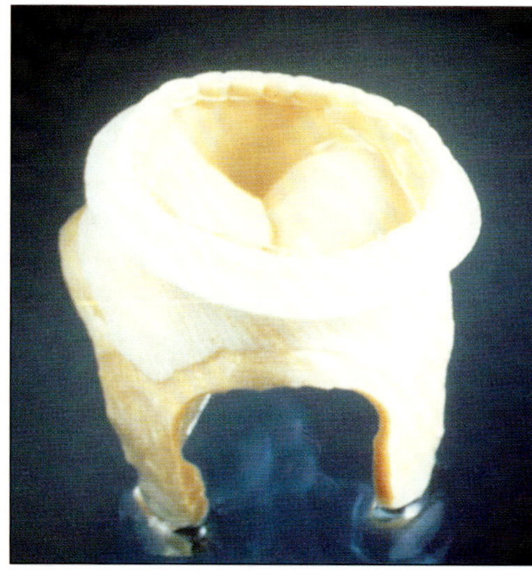

Figure 7.3. Prototype with entire inflow rim reinforced with polyester cloth.

commissures. A minimal amount of tissue has been left at the inflow annulus for the first suture line. This prototype was also designed to be invertible. Unfortunately, the handleability of this device was poor compared with that of stented valves.

Figure 7.2 shows a successive modification of the device, where the inflow rim in the area of the muscle shelf is also reinforced with polyester cloth. The supra-annular ridge is covered with cloth as well and is intended to support the second suture line. With this placement of the second suture line, the commissure tips had to be tacked down separately.

The design in Figure 7.3 includes reinforcement of the entire inflow rim with polyester cloth that has been shaped into a pliable sewing ring to facilitate placement of the first suture line. The cloth covers the muscle shelf and supports the commissures externally. Dr Cosgrove experimented with this device *in vitro* and determined that it felt too much like a stented valve.

Several design iterations incorporated a scalloped inflow rim, but it was decided that this could make implantation difficult in patients with a diseased or bicuspid native valve, where the contours of the valve might not match the contours of the patient.

Figure 7.4 shows the last prototype before our current design. On the basis of his observation that homografts frequently become distorted during implantation, Dr Cosgrove suggested that the outflow end of the valve should be left essentially intact, with removal of the ligated coronary arteries only. The 'tabs' of tissue distal to the coronary ostia could then be removed (if desired) before placement of the second suture line.

Figure 7.5 shows Baxter's current design, incorporating the best design charcteristics of the various prototypes. The inflow rim and muscle shelf are reinforced with polyester cloth. A line of green suture is used to attach the cloth to the inflow rim of the valve and provides a visual guide to surgeons for safe suture placement. The integral holder, shown attached to the valve in Figure 7.6, provides added stability during the first suture line and assists in orientating the valve. Green marking sutures, shown in Figure 7.7, are placed on the inflow rim, midway between each set of commissures. These sutures help the surgeons to place their sutures evenly around the annulus.

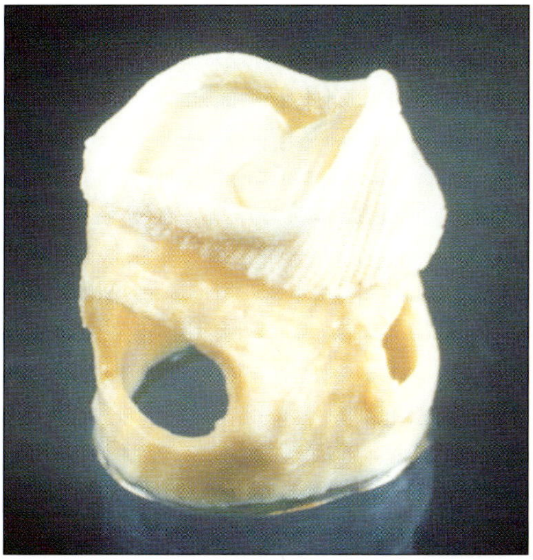

Figure 7.4. Prototype showing intact outflow end of the valve, with removal of coronary arteries only.

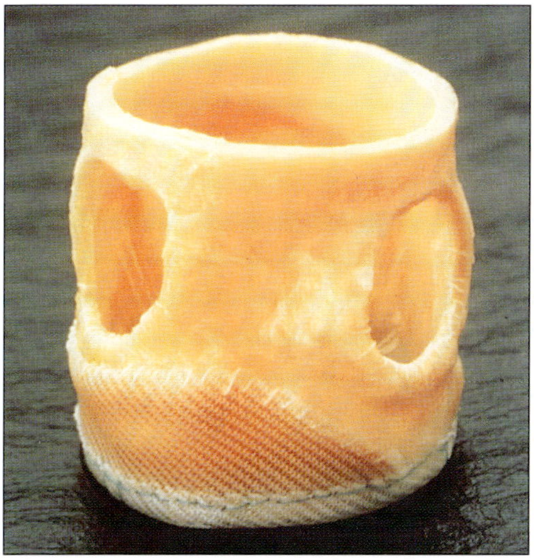

Figure 7.5. Current design of the Edwards Prima™ stentless bioprosthesis.

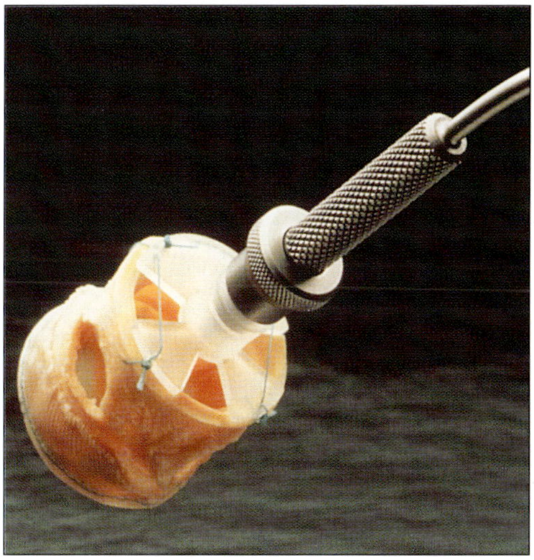

Figure 7.6. Current design, showing integral holder attached to the valve.

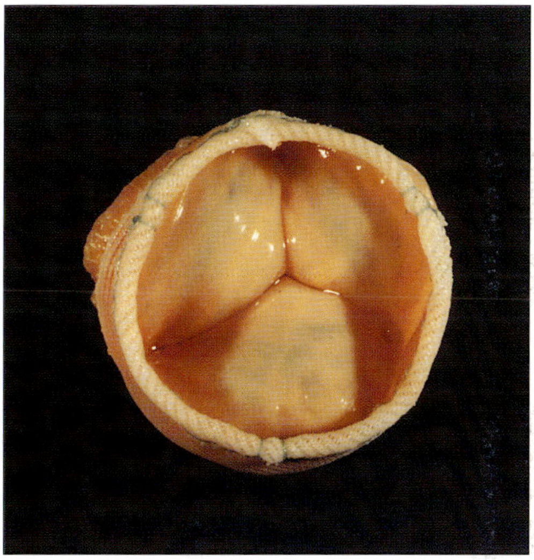

Figure 7.7. Current design showing green marking sutures.

Valve testing

Preclinical testing, both *in vitro* and *in vivo*, has been conducted and a 200-valve clinical trial is in progress. A valve explanted after 5 months in the aortic valve position in a sheep is shown in Figure 7.8; the radiograph is shown in Figure 7.9. It is interesting to note that there is no calcification of the leaflets and only very early signs of calcification of the aortic wall. This would suggest transference of the stress from the leaflets to the aortic wall.

The Edwards Prima™ stentless bioprosthesis is offered in sizes 17–29 mm, sized by outer diameter. Specially designed sizers are available that show how the valve will fit both in the annulus and in the aorta.

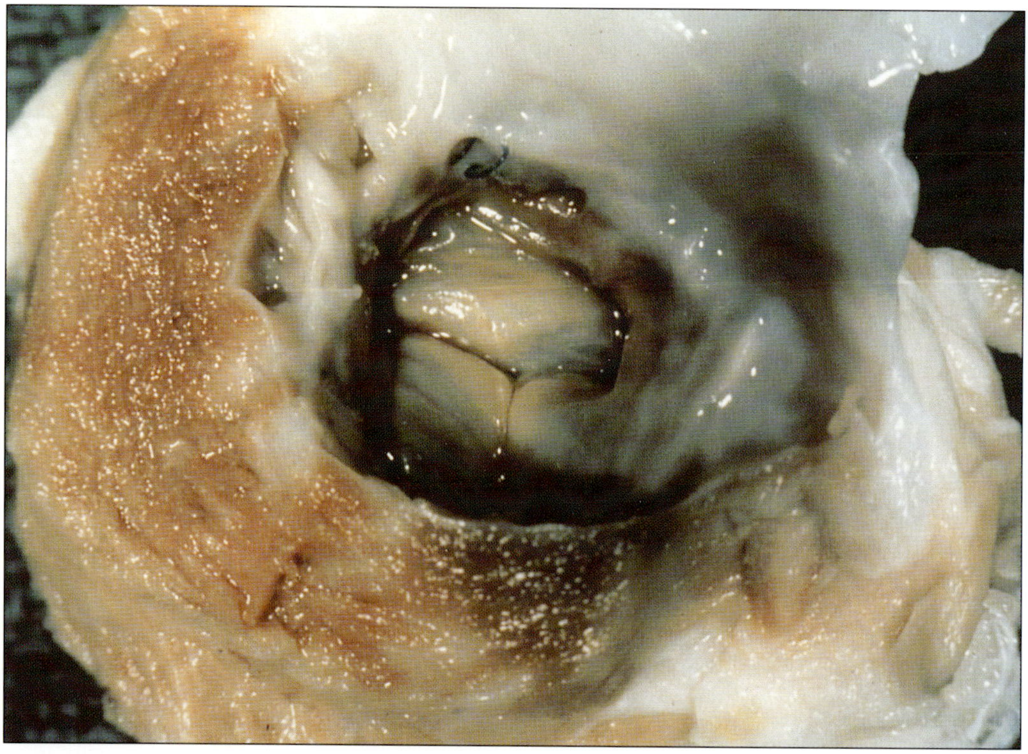

Figure 7.8. Edwards Prima™ stentless bioprosthesis explanted after 5 months in the aortic valve position in a sheep.

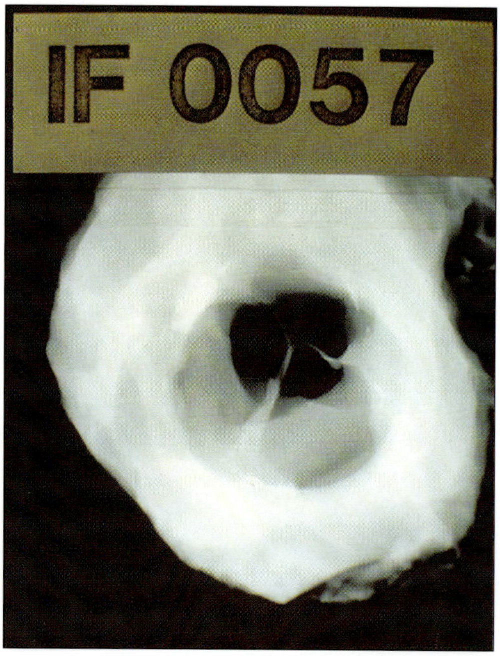

Figure 7.9. Radiograph of valve shown in Figure 7.8.

References

1. Solymar L, *et al* 1991 Prosthetic valves in children and adolescents. Am Heart J 557–568
2. Cipriano PR, *et al* 1982 Calcification of porcine prosthetic heart valves: a radiographic and light microscopic study. Circulation 66(5): 1100–1104
3. Williams W G 1988 Valve surgery in children. Can J Cardiol 4(6): 311–313
4. Kutsche L M *et al* 1979 An important complication in Hancock mitral valve replacement in children. Cardiovasc Surg (Suppl 1) 60(2): I–98–I–103

5. Jamieson W R E, Allen P, Miyageshima R T, Gerein A N, Munro A I, Burr L H, Tyers G F O 1990 The Carpentier–Edwards standard porcine bioprosthesis: a first-generation tissue valve with excellent long-term clinical performance. J Thorac Cardiovasc Surg 99: 543–561
6. Finkelmeier B A, *et al* 1989 Implications of prosthetic valve implantation: an 8-year follow-up of patients with porcine bioprostheses. Heart Lung 18: 565–574
7. Bernal J M, *et al* 1991 Valve-related complications with the Hancock I porcine bioprosthesis: a twelve- to fourteen-year follow-up study. J Thorac Cardiovasc Surg 101: 871–880
8. Hasenkam J M, *et al* 1989 Turbulent stresses downsteam of porcine and pericardial aortic valves implanted in pigs. J Card Surg 4(1): 74–78
9. Rashtian M Y, *et al* 1990 Flow characteristics of bioprosthetic heart valves. Chest 98(2): 365–375
10. Weerasena N, *et al* 1992 The hydrodynamic function and leaflet dynamics of aortic and pulmonary roots and valves: an *in vitro* study. Eur J Cardio-Thorac Surg 6(7): 350–356
11. Maxwell L, Gavin J B, Barratt-Boyes B G 1989 Differences between heart valve allografts and xenografts in the incidence and initiation of dystrophic calcification. Pathology 21: 5–10
12. All information on design history received from Diana Nguyen, The Baxter Company, Irvine, CA, USA and Dr Delos Cosgrove

Chapter 8

Pericarbon™ stentless valve

S. Rinaldi, M. Curcio, E. Pasquino, C. Stacchino and F. Vallana

The concept of a stentless aortic bioprosthetic valve has recently received renewed and increasing attention owing to the expectation of a better clinical performance compared with the standard stented bioprostheses. These expectations are derived from technical motivation as well as from an extrapolation of the good clinical results obtained using homografts.

Advantages of stentless bioprostheses

The first advantage of using a stentless bioprosthesis is better haemodynamic performance, particularly in terms of reduced transvalvular pressure gradient;[1–3] this is an obvious consequence of the significant increase in the valve orifice diameter attainable, for the same implant diameter, by eliminating the obstructing stent and the sewing ring.

Increased durability and reduced rate of calcification are other expected advantages of stentless prostheses;[3–5] these expectations are based on the fact that stentless valves are sutured directly into the aortic root, which is much more elastic and compliant than the stent. As a consequence one may expect a reduction in the stress and in the stress concentration, particularly in the cusps; the influence of these phenomena on the mechanical durability as well as on the rate of calcification is well known.

These considerations are further strengthened by experience with the aortic homograft, which has shown much greater durability when free-sewn in the aortic root than when premounted on stents: Angell and others,[5] for example, reported that, in their experience, a fresh aortic allograft lasted 12.1 years (median years to structural deterioration) when free-sewn and only 6.6 years when premounted on stents, calcification and mechanical rupture being, in the latter case, the more common causes of structural deterioration.

Another possible advantage of stentless bioprostheses that can be expected, by analogy with the homograft, is a greater resistance to recurrent infection,[3] particularly when the valve replacement is made for active endocarditis. Indeed, the better results obtained using homografts than with prosthetic valves[6] have been attributed to, and are likely to be at least related to, the complete absence of prosthetic material in the replacement device. This characteristic can be maintained in the design of a stentless bioprosthesis, although this is not always the case and many incorporate a fabric cuff.

For the above reasons, and with the aim of best exploiting the potential advantages of the stentless aortic bioprosthesis, a new valve made of pericardium was developed (the Pericarbon™ stentless valve), as described below.

The Pericarbon™ stentless valve

Design

In designing a stentless valve, advantage was taken of previous experience in bioprosthetic valve design and biological material treatments developed for the Pericarbon™ stented valve (Figure 8.1), which has been in clinical use for 8 years.

Figure 8.1. Pericarbon™ stented valve.

For both valves the same biological material (pericardium) was used, treated with an original low-pressure, progressive glutaraldehyde process; the same unique design was adopted, based on a two sheet structure.

The decision to maintain the choice of bovine pericardium as the raw material for manufacture of the bioprosthesis followed confirmation that the mechanical failures encountered in the first-generation pericardial valves resulted from shortcomings in design, manufacture and quality control, rather than from the intrinsic properties of the material itself.[7–10] Indeed, the last generation of pericardial valves, with basic changes in valve design and tissue preservation have shown better long-term durability.[10–14]

The pericardium offers also, from the manufacturing point of view, the significant advantage of allowing standardization, which results in a high degree of reproducibility; the adequacy of its biological characteristics has been again recently confirmed.[15]

Construction

After harvesting in sterile conditions, the membrane treatments start within 10 h in order to prevent tissue contamination and degradation. Careful selection for appropriate thickness, freedom from vascularization and absence of inflammatory processes ensures uniform material characteristics.

Tissue processing is based on the simultaneous fixation and shaping obtained with a fluid system in which increasing glutaraldehyde concentrations gradually and slowly (72 h are required) displace the collagen fibres to the final configuration, avoiding structural disruption. Furthermore, the solution diffusion through the membrane is flow controlled, thus ensuring in-depth fixation.

The unique two-sheet design developed for the Pericarbon™ valve (Figure 8.2a) involves the suturing of an internal tricusp-shaped pericardial membrane to an external flat membrane by means of a pattern studied and tested in order to optimize the stress distribution, ensuring that the stitches themselves do not become the origin of fatigue fracture lines and that no excessive stress concentration occurs.

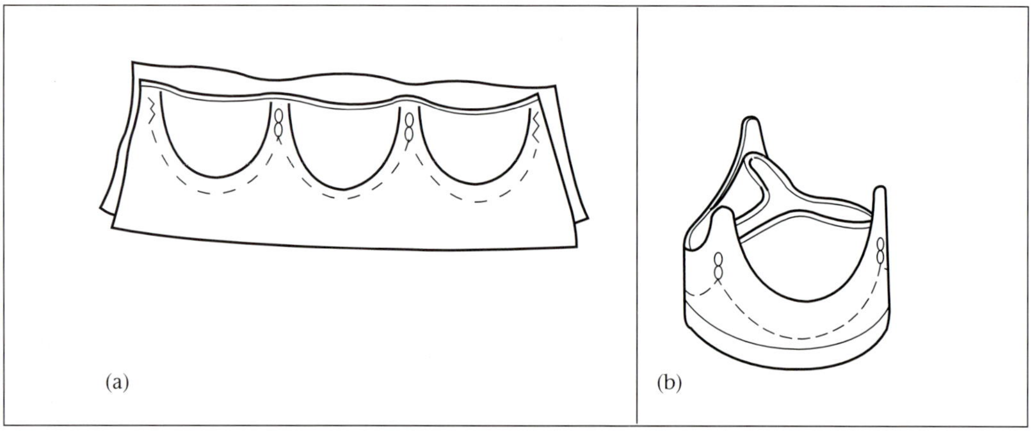

Figure 8.2. (a) Pericarbon™ valve two-sheet assembly; (b) final cylindrical structure.

In the commissural areas, which clinical experience has shown to be points of maximum stress, the suturing is in a cross-stitch pattern, superimposing the two sheets in a flat position: the leaflet opening allows the leaflets to assume their original configuration and they are thus completely stress free.

The resulting structure is a cylinder that can be trimmed to match the shape of the stent posts and its inflow profile in the case of the stented Pericarbon™ valve (Figure 8.2b). The Pericarbon™ stentless valve incorporates the same structure, resulting in an external scalloped pericardial cylinder that includes the internal cylinder forming the three cusps (Figure 8.3). The external cylinder is trimmed at the inflow side in such a way that at this level, there is only one sheet of biological tissue, in order to allow the best ratio between internal orifice and outer diameter. No fabric reinforcement is added, because the pericardium offers the proper mechanical resistance with the minimum thickness; this enables the use of non-biological materials to be avoided completely.

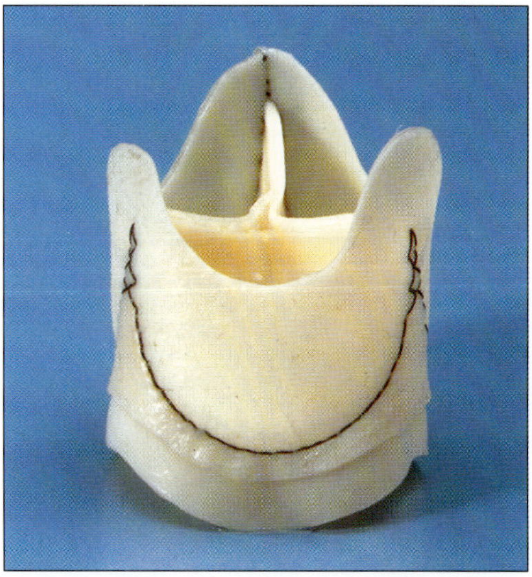

Figure 8.3. Pericarbon™ stentless valve.

The external cylinder extends on the outflow side where it is scalloped, in order to match the shape of the commissure but not to obstruct the Valsalva sinuses; some extra tissue is, however, left to be trimmed at the time of implantation, to adapt the valve to the anatomy of the individual patient.

Testing

The above-mentioned process and design characteristics have been validated through 8 years' clinical experience with the Pericarbon™ valve, in which no tissue tear has been reported.[13,14]

Advantages

The main advantages of the Pericarbon™ stentless valve design can be summarized as follows:

1. The morphology and configuration permit exploitation of all the expected advantages of the stentless aortic valve concept — optimal haemodynamics, improved mechanical durability and reduced calcification rate.
2. The totally biological configuration (absence of polymeric materials) justifies expectation of a very high resistance to infection and very low thrombogenicity.
3. The valve is adaptable to most of the surgical techniques used with homografts. The valve can be inserted with the inversion technique without consistently increasing the tissue bulk in the outflow tract, or can be inserted without inversion with interrupted sutures. The distal suture to the aortic wall may be tailored by means of intraoperative trimming of the external cylinder to match the coronary ostia.
4. The material and techniques chosen for the manufacturing process permit a high degree of reproducibility and the availability of a range of implant diameters from 15 to 29 mm.

Early clinical experience[16,17] in aortic valve replacements has confirmed the above-mentioned characteristics, in particular the excellent haemodynamic function and the low rate of complications.

References

1. David T E, Pollick C, Bos J 1990 Aortic valve replacement with stentless porcine aortic bioprosthesis. J Thorac Cardiovasc Surg 99: 113–118
2. Jaffe W M, Coverdale A, Roche A H G, Whitelock R M L, Neutze J M, Barratt-Boyes B G 1990 Rest and exercise hemodynamics of 20–23 mm allograft, Medtronic Intact (porcine) and St. Jude Medical valves in the aortic position. J Thorac Cardiovasc Surg 100: 167–174
3. Barratt-Boyes B G, Christie G W, Raudkivi P J 1992 The stentless bioprosthesis: surgical challenges and implications for long-term durability. Eur J Cardio-Thorac Surg 6 (Suppl 1): S39–S43
4. Angell W, Oury J, Lambert J, Koziol J 1989 Durability of the viable aortic allograft. J Thorac Cardiovasc Surg 98: 48–56
5. Angell W, Pupello D, Bessone L, Hiro S, Brock J 1991 Effect of stent mounting on tissue valves for aortic valve replacement. J Card Surg 6: 595–599
6. Haydock D, Barratt-Boyes B, Macedo T, Kirklin J, Blackstone E 1992 Aortic valve replacement for active infectious endocarditis in 108 patients. J Thorac Cardiovasc Surg 103: 130–139
7. Thiene G, Bortolotti U, Valente M, Milano A, Calabrese F, Talenti E, Mazzucco A, Gallucci V 1989 Mode of failure of the Hancock pericardial xenograft. Am J Cardiol 63: 129–133
8. Walley V M, Rubens F D, Campagna R T, Pipe A L, Keon W L 1991 Pattern of failure in Hancock pericardial bioprostheses. J Thorac Cardiovasc Surg 102: 187–194
9. Schoen F J, Fernandez J, Gonzales-Lavin L, Cernaianu A 1987 Causes of failure and pathologic findings in surgically removed Ionescu–Shiley standard bovine pericardial bioprostheses: emphasis on progressive structural deterioration. Circulation 76: 618–627
10. Valente M, Minarini M, Ius P, Talenti E, Bortolotti U, Milano A, Thiene G 1992 Durability of glutaraldehyde-fixed pericardial valve prostheses: clinical and animal experimental studies. J Heart Valve Dis 1(2): 217–223
11. Thiene G, Laborde F, Valente M, Gallix P, Talenti E, Calabrese F, Piwnica A 1989 Morphological survey of a new pericardial valve prosthesis (Pericarbon): long term experimental model. Eur J Cardio-Thorac Surg 3: 65–74

12. Perier P, Mihaileanu S, Fabiani J N, Deloche A, Chauvaud S, Jindani A, Carpentier A 1991 Long-term evaluation of the Carpentier–Edwards pericardial valve in the aortic position. J Card Surg 6(Suppl 4): 589–594
13. Van-Cao A N, Kucharsky K, Veyssie L, Faris B, Manasche P, Bloch G, Piwnica A 1993 The five year survival analysis of a new bioprosthesis: the Pericarbon. J Am Coll Cardiol 21 (2A): 342
14. Waunberger E, Jault F, Fontanel M, Nataf P, Cantoni M, Parie A, Cabrol G, Gandjbakhch I 1994 Bioprothèse péricardiaques de Sorin – Péricarbon. Arch Mal Coeur 87: 325–328
15. Liao K, Seifter E, Hoffman D, Yellin E L, Frater R W M 1992 Bovine pericardium versus porcine aortic valve: comparison of tissue biological properties as prosthetic valves. Artif Organs 16: 361–365
16. Casabona R, De Paulis R, Zattera G, di Summa M, Bottone W, Stacchino C, Vrandecic M O P, Morea M 1992 Stentless porcine and pericardial valve in aortic position. Ann Thorac Surg 54: 681–685
17. De Paulis R, Zattera G F, Parisi F, Leotta E, Casabona R, Galloni M, Pugliese G, di Summa M, Morea M 1995 Indications, implantation techniques and results of the various types of stentless bioprostheses: the surgeon's standpoint and the Sorin Pericarbon™ stentless valve. In: Piwnica A, Westaby S (eds) Stentless bioprostheses. Isis Medical Media, Oxford, ch 25, pp 193–198

Chapter 9

Labcor stentless aortic valve: the engineer's standpoint

I.S.J. Casagrande, H.G. Menezes, M.L.G.M. Coelho, A.N.J. Casagrande and P. Cardoso Neto

Labcor Laboratories is a Brazilian company located in Belo Horizonte, Brazil. Experience in the manufacturing of bioprostheses since 1983, with more than 20 000 implants in Brazil and other countries (Argentina, Uruguay, Portugal, Spain, France, Italy, Austria, Bulgaria, Israel, United Kingdom), added to research activities, with laboratory tests *in vitro* and *in vivo*, has allowed LABCOR to succeed in improving its manufacturing process.[1]

Processing

Porcine valves are obtained from selected slaughterhouses by specially trained personnel. Immediately after slaughter the valves are placed in a balanced saline

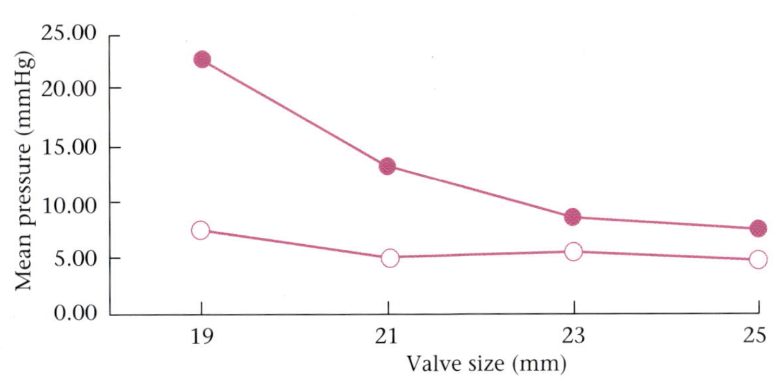

Figure 9.1. Hydrodynamic performance of the Labcor aortic prosthetic valve: results of VSI heart valve function test (mean transvalvular pressure) of stented (●) and stentless (○) valves.

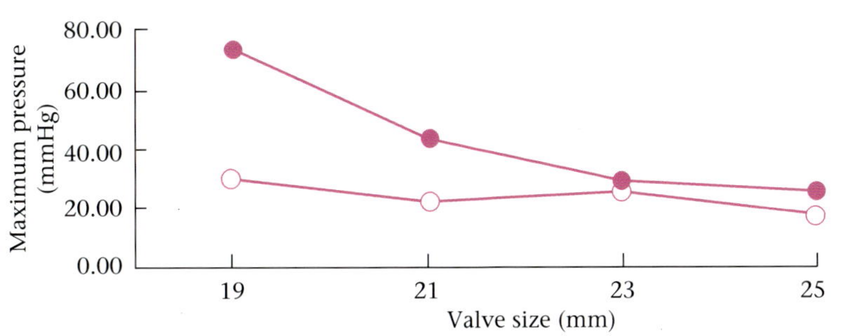

Figure 9.2. Results of VSI heart valve function test (maximum transvalvular pressure) for stented (●) and stentless (○) valves.

solution, buffered with phosphate at pH 7.4 at 4°C and rapidly transported to the laboratory. As the slaughterhouse is near the laboratory, the average time between collection and beginning of preservation in glutaraldehyde is 6 h.

After the surrounding tissue has been cleansed, the selected valves are placed in a glutaraldehyde solution buffered with phosphate at pH 7.4 at 8–10°C and no pressure is exerted over the leaflets (zero pressure fixation) for 8 weeks.[2–8]

Testing

Tests *in vitro* (Heart Valve Function Test System, VSI Vivitro Systems Inc., Victoria BC, Canada) have permitted evaluation of hydrodynamic performance (which was superior to results with the Labcor stented bioprostheses results, mainly for valves of smaller diameters – 19 and 21mm) and have demonstrated that symmetry of the leaflets and suture are essential for good performance (Figures 9.1–9.5).

Impants have been performed in sheep, which have been killed after 80–90 days for precise evaluation of prosthesis dimensions and diameter, as well as of failures due to implantation techniques. Such failures have included suture dehiscence (Figure 9.6), leaflet tearing (Figure 9.7), leaflet perforation (secondary to the suture; Figure 9.8) and paravalvular leakage (Figure 9.9), supplying essential data to aid development of the manufacturing process and the implantation technique.[9–11]

Assembly

The Labcor stentless aortic valve is constructed from three different leaflets (Figure 9.10) (composite valve) selected from porcine aortic valves, rejecting all septal

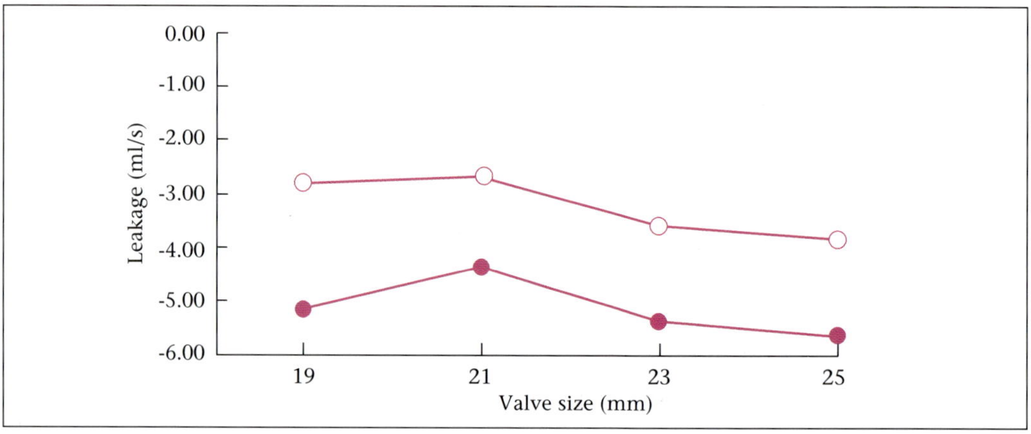

Figure 9.3. Results of VSI heart valve function test (leakage) of stented (●) and stentless (○) valves.

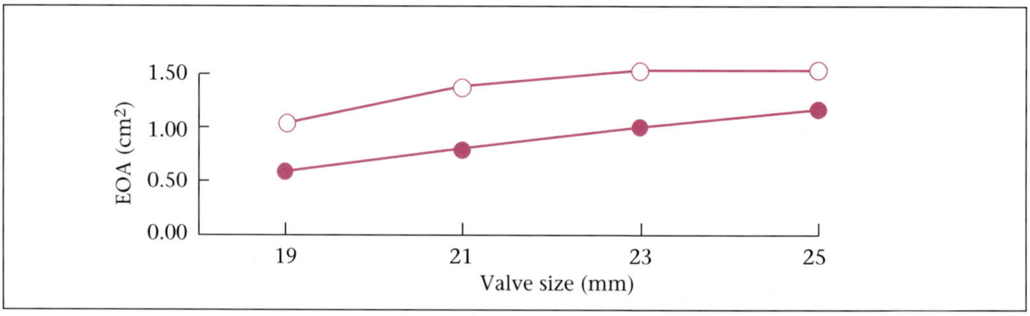

Figure 9.4. Results of VSI heart valve function test (effective orifice area; EOA) of stented (●) and stentless (○) valves.

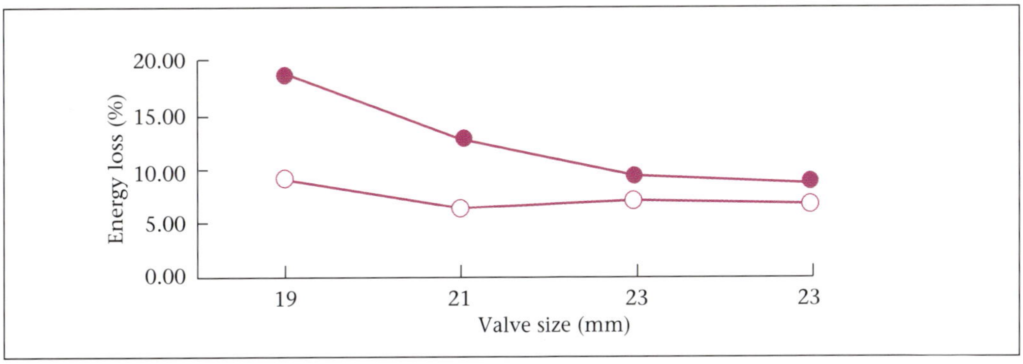

Figure 9.5. Results of VSI heart valve function test (energy loss) of stented (●) and stentless (○) valves.

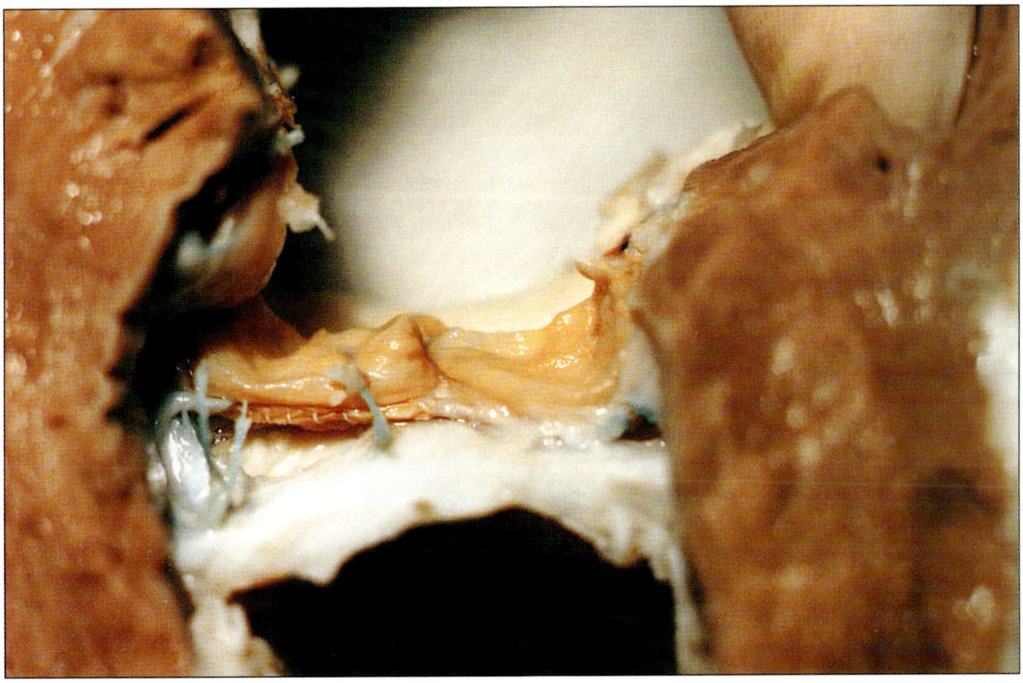

Figure 9.6. Aortic implant of 12 mm Labcor stentless aortic valve, showing dihiscence between the prosthesis and aortic root.

leaflets with a muscle shelf, or leaflets that present evidence of structural damage (e.g. myxomatous degeneration) or large areas of lipid infiltration (Figures 9.11–9.13).

During construction of the prostheses the symmetry between the leaflets is considered based on circumferential diameter, radial diameter, site of leaflet implantation, coaptation area and dimensions of the aortic wall.[12]

Once assembled, the prostheses are inspected under polarized light with magnification, which permits verification of the collagenic structure, as well as detection of lesions originating during manufacture[13] (Figures 9.14 and 9.15).

Presentation

When they have passed inspection, the prostheses are individually sterilized and packed in a 0.6% glutaraldehyde/20% ethanol solution and packed in a 0.2% glutaraldehyde solution. Serial cultures in thioglycolate, Casoy, Saboraud–Dextrose, Micosel and Lowenstein–Jensen media are obtained and the prostheses

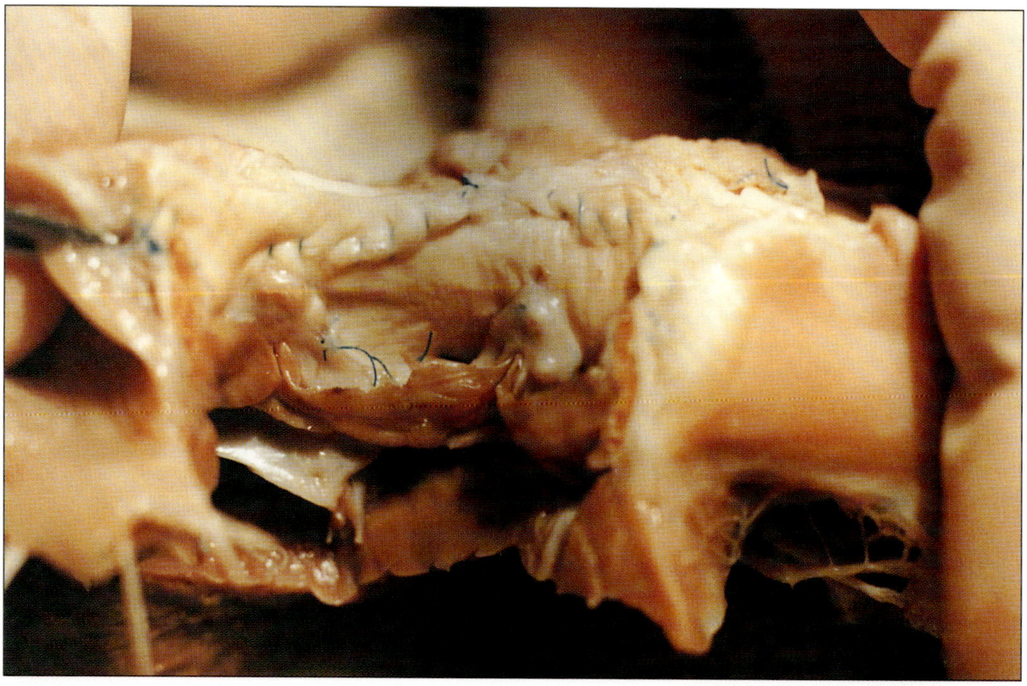

Figure 9.7. Aortic implant of 12 mm Labcor stentless aortic valve, showing tearing of leaflet by a polypropylene suture.

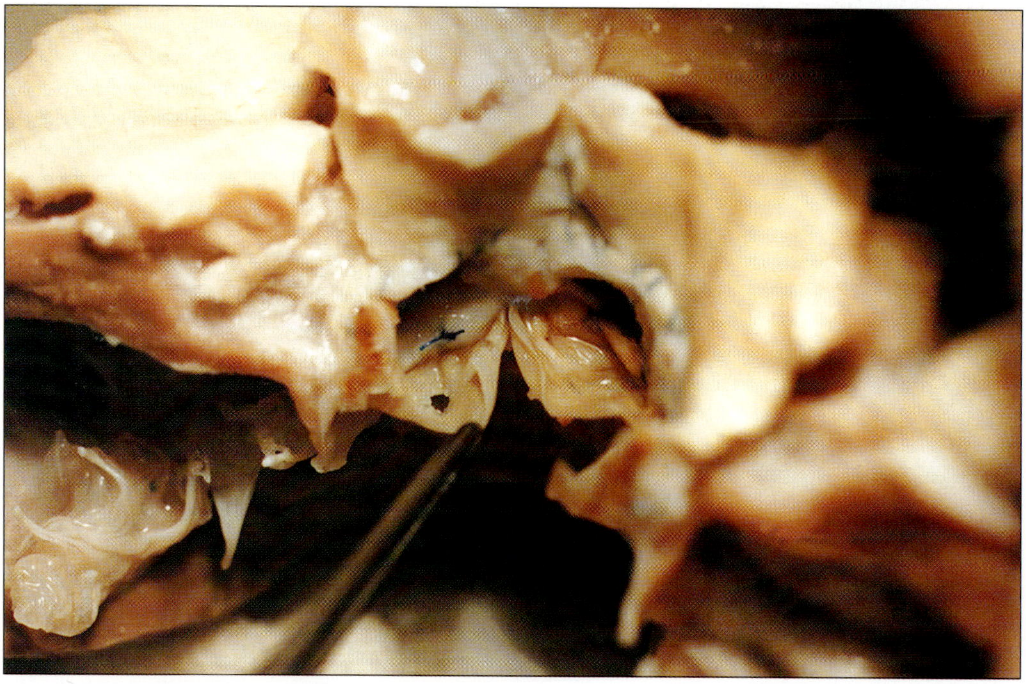

Figure 9.8. Aortic implant of 12 mm Labcor stentless aortic valve, showing perforation of leaflet secondary to polypropylene suture.

are kept in isolation. Once their sterility has been confirmed they are ready for clinical use, packed in a glutaraldehyde solution (Figures 9.16–9.18).

The prostheses are presented in dimensions corresponding to their internal diameters in millimeters, which correspond to the sizers utilized during implantation.

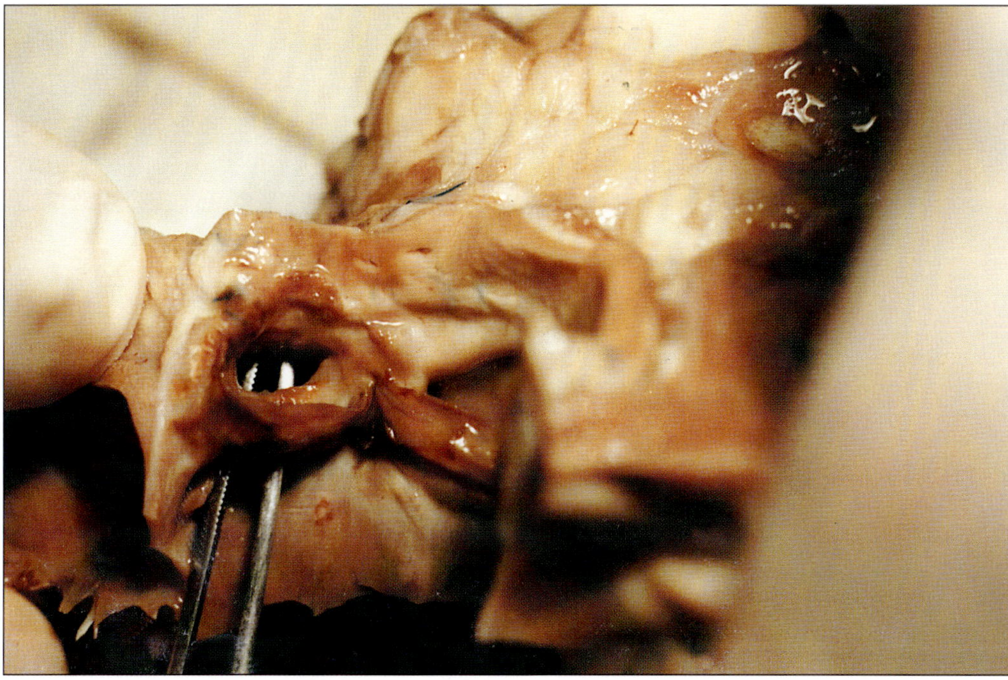

Figure 9.9. Aortic implant of 12 mm Labcor stentless aortic valve, showing paravalvular leakage.

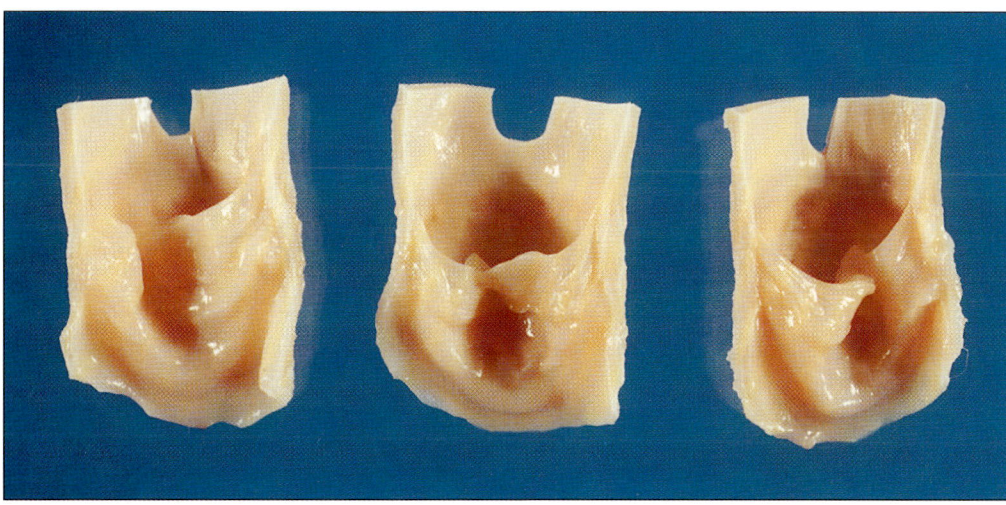

Figure 9.10. Glutaraldehyde-preserved non-septal leaflets of aortic porcine valve.

Performance

Labcor stentless aortic valve implants have been used since April 1993 in seven hospitals in six Brazilian cities, with good haemodynamic performance.[1]

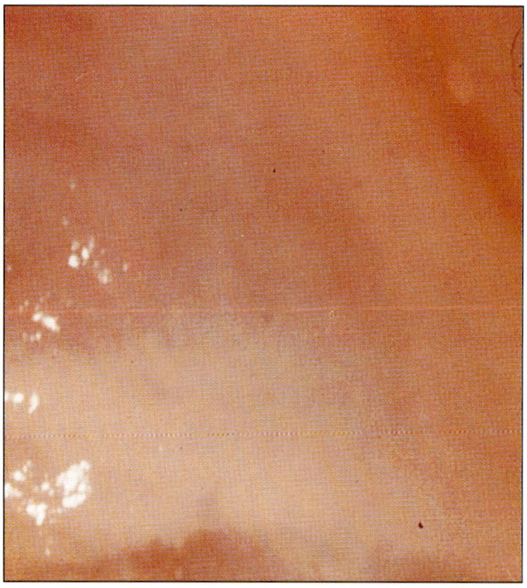

Figure 9.11. Macroscopic view of myxomatous degeneration of porcine aortic leaflet.

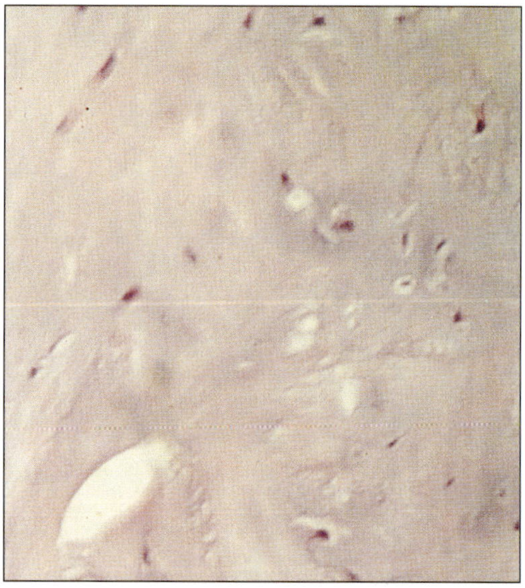

Figure 9.12. Myxomatous degeneration of porcine aortic leaflet, stained with haematoxylin and eosin, × 160.

Figure 9.13. Lipid infiltration of porcine aortic leaflet, shown by polarized light with magnification.

Figure 9.14. Photomicrograph of transverse cross-section of porcine aortic valve, preserved in glutaraldehyde by zero-pressure fixation, showing preservation of the entire tissue, as demonstrated by a birefringence pattern within the collagen fibres , × 630.

Figure 9.15. Photomicrograph of porcine aortic valve preserved in glutaraldehyde by zero-pressure fixation. Magnification (×40) under polarized light permits non-destructive evaluation of collagen morphology: the bent strips are seen as the natural valve surface, and the circumferentially orientated birefringent bundles of collagen fibres display a banding pattern, indicative of collagen crimp retention.

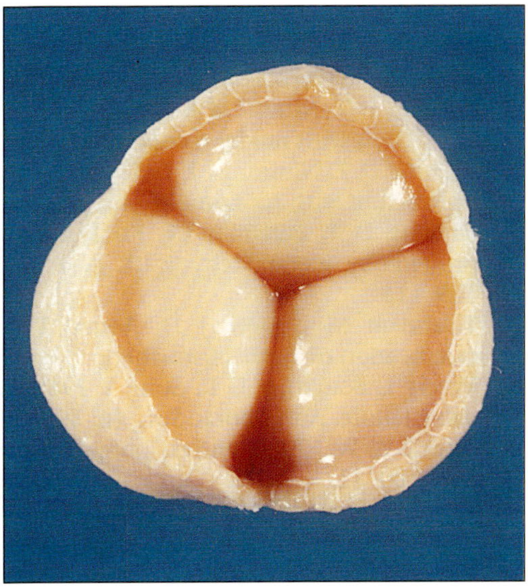

Figure 9.16. Labcor stentless aortic valve, inflow view.

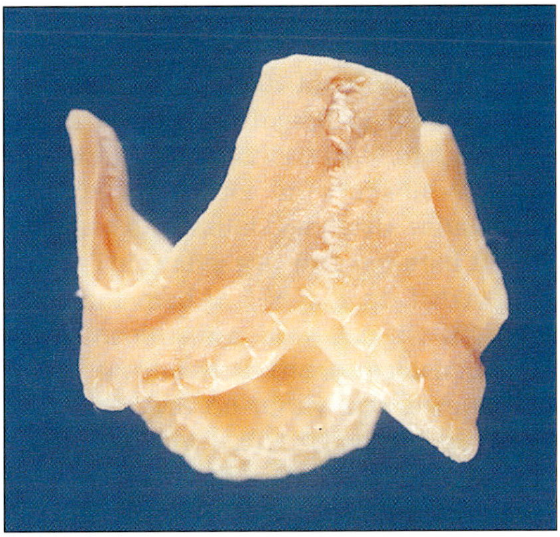

Figure 9.17. Labcor stentless aortic valve, lateral view.

Figure 9.18. Labcor stentless aortic valve packed ready for use.

References

1. De Souza J M, Oliveira S A 1995 Early experience with the Labcor stentless porcine aortic bioprosthesis. In: Piwnica A, Westaby S (eds) Stentless bioprostheses. Isis Medical Media, Oxford, ch 26, pp 199–203
2. Broom N, Thompson F J 1979 Influence of fixation conditions on the performance of glutaraldehyde-treated porcine aortic valves: towards a more scientific basis. Thorax 34: 166–176

Valve design

The O'Brien–Angell Model 300 valve is a composite design constructed from the non-coronary leaflets of three porcine aortic valves. After fixation has been completed, initial scalloping of the sinuses is performed, followed by careful removal of the non-coronary leaflet. Each leaflet is examined in detail to determine leaflet structure and integrity. Examination of the leaflet after removal from the conduit allows the technicians access to the entire leaflet surface area. Any leaflet that does not meet the critical specifications is rejected. Further dissection is necessary to remove the remnant ventricular muscle and create the unique low-profile inflow surface of the valve.

The leaflets are critically matched for size and symmetry with a proprietary template to assure competence, maximum leaflet coaptation and synchronous opening.

Valve assembly

Assembly of the valve is begun by matching three non-coronary leaflets, the total surface area of which will equal the desired valve size (Figure 10.1). Initial interrupted, and then running, sutures are placed along the free edges of the aortic wall, carrying the sutures the full length of the leaflet commissures. A blanket stitch is placed to secure the inflow surface of the valve and final sculpting of the sinuses is completed.

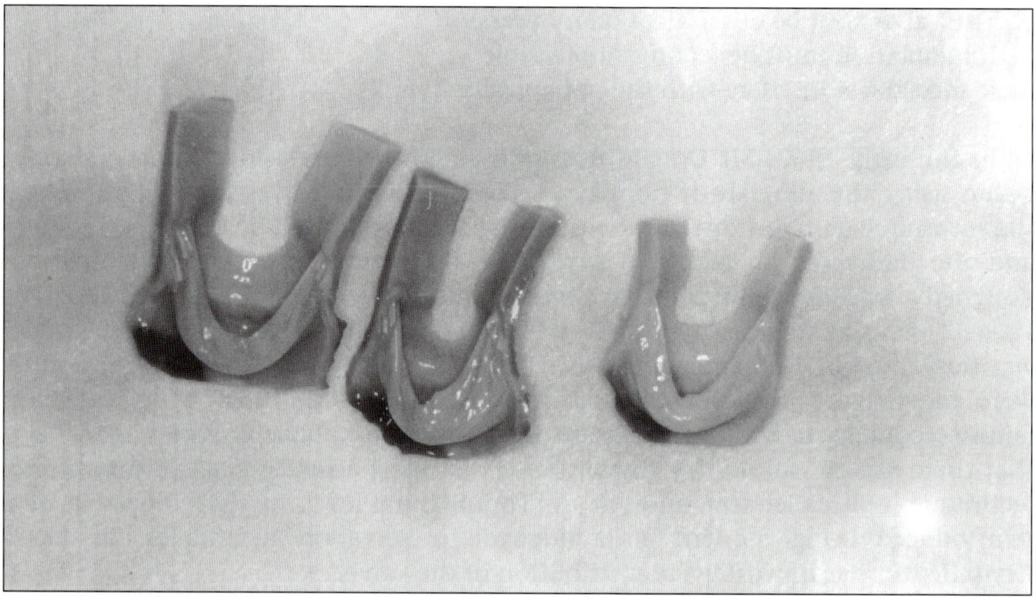

Figure 10.1. Three non-coronary leaflets of porcine aortic valves, as used in the O'Brien–Angell Model 300 stentless valve.

At the completion of assembly, the valve is grossly examined to ensure that construction meets Bravo's strict guidelines (Figure 10.2). Each valve is sized using calibrated sizers.

Finally, each valve undergoes a Quality Assurance microscopic analysis. The examination includes the leaflet surface area, coaptation, and all suture lines.

This unique design provides optimal haemodynamics as a result of removing all ventricular muscle, with an end product of gradients approaching those of allograft heart valves.

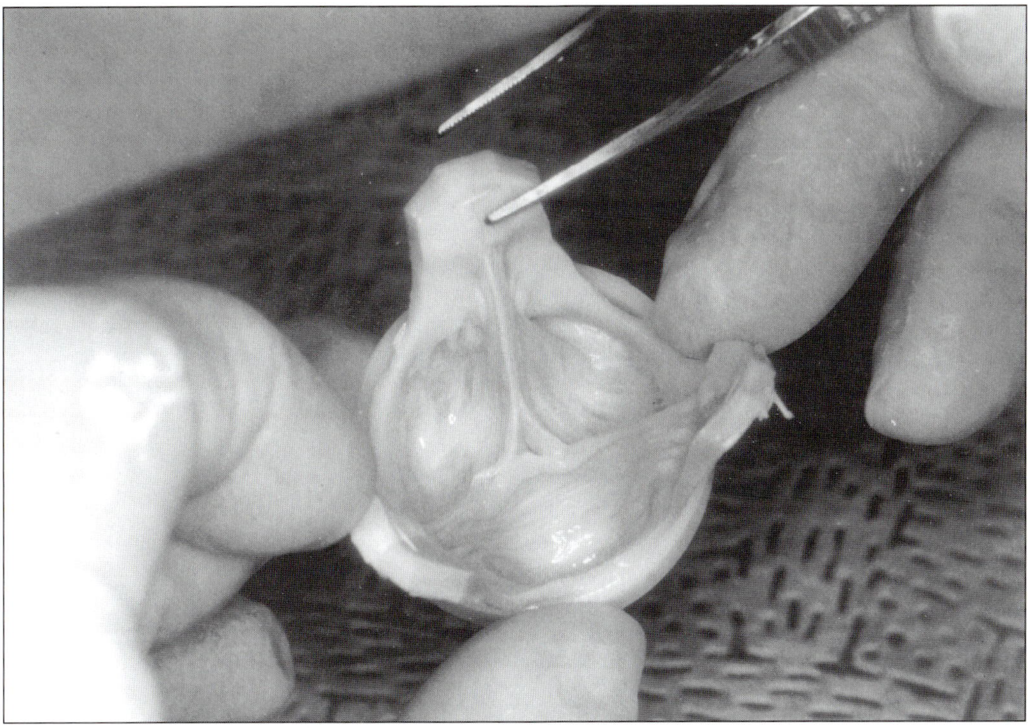

Figure 10.2. Examination and sizing of the assembled valve.

Properties of the O'Brien–Angell Model 300 stentless valve

Testing performed *in vitro* on the O'Brien–Angell valve comparing it to stented bioprostheses illustrates the excellent gradient qualities of the valve (Table 10.1). All three valves are sized at 25 mm. Cardiac output is 4.9 l/min at a pulse rate of 75 bpm. The peak transvalvular pressure gradients across the valve in systole are 30% less than those of stented valves, with an effective orifice area approximately 20% greater than that of the stented test valves. Closing volumes are about one-half those of the stented valves.

*Table 10.1 Hydrodynamics of the O'Brien–Angell Model 300 stentless valve**

Valve type	Peak Δ p† (mmHg)	EOA†† (cm^2)	Closing volume (ml)	Leakage (ml)
O'Brien–Angell	7	2.82	1.2	0
Angell–Duran	10	2.17	2.1	0.2
Hancock	10	2.24	2.0	0.2

*Valve size 25 mm; †pressure drop; ††effective orifice area.

The absence of additional biological or synthetic material eliminates excess bulk at the annulus (Figure 10.3). The supra-annular design of the valve allows for use of one size larger valve than measured, i.e. a 25 mm annulus will accept a 27 mm O'Brien–Angell valve. Implantation times are relatively brief with the use of the single suture line technique utilizing a double-armed 3/0 polypropylene suture. The sutures must be triangulated deep in the sinus, incorporating as much xenograft aortic wall as possible without compromising the leaflet. The sutures are placed at the centre, or nadir, of each sinus and tied, fixing the valve at these three

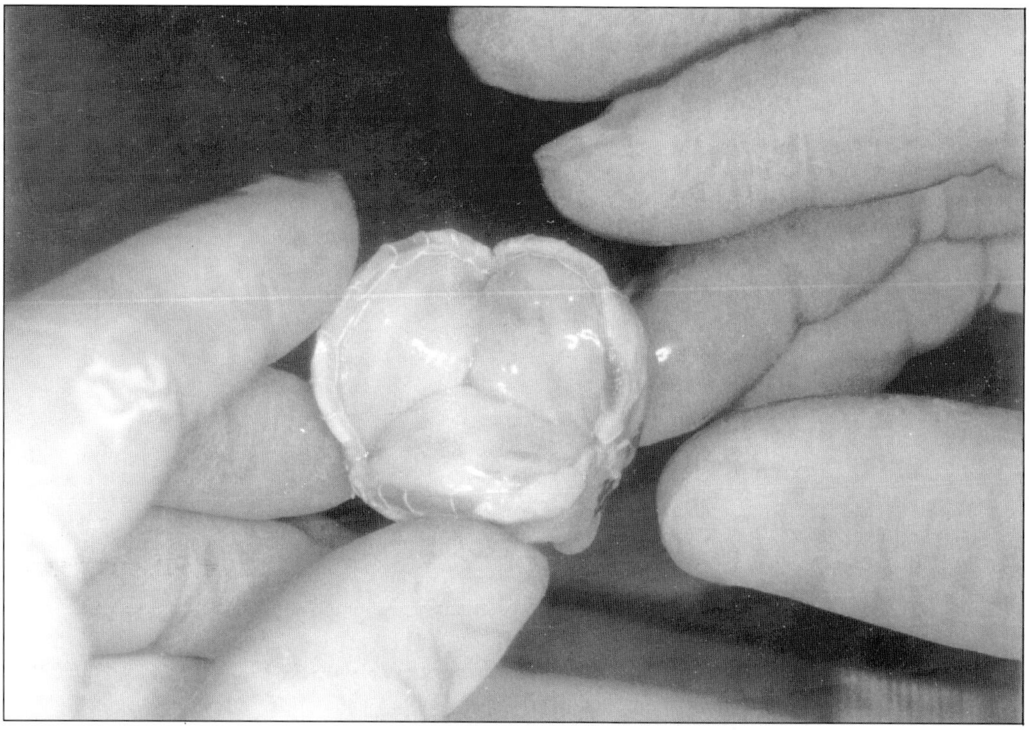

Figure 10.3. Assembled valve, showing lack of excess bulk at the annulus.

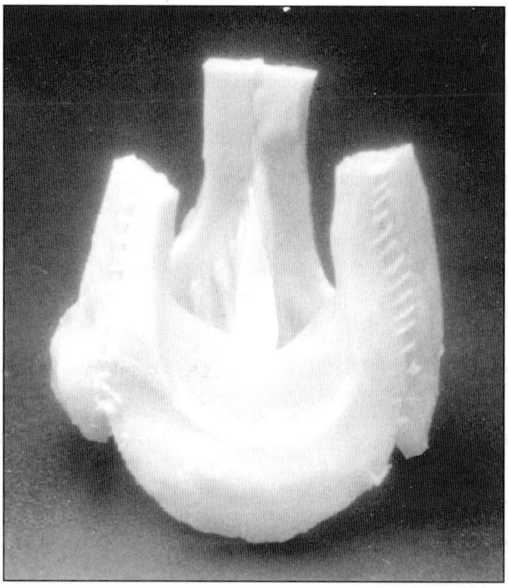

Figure 10.4. The O'Brien–Angell Model 300 stentless valve.

points. Each arm of the suture attaches the xenograft to the recipient aorta, terminating at the top of the commissure where the suture is taken out through the aortic wall and secured to the adjacent arm, resulting in six fixation points. Proper suture placement is critical to a successful implantation.

The lack of any synthetic material significantly lessens the risk of postoperative infection. In the patient with endocarditis, and in the absence of a homograft, the O'Brien–Angell valve could well be the valve of choice.

The O'Brien–Angell Model 300 stentless valve (Figure 10.4) is designed after the valve that proves itself hundreds of millions of times each day — the homograft valve. Clinical results from around the world indicate implant times of 30–60 min, with gradients approaching those of homografts, and excellent short-term performance.

Chapter 11

Wear-tested pulmonary porcine valve: mechanical properties and ultrastructural findings

G. Gerosa, H. David, M. Valente, D. Boughner, I. Vesely, G. Thiene, J. Davies, C. Muneretto and D. Casarotto

The potential value of the pulmonary valve for use as an aortic valve replacement has been suggested by Ross *et al.*[1,2] In their experience pulmonary autografts have performed extremely well over the long term[2,3] and pulmonary homografts have shown satisfactory and promising short-term results,[1] proving in the clinical setting that the natural pulmonary valve can readily sustain diastolic pressure at systemic levels without tissue disruption. Thus, the porcine pulmonary valve might also be an alternative worth considering as an aortic bioprosthetic device.

The stentless aortic porcine bioprosthesis is emerging as a promising tool to improve the durability of glutaraldehyde-fixed porcine aortic xenografts. Indeed, the porcine pulmonary valve as an unstented replacement device may be anatomically and biochemically superior to the porcine aortic valve. The latter has the structural disadvantage of a muscular shelf in the right coronary cusp, which reduces the functional valve orifice size. In the pulmonary porcine valve this asymmetrical muscle bar is absent and the total content of calcium in the leaflets and wall is lower than in its aortic counterpart.[4]

Although clinical experience has shown that pulmonary autografts and homografts can perform well, there is still great concern regarding the ability of the pulmonary valve to withstand systemic pressure. The recent demonstration that fresh human, as well as fresh and fixed porcine, pulmonary and aortic valve tissues have similar tensile properties gave experimental evidence that the pulmonary valve is mechanically suitable for use as an aortic replacement device.[1,5] Nevertheless, it is current opinion[6] that in the long term the thinner pulmonary leaflets would fail, thus harming valve durability.

It was, therefore, decided to examine the mechanical properties and ultrastructural characteristics of pulmonary valves after they had been tested in a wear test system. A comparison was then made with glutaraldehyde-fixed wear-tested porcine aortic valves, for which the projected durability is known.

Materials and methods

Valve preparation

Low-pressure fixation (less than 4 mmHg), using 0.2% glutaraldehyde in pH 7.4 buffer, was used for treatment of pairs of aortic and pulmonary valves harvested from similar pigs. Similar-sized valves, 23 or 25 mm, were sewn into appropriate sizes of porcine bioprosthesis stents. Three aortic and three pulmonary valves were mounted into the appropriate clamping systems of six independent test chambers in a Rowan Ash Accelerated Heart Valve Fatigue Test apparatus. Individual test chambers were numbered 1 to 6 and details were recorded of each valve mounted into its chamber. The Rowan Ash apparatus was 'balanced' in that three size 23 mm valves were mounted in positions 1, 3 and 5, with three size 25 mm valves in positions 2, 4 and 6.

After all the valves had been mounted, the apparatus was switched on and each valve viewed with a stroboscope lamp. The amplitude of the driving plate was increased until the least mobile of all the valve leaflets was seen to open fully and close completely during each cycle. The apparatus was operated at 1420 vibrations/min, equivalent to 20 times the normal heart rate of 70 bpm.

A Gael-Tech differential pressure transducer was used to measure the pressure cycle experienced by each valve on test. A pressure cycle showing a difference of 75–100 mmHg was required for all valves on test and, if necessary, the amplitude of the driving plate was increased further to achieve this pressure differential cycle together with complete opening and closing of each valve. The cycle counter was set to zero, the 0.2% glutaraldehyde test solution was degassed and the system was left to run.

The apparatus was examined every working day by stroboscope lamp to detect gross morphological damage. In accordance with I.S.O. 5840, the system was stopped every 38 × 10 cycles (1 year equivalent) and all the valves were removed in turn from the apparatus and examined with a ×10 stereomicroscope. Observations and cycle numbers were recorded and the system was restored to use, degassed and left to run for a further 38 × 10 cycles.

At least 380 × 10 cycles (10 years equivalent) were used to assess all six valves. Independent test chambers permitted continued testing of undamaged valves in the presence of defects on other valves.

Morphological study

Once they were fatigued, all six prostheses were examined grossly and photographed. All three leaflets were then removed from each valve for a total of 18 cusps. One leaflet from each valve was put aside for histological and ultrastructural studies. For histology, blocks of leaflets were paraffin embedded, sectioned 5 mm thick and stained with haematoxylin and eosin, Weigert–Van Gieson and Azan stains. For transmission electron microscopy, small blocks of leaflets were fixed in 2% glutaraldehyde, post-fixed in osmium tetroxide and embedded in Epon. Semithin sections were stained with toluidine blue and examined by light microscopy. Ultrathin sections of selected areas were stained with uranyl acetate and lead citrate and examined by transmission electron microscopy. For scanning electron microscopy, small blocks of leaflets were dehydrated at critical-point drying, coated with gold and examined by scanning electron microscopy.

As a control group, three aortic and three pulmonary valves, removed from pigs soon after slaughter, were studied in the same manner.

Mechanical study

The remaining 12 leaflets were cleaned in modified Hanks' solution (pH 7.4, 31 mOs), inspected for any abnormalities and then stored in chilled Hanks' solution until testing. Samples were cut with a scalpel from the central portion of each cusp, producing either a 5 mm wide circumferential or 10 mm wide radial strip; thus, four test groups of three specimens each were created. The test groups comprised aortic and pulmonary circumferential strips and aortic and pulmonary radial strips. The test strip widths were chosen to exceed the minimum needed to achieve reproducible results, as shown by Lee *et al.*[7] The precision in cutting the test strips to the desired widths was estimated to be ± 0.2 mm (2% for a 10 mm wide radial strip). The thickness of each leaflet was measured to an estimated accuracy of ± 0.05 mm using a specially modified microscope.[8] Mean specimen thicknesses are shown in Table 11.1 (see p. 87).

All tensile stress/strain and stress relaxation tests were performed with a screw-driven Instron tensile testing machine (Model 1125, Instron Corp., Canton, Massachusetts, USA). A 500 g load cell was used for all tests. Test strips were clamped between stainless steel jaws covered with 320 emery paper to prevent slippage. The whole clamping apparatus was attached to the base of a Plexiglass tub which was itself fixed to the lower, movable crossbar of the Instron. The tub was filled with modified Hanks' solution at 37°C. The gauge length of the tissue strip was estimated at zero load as transition occurred from wrinkled to straight during elongation, as previously described.[8] Each specimen was strained repeatedly by cyclically loading it to 100 ± 1g (0.98 N) at an elongation rate of 10 mm/min. This preconditioned the tissue and resulted in reproducible load/elongation curves. At the point of zero loading, the gauge length was measured from a calibrated scale on the tissue clamping apparatus.

The test extension rate was then set at 200 mm/min and the tissue loaded to 200 ± 1g (1.96 N) for circumferential strips and 167 ± 1g (1.64 N) for radial strips. At maximum load, the Instron crosshead was stopped and the tissue allowed to stress relax for 100 s. The specimen was then unloaded and then retested at extension rates of 50, 10 and 2 mm/min. Finally, each specimen was strained to fracture at an extension rate of 10 mm/min.

Data analysis

The resulting load/extension (stress/strain) and load/time (stress relaxation) curves were digitized using a Summagraphics tablet with an accuracy of ± 0.25 mm. Thirty points from both the load/extension and load time curves were recorded. Thirty points were also recorded from the fracture curves, with digitizing ceasing at the first significant drop in stress. This final point was later recorded as the fracture stress and strain. A Lotus-123 worksheet was used to calculate the resulting stress/strain, stress relaxation and fracture curves. The initial cross-sectional area was used to calculate the stress. This avoided the added error of attempting to estimate the Poisson ratio and the contraction of the sides of the specimen during testing. Thus, the stress calculated was the 'engineering' stress and not the 'true' stress, resulting in underestimation of the stress value. Owing to the comparative nature of this study and the identical testing methods, engineering stress was a valid measure of the tissue properties. Three constants were also calculated for statistical evaluation: these were (1) the elastic modulus of the stress/strain curve at 200 kPa, (2) the low modulus length of the stress/strain curve, or the leaflet extensibility, and (3) the slope of the stress relaxation curve. The elastic modulus was calculated at 200 kPa because this falls within the physiological range experienced by the leaflet. The extensibility of the tissue was calculated as the intersection point of the tangent line to the stress/strain curve at 200 kPa and the *x*-axis. The slope of a regression line fit of the relaxation curve was used as the third constant, the slope of the stress relaxation curve. Measuring the relaxation slope gave a measure of the viscosity of the tissue.

The three individual stress/strain and stress relaxation curves for each test group and extension rate were averaged to obtain mean and standard error curves. This enabled visual and statistical comparison to be made between test groups and extension rates. Lotus-123 software was used to perform paired and independent *t*-tests at the 99% confidence level on the three constants. Paired *t*-tests were used to compare the different strain rates for the data from the same test group, as the samples were used. Independent *t*-tests were used to make all statistical comparisons between test groups. Finally, the data on the fatigued cusps in this study were compared with the fresh and fixed cusps tested in an identical manner and described in a previous paper by this group.[5] As the testing and data evaluation methods were identical in both studies, a direct comparison was valid.

Results

Morphological study

At gross examination two of three pulmonary prostheses showed cusp perforations: one had two rounded holes in two different cusps, 5 and 2.5 mm in length, respectively; the second prosthesis had an initial tear 3 mm in length. One of three aortic prostheses had an initial T-shaped perforation at the right cusp, 2 mm in length. All the tears were located in the middle of the body of the semilunar leaflets.

Histological examination disclosed, at the site of laceration, disruption of the tunica fibrosa with swelling of the underlying spongiosa.

With transmission electron microscopy, both aortic and pulmonary prostheses in areas with intact leaflets on gross and histological examination, showed regular waviness of collagen bundles, normal elastic fibres and fairly well preserved cells, surrounded by a halo of empty spaces suggesting interstitial oedema (Figure 11.1).

Borders of tears featured severe disruption of connective tissue fibres, with loss of crimping of the collagen scattered within abundant amorphous electrolucent material, probably consisting of insudation of medium; cusp cells, on the other hand, were fairly well preserved (Figure 11.2).

With scanning electron microscopy, partial detachment of the endothelial lining was observed in both aortic and pulmonary prostheses on each side.

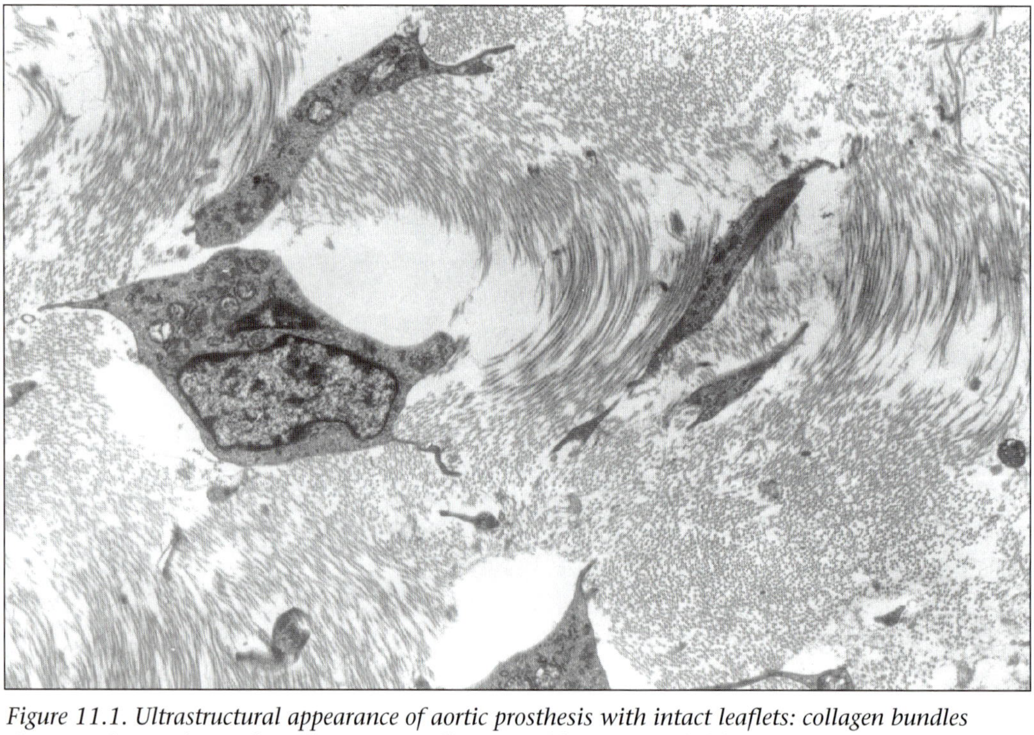

Figure 11.1. Ultrastructural appearance of aortic prosthesis with intact leaflets: collagen bundles show regular waviness; the cells are very well preserved but surrounded by an empty halo. Uranyl acetate and lead citrate (× 6000).

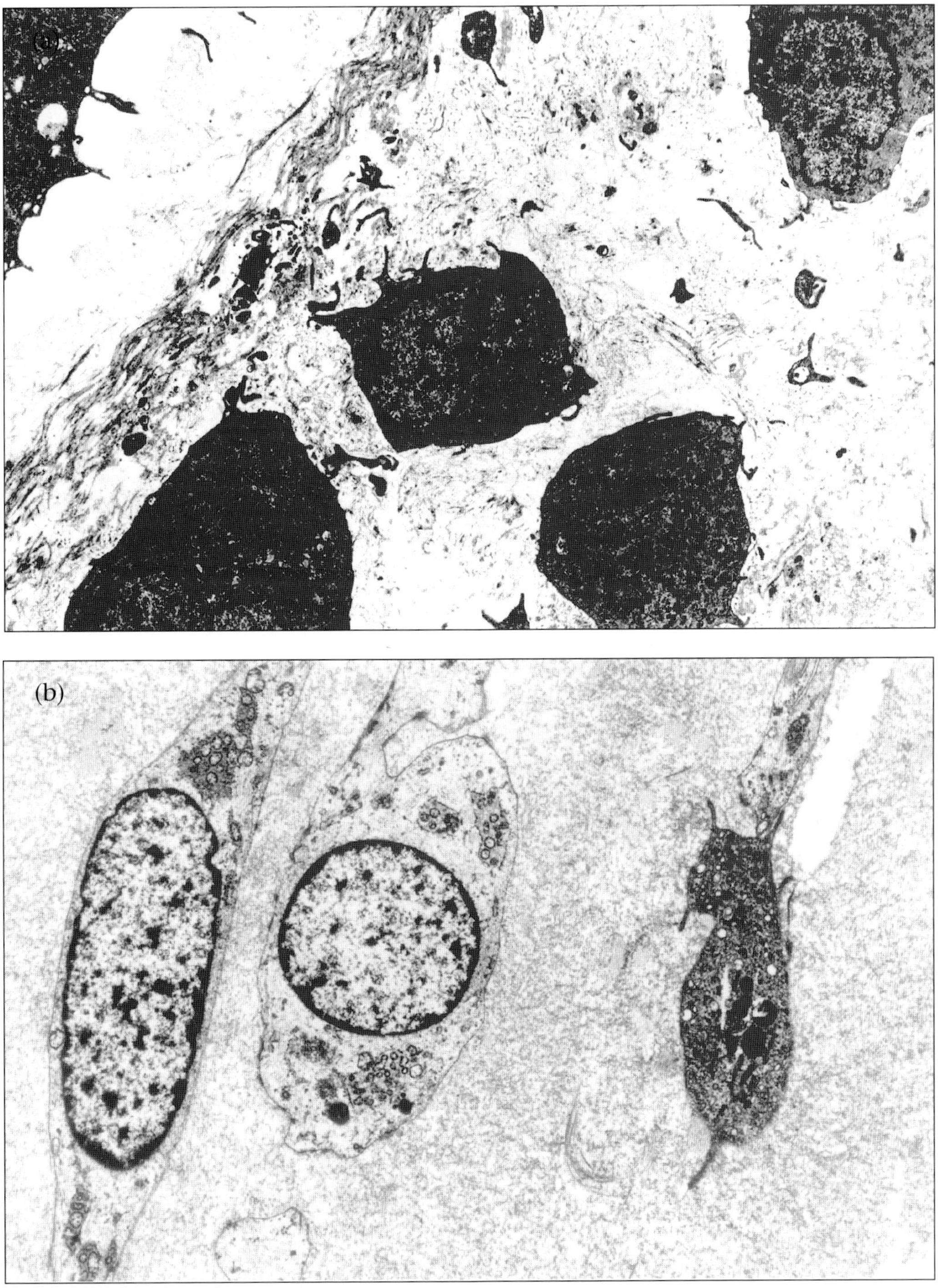

Figure 11.2. Ultrastructural appearance of aortic (a) and pulmonary (b) leaflets at the borders of the tears: collagen fibres are disrupted, whereas cusp cells appear fairly well preserved. Uranyl acetate and lead citrate (a × 4800; b × 5000).

Histological and ultrastructural studies performed on porcine valves did not show significant differences between aortic and pulmonary valvular tissue (Figure 11.3).

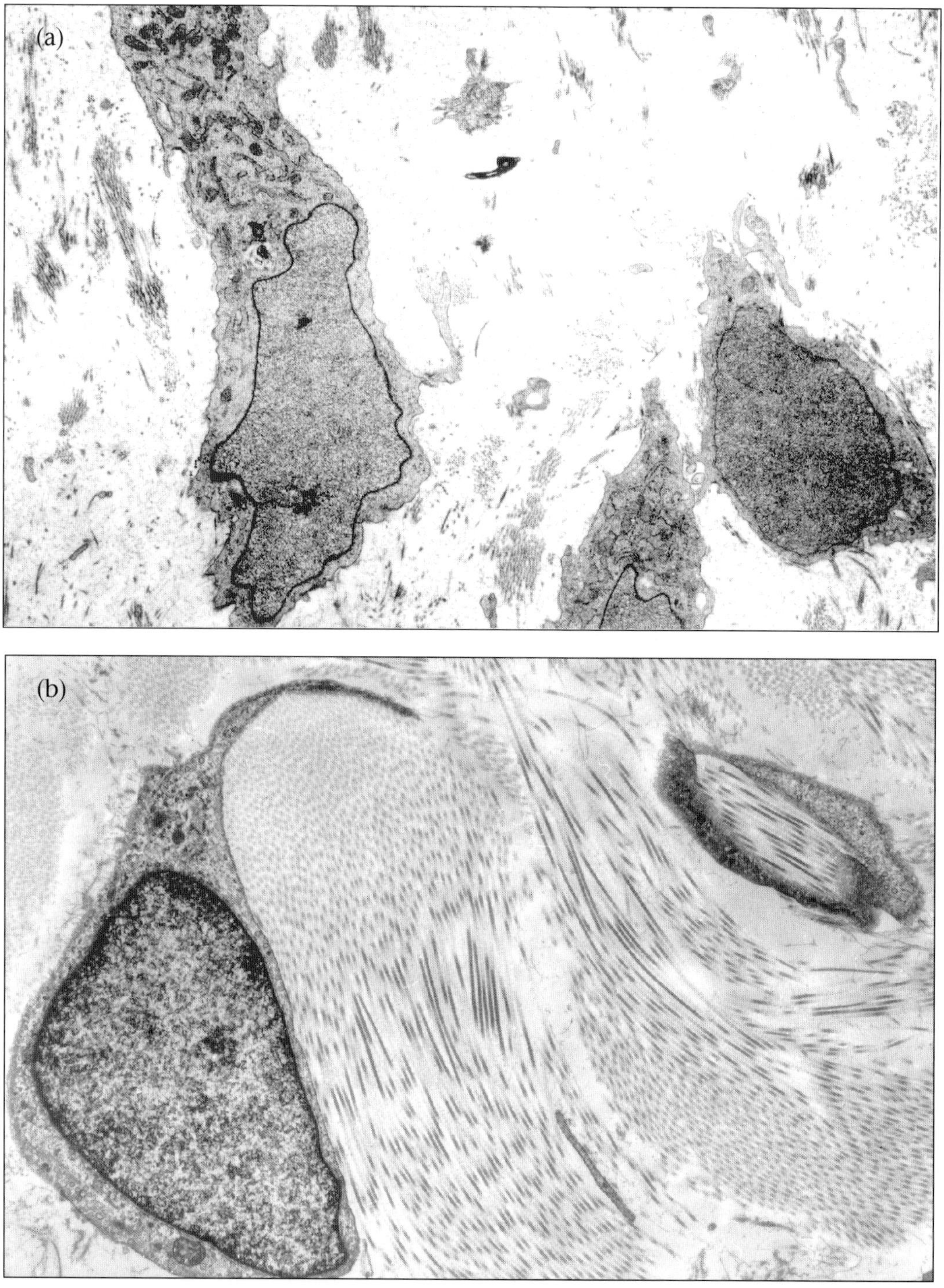

Figure 11.3. Ultrastructural appearance of aortic (a) and pulmonary (b) valves removed from pigs soon after slaughter: note very well preserved cells, without any empty halo, ground substance and scattered bundles of collagen fibres in aortic spongiosa (a) and variably orientated collagen fibres in the fibrosa of pulmonary valve (b). Uranyl acetate and lead citrate (a × 6000; b × 9000).

Mechanical study

Leaflet thickness

Fatigued pulmonary valve leaflets had similar thicknesses whether cut radially or circumferentially and were noticeably thinner than their aortic counterparts. For the fatigued aortic strips the thickness differed, with the tissue being slightly thicker in the circumferential direction (0.59 mm) than in the radial direction (0.48 mm) (Table 11.1).

*Table 11.1 Measured thickness and unloaded 'gauge' length of fatigue-tested circumferential or radially cut strips of aortic and pulmonary valve tissue**

Fatigued strips	Thickness (mm)	Gauge length (mm)
Circumferential aortic	0.59 ± 0.06	7.1 ± 1.01
Circumferential pulmonary	0.31 ± 0.07	8.2 ± 0.58
Radial aortic	0.48 ± 0.02	2.83 ± 0.29
Radial pulmonary	0.32 ± 0.09	2.42 ± 1.01

*Values are means ± SD

When these results were compared with previous measurements of fresh and unstressed glutaraldehyde-fixed valves,[5] both pulmonary and aortic cusps were thinner after the fatigue testing. Fresh and fixed aortic circumferential strips measured 0.67 and 0.70 mm respectively, compared with 0.59 mm after cycling in the fatigue test system. Similarly, fresh and fixed pulmonary valve circumferential strips measured 0.55 mm and 0.54 mm, respectively, compared with 0.31 mm for this study. The radial strips from both valves showed a similar decrease in thickness. This decreased thickness probably results from tissue elongation and thinning (stress relaxation) during cycling in the fatigue testing system.

Elastic moduli

The stress/strain diagram and elastic moduli measured showed both pulmonary and aortic cusps to be highly anisotropic. Circumferential strips from both valves had a much higher elastic modulus than was evident for the radial strips (Table 11.2). This result could be anticipated, owing to the obvious preferential alignment of load-bearing collagen fibres in the circumferential direction.

The elastic moduli of fatigued circumferential pulmonary cusps, at all four extension rates and within both the circumferential and radial direction groups, usually differed, with pulmonary cusps being stiffer (Table 11.2). There was less of a difference at high extension rates. The opposite was true with fatigued radial specimens, where the aortic specimens were stiffer (Table 11.2).

*Table 11.2 Three calculated constants (elastic modulus, low-modulus phase and stress relaxation) obtained from four types of fatigued strips at a strain rate of 10 mm/min**

Fatigued strips	Elastic modulus† (kPa)	Length of low-modulus phase (%)	Slope of stress relaxation curve
Circumferential aortic	10.4 ± 1.2	18.1 ± 3.4	–2.6 ± 0.5
Circumferential pulmonary	16.4 ± 1.4	20.3 ± 2.6	–4.9 ± 3.0
Radial aortic	3.0 ± 0.8	62.3 ± 12.8	–3.9 ± 0.3
Radial pulmonary	2.1 ± 0.2	75.9 ± 14.5	–5.3 ± 1.4

*Values are means ± SD; † at 200 kPa

Although the elastic moduli of aortic and pulmonary cusps were similar at each extension rate when the highest extension rate (200 mm/min) was compared with the lowest extension rate (2 mm/min), a lower elastic modulus was observed at the higher strain rate (i.e. 8.0 kPa vs 12.0 kPa). This observation was the same as that noted for fresh and unstressed glutaraldehyde-fixed valve tissue.[5]

When comparing the fatigued samples with the fresh and glutaraldehyde-fixed valves that were tested previously,[4] it is interesting that the elastic moduli of the fatigued cusps have greater similarity to those of the fresh cusps (i.e. 10.4 kPa vs 11.6 kPa) than those of the fixed cusps (i.e. 10.4 kPa vs 7.7 kPa). Although the elastic moduli of the fatigued samples were closer to those of the fresh cusps, there were still significant differences between most of the samples, especially in the circumferential direction. This may indicate that the apparent lowering of the elastic moduli during fixation is only temporary. In addition, the same relationship of decreasing elastic moduli with increasing extension rate held for all groups. Fatigued pulmonary strips had radial elastic moduli similar to those of fresh aortic strips (2.1 kPa vs 3.1 kPa), whereas circumferential pulmonary specimens had higher elastic moduli than similar fresh aortic strips (16.4 kPa vs 10.7 kPa).

Extensibility

The extensibilities of the fatigued aortic and pulmonary cusps, when measured as the length of the low modulus phase, were similar (Table 11.2). For both aortic and pulmonary cusps, the circumferential strips were much less extensible than radial strips (i.e. for aortic cusps, 18.8 vs 62.3%). The much less extensible collagen fibres that are preferentially orientated in the circumferential direction prevent the very extensible elastin fibres from stretching circumferentially.

Statistically, there was little change in the extensibility of the cusps as the strain rate increased. Comparison of the extensibilities of circumferential strips of both aortic and pulmonary valves at the lower and upper strain rates of 2 and 200 mm/min showed little change (18.1 vs 17.9% strain and 20.3 vs 18.6% strain, respectively). When the same extension rates were compared for radial strips there again were no statistically significant differences between them, but a trend towards lower extensibility at higher strain rates was observed (i.e. 62.0 vs 63.7% strain). At higher extension rates, the tissue would have less time to stress relax through viscous flow within the tissue, resulting in a more rapid onset of higher stresses. It is expected that even higher extension rates, closer to actual conditions *in vivo*, would result in even lower extensibility.

Again, comparison of the fatigued-cusp data with previous fresh and glutaraldehyde-fixed data[5] showed that the same anisotropic and strain rate relationship applied. In all three cases, there was little dependence of extensibility on whether the cusp was aortic or pulmonary, or the extension rate at which it was tested. For radial strips, the same greater extensibility was observed when compared with circumferential strips, as well as the same trend towards less extensibility at higher extension rates. Whereas the fatigued cusps exhibited extensibilities similar to those of the fresh cusps, there was a statistically significant difference from those of the fixed cusps: for example, fatigued circumferential pulmonary strips had an extensibility of 20.3% strain, whereas the same strips of fresh and fixed tissue had extensibilities of 22.9% and 46.1% strain, respectively. Fatigued pulmonary strips proved to have extensibilities similar to those of fresh aortic strips (Table 11.2).

Viscosity

The viscosity of the cusp was measured as the slope of the stress relaxation curve. Statistically, the fatigued radial pulmonary strips relaxed more quickly than aortic strips (–5.3 vs –3.9) and radial aortic strips relaxed more quickly than circumferential strips (–3.9 vs –2.6). The radial pulmonary strips also showed a tendency, although not a statistically significant difference, towards more rapid relaxation (–5.3 vs –4.9). Owing to a lack of aligned collagen, the radial strips would be more dependent on the viscous component for load bearing and would therefore be expected to stress relax more quickly.

Comparison of the relaxation slopes at the various extension rates showed no clear trends or differences. For example, fatigued radial aortic specimens tested at the four extension rates of 2, 10, 50 and 200 mm/min had relaxation slopes of –3.2, –3.9, –3.6 and –3.3, respectively. However, visual examination of the stress relaxation curves showed that as the extension rates increased, a rapid initial fall in percentage stress remaining developed within the first 0.1 s of the test. The result was that at the end of a 100 s test, the percentage stress remaining for a 200 mm/min extension rate test was 15–20% lower than that for a 2 mm/min test. The stress remaining 100 s after the 10 mm/min test was between 80 and 90%, which is comparable to that found in the authors' previous studies.[5] It is probable that, as the extension rates increased, the tissue became incapable of straining fast enough to accommodate the stress applied to it. As a result, the stress on the strip was temporarily artificially increased and this excess stress rapidly dissipated once the stress/strain test was stopped and the tissue began the stress relaxation test.

Fatigued aortic relaxation slopes were less than those of their fresh or fixed counterparts[5] (i.e. –2.6 vs –5.8), with the exception of the glutaraldehyde-fixed radial specimens (–3.9 vs –4.1). Fatigued pulmonary relaxation slopes were statistically similar to those of the fresh and fixed slopes (i.e. –4.9 vs –4.8); in addition, the fatigued pulmonary strips had relaxation slopes similar to those of both fresh and fixed aortic strips.

Fracture

All specimens were strained to fracture at an extension rate of 10 mm/min. The fatigued aortic and pulmonary cusps had similar fracture stresses and exhibited the same marked anisotropy as evidenced previously (Table 11.3). The radial strips proved to be much weaker, fracturing at 594 kPa vs 2148 kPa for circumferential strips. In both the aortic and pulmonary groups, the radial strips were also much more extensible than circumferential strips (i.e. 87.6% strain at fracture vs 34.1% strain at fracture). Circumferential aortic and pulmonary specimens had similar fracture strains of 34.1 and 37.3%; however, radial aortic specimens fractured at a much lower strain than radial pulmonary specimens (87.6 vs 117.7% strain). The reason for the greater extensibility of the radial pulmonary strips is not known.

Table 11.3 Failure stress and failure strain for four types of fatigued strips[†]

Fatigued strips	Failure stress (kPa)	Failure strain (%)
Circumferential aortic	2148 ± 431	34.1 ± 6.5
Circumferential pulmonary	2654 ± 361	37.3 ± 4.0
Radial aortic	594 ± 146	87.6 ± 2.3
Radial pulmonary	595 ± 160	117.7 ± 27.3*

[†]Values are means ± SD; *significant difference ($p<0.05$; independent t-test)

Fatigued specimens of all groups exhibited fracture stresses similar to those of the fresh and glutaraldehyde-fixed samples tested previously.[5] Fatigued circumferential strips fractured at a lower strain than either the fresh strips (i.e. 34.1 vs 48.0% strain) or fixed strips (i.e. 34.1 vs 60.3% strain). Fatigued radial strips fractured at strains similar to those of fresh strips (i.e. 87.6 vs 82.2% strain) but lower than those fixed strips, which again illustrates that the increase in extensibility given by glutaraldehyde fixation appears to be only temporary. Fatigued pulmonary circumferential specimens fractured at a lower strain than fresh aortic specimens (37.3 vs 48.0%). However, radial fatigued specimens failed at higher strain (117.7 vs 82.2%).

Discussion

The search continues for bioprosthetic alternatives to the widely used stented xenografts. The hypothesis that stress might play an important role in the onset of calcification[9] is forcing surgeons to turn to the use of stentless bioprostheses. Indeed, additional reasons leading to their use are the satisfactory mid-term clinical results coupled with the excellent haemodynamic performance in the smaller annular sizes.[10]

The pulmonary porcine valve, with its symmetrical orifice and the absence of a muscle bar beneath the right coronary cusp, might even improve the haemodynamic characteristics of present aortic stentless valves. The authors have already shown in a recent study that the calcium content of pulmonary porcine leaflets and artery wall is lower than that in aortic valves;[4] if this might reduce the onset of calcification is still an open question but it deserves constant attention.

Previous comparative studies of porcine aortic and pulmonary cusps have increased the understanding of both their mechanical properties and their behaviour as xenografts.[5] In this paper, the effect of actual use on ultrastructure and mechanical characteristics of the chemically treated tissues has been examined.

As previously shown in human valves,[1] ultrastructural differences between aortic and pulmonary porcine tissues are minimal, relating more to 'quantity' than to 'quality'. As far as morphology is concerned, the fibrosa, spongiosa and ventricularis have basically a similar structure. The most striking difference relates to thickness: pulmonary cusps are thinner than aortic cusps in both the circumferential and radial directions. Only the aortic cusps examined here showed a difference in thickness between their radial and circumferential directions, with the radial specimens being thinner. The fatigued cusps were found to be thinner than either their fresh or glutaraldehyde-fixed counterparts, probably owing to stretching during use. The only effect of a thinner valve would be on the failure load. Although a cusp that is thinner, and therefore has a lower cross-sectional area, might fail at a stress similar to that for a thicker valve, it would fail at a lower load. However, this should be of little consequence, since expected loads *in vivo* would be much lower than the load needed to cause tissue disruption.

After cycling in the fatigue testing system, one aortic and two pulmonary valves presented with cusp perforation. Ultrastructural examination revealed normal tissue in the well-functioning cusps and similar degenerative changes of the collagen structure both in the aortic and the pulmonary cusps presenting with perforations.

The extreme variability already observed when comparing bioprosthetic valves in a fatigue testing system[9] does not enable a definitive conclusion to be drawn as far as predicted durability *in vivo* is concerned.

The elastic moduli of the circumferential pulmonary fatigued cusps were higher than those of similar aortic cusps; the opposite proved to be true for radial specimens. All circumferential specimens had higher elastic moduli than the radial specimens. The elastic moduli of the fatigued cusps were found to be a closer match to those of fresh cusps than to those of glutaraldehyde-fixed cusps. This indicates that during use the fixed leaflets appear to revert to their 'fresh' properties. When fatigued pulmonary and fresh aortic specimens were compared, the elastic moduli in the radial direction proved to be similar, whereas a difference was evident between circumferential groups. Perhaps during use, when the heavily crimped fixed collagen fibres begin to strain permanently and therefore to straighten out, the cross-links serve to stiffen the cusp. Both exhibited much greater extensibility in the radial direction.

As previously reported,[5] an unexpected fall in the elastic modulus was found at the highest extention rate of 200 mm/min.

Similar extensibilities were found for the aortic and pulmonary fatigued cusps. Both exhibited much greater extensibility in the radial direction. There was also a trend towards lower extensibilities at higher extension rates (20 mm/min). This apparently results from the cusps having a short reaction time at the higher extension rates (closer to conditions *in vivo*), with resultant decrease in extensibility.

The increased extensibilities of glutaraldehyde-treated cusps found in previous studies[5] were not present for the fatigued cusps. Continued use apparently removes the 'crimping' effect of fixation on the collagen fibres. The result was that the fatigued cusps demonstrated extensibilities similar to those of fresh cusps.

Pulmonary fatigued cusps stress relaxed more rapidly than similar aortic cusps. The radial strips relaxed more rapidly than circumferential strips, which would be expected because of the greater resiliance of the radial strips on viscous components for load-bearing rather than on collagen. Although the relaxation slopes did not alter as the extension rates changed, a rapid initial drop was observed in percentage stress remaining within the first 0.1 s as the extension rates increased. The greater strain rates at the higher extension rates did not allow the tissue to strain sufficiently to accommodate the increased stress upon it. An artificially high stress was then introduced, which rapidly dissipated once further straining ceased.

In general, the stress relaxation results of the fatigued pulmonary cusps conformed to the results for fresh and fixed cusps, whereas those of fatigued aortic cusps did not, usually being less. Fatigued cusps did not show a clear trend towards higher stress relaxation rates at higher extension rates, which perhaps would indicate that the viscous component plays less of a role in the fatigued cusps.

Both aortic and pulmonary fatigued cusps showed marked anisotropy during the fracture tests. The circumferential strips failed at higher loads and lower strains than radial strips. Aortic and pulmonary failure strains and stresses were similar, except that aortic radial specimens were less extensible than similar pulmonary strips. The thinner pulmonary cusps failed at lower loads but this should be of little importance since they were well below loads *in vivo*. Fatigued cusp failure stresses were similar to previous results for fresh and fixed tissues. However, the fatigued leaflets failed at similar (but slightly lower) strains than fresh leaflets and at much lower strains than fixed leaflets. This was consistent with the elastic moduli and extensibility results. Pulmonary specimens were fractured at lower strains for circumferential and higher strains for radial specimens when compared with fresh aortic strips. Again, loads and strain *in vivo* would be expected to be much lower than the fracture loads and strains observed and should pose no problems.

If these results can be extrapolated to the actual working conditions of the leaflets themselves, then glutaraldehyde-treated porcine pulmonary valves should prove suitable for long-term implant use. The factors influencing valvular calcification and structural deterioration remain unclear and the potential benefit available from pulmonary tissue over aortic tissue for unstented freehand-inserted valves has yet to be investigated. Further experimental evaluation using calf and growing sheep models are mandatory in order to investigate the possible reduction in calcification and the potential durability of the pulmonary porcine stentless valve. For a better understanding of the behaviour of pulmonary human and porcine valves implanted in the systemic circulation, it is essential to investigate the differential value of 'active' and 'passive' remodelling of pulmonary valve leaflets when exposed to diastolic systemic pressure.[11] For instance, fibroblasts of pulmonary autografts can play an active role in producing more collagen fibres and thus active remodelling of the leaflets might take place. Homografts and glutaraldehyde-treated pulmonary tissues are virtually unable to do so and therefore architectural and morphological variation would be unlikely. To clarify this key point further, the authors are investigating pulmonary and aortic valve leaflets, harvested from adults with primary pulmonary hypertension who underwent heart and lung transplant procedure. Histological and electron mircroscopy studies should enable the significance of pulmonary leaflet thickness, when adapting to systemic diastolic pressure, to be verified.

References

1. Gerosa G, Ross D N, Brucke P E, Dziatkowiak A, Mohammad S, Norman D, Davies J, Sbarbati A, Casarotto D 1994 Aortic valve replacement with pulmonary homografts: early experience. J Thorac Cardiovasc Surg 107: 424–437
2. Ross D N, Jackson M, Davies J 1992 The pulmonary autograft — a permanent aortic valve. Eur J Cardio-Thorac Surg 6: 113–117
3. Gerosa G, McKay R, Davies J, Ross D N 1991 Comparison of the aortic homograft and the pulmonary autograft for aortic valve or root replacement in children. J Thorac Cardiovasc Surg 102: 51–61
4. Gerosa G 1993 The pulmonary valve (autograft, homograft, xenograft) for aortic valve replacement. A clinical, ultrastructural and biomechanical evaluation. I1 Cuore 10(4): 349–364
5. David H, Boughner D R, Vesely I, Gerosa G 1994 The pulmonary valve: is it mechanically suitable for use as an aortic valve replacement? ASAIO J. 40: 206–212
6. Barratt-Boyes B 1994 Discussion of reference 1
7. Lee J M, Courtman D W, Boughner D R 1984 The glutaraldehyde stabilized porcine aortic valve xenograft. I. Tensile viscoelastic properties of the fresh leaflet material. J Biomed Materials Res 18: 61–67
8. Vesely I, Lozon A, Talman E 1994 Is zero pressure fixation truly stress free? J Thorac Cardiovasc Surg. 106: 288–298
9. Gabbay S, Kadam P, Factor S, Cheung T K 1988 Do heart valve bioprostheses degenerate for metabolic or mechanical reasons? J Thorac Cardiovasc Surg 95: 208–215
10. David T E, Pollick C, Bos J 1990 Aortic valve replacement with stentless porcine aortic bioprostheses. J Thorac Cardiovasc Surg 99: 113–118
11. Vesely I, Boughner D R 1989 Analysis of the bending behaviour of porcine xenografts and of natural aortic valve material: bending stiffness, neutral axis and shear measurements. J Biomech 22: 655–671

Chapter 12

Stentless heart valve concept: new anticalcificant tissue tanning — preclinical trial

M. Vrandecic, B. F. Gontijo, F. A. Fantini, S. Gabbay, E. A. Vrandecic and E. Vrandecic

The demonstration by Hufnagel[1] in 1951, using a caged-ball prosthesis implanted in the human descending aorta, that man-made materials were tolerated in the bloodstream, heralded the beginning of a new era, with the possibility of replacement of diseased heart valves. In 1954, Gibbon,[2] with the development of the extracorporeal circulation, also contributed to making heart valve replacement surgery a reality. Since then, a wide range of research projects have led to the extensive development of heart valve substitutes, involving medical research from many different scientific backgrounds. Close collaboration between medical and non-medical scientists is essential in order to achieve the ideal heart valve substitute. Of prime importance are the long-term clinical results with the use of the heart valve substitutes. In this regard, there is a substantial database available of thousands of patients studied by multicentre valve trials; these have shown that the current models, mechanical and tissue valves, lead to similar rates of survival and complications.[3]

Further research is necessary to improve durability and reduce complications. The shortcomings encountered in the literature available on the short- and long-term fate of heart valve substitutes can be attributed to their design and the nature of the materials used. This chapter describes the authors' contribution to the search for an ideal heart valve substitute: their daily observations in the operating room, while constructing or replacing heart valves using the models currently available, have triggered their conviction that man cannot do better than nature. The deleterious effect of the stent on the long-term performance of treated tissue is well known; the basic design and materials should therefore be as close as possible to the native heart valve.

Anatomy and function

In 1487 Leonardo da Vinci,[4] describing the structure of the heart and specifically of the mitral and aortic valves, correlated their anatomy and function as well as can be done today. As stated by Robicsek,[5] the aortic valve consists of the aortic annulus, three leaflets, three commissures and three sinuses. The functional significance of the sinuses of Valsalva in circulatory dynamics is today well accepted. The importance and the role of the eddying currents of the sinus of Valsalva in relation to the closure of the aortic valve is a masterpiece of engineering, enabling tissue stress to be reduced and optimizing its function (Figures 12.1 and 12.2). The mitral heart valve consists of the mitral annulus, two true leaflets (anterior and posterior), two free leaflets, and a chordal support

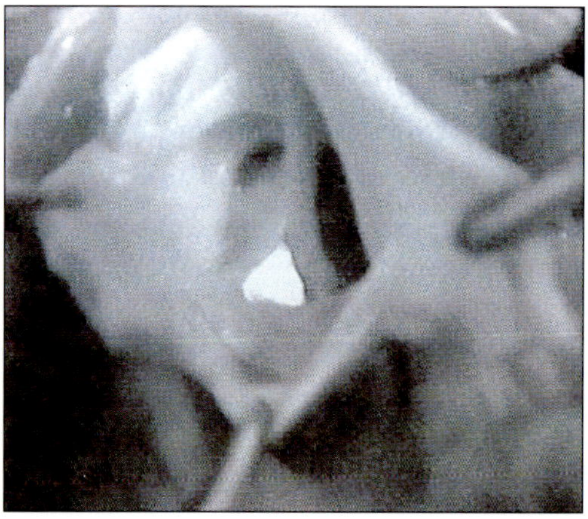

Figure 12.1. *Fresh porcine aortic valve showing anatomical landmarks.*

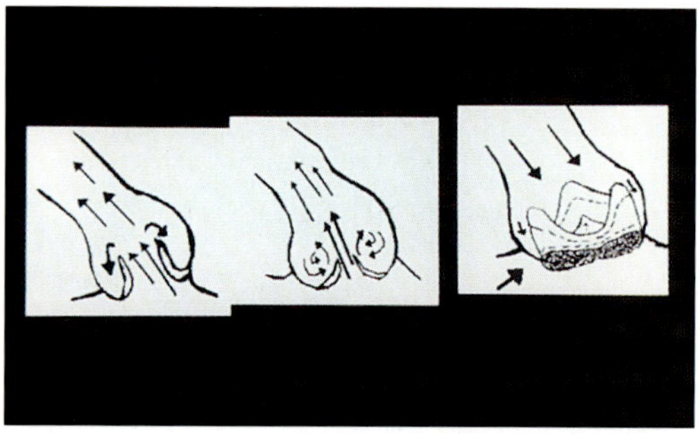

Figure 12.2. *Eddy currents are important determinants of leaflet and commissural stress.*

derived from the anterolateral and posteromedial papillary muscles. This provides the annular–ventricular continuity essential for good left ventricular function (Figure 12.3). The stentless concept has therefore a natural application to the design and manufacturing of the heart valve substitutes.

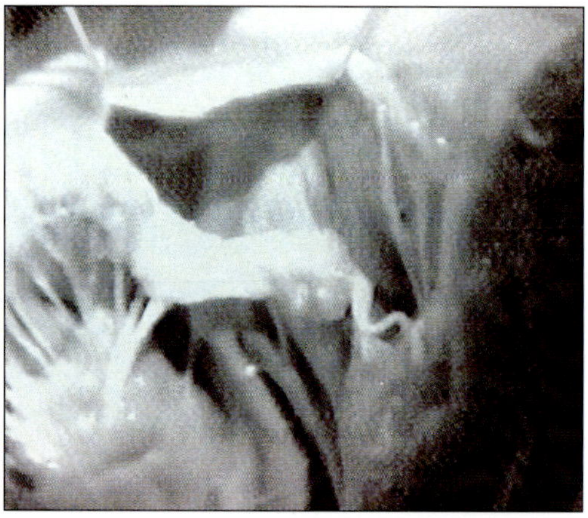

Figure 12.3. *Fresh porcine mitral valve showing the annuloventricular continuity; note the left ventricular outflow tract or intertrigonal area.*

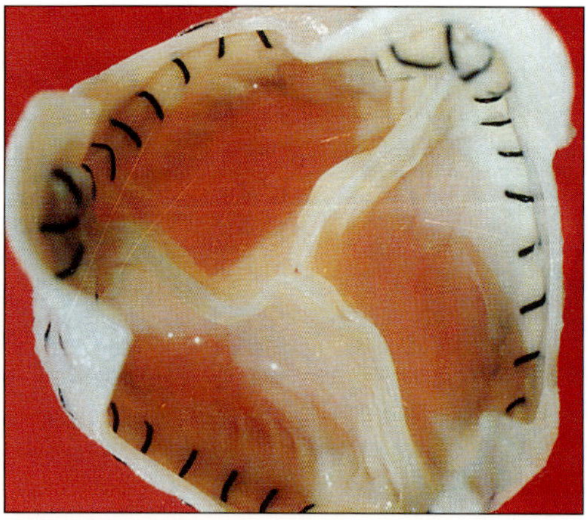

Figure 12.4. *Porcine stentless aortic valve.*

The porcine stentless aortic valve model was based on the well-established anatomofunctional concept and clinical experience with the use of the aortic homograft.[6] Its exclusive design mimics the shape and form of the native aortic heart valve, addressing such issues as differences in shape of the ascending aorta, thus avoiding aortic valve insufficiency. Handling of the stentless aortic valve during implantation is greatly improved, owing to its pliability. As the stentless wall is fairly thin, there is a larger internal valve surface area (Figure 12.4).

The porcine mitral stentless valve includes all the native anatomical structures, providing annular–ventricular continuity. Pericardial reinforcement of its annulus and the anatomical–functional way to secure the new origin of the chordae restore the specific features of flow (vortex) and left ventricular performance. The mitral heart valve, created by the Great Master, has been shown throughout the history of man to be the best-engineered valve at the left atrial–ventricular level. Its excellent function as assessed by pulse duplicator and fatigue testing of the stentless mitral valve, and the insight obtained by the acute animal studies, have been the basis of its clinical feasibility.

Both products also have the advantage of being available in sizes ranging from 10 to 31 mm, for the aortic and pulmonary valves, and from 11 to 35 mm for the mitral and tricuspid valves (Figure 12.5).

Figure 12.5. Porcine stentless mitral valve.

Tanning of biological tissues

Having addressed one of the drawbacks of current tissue heart valves regarding design, the next question is how to obtain better 'tanning' (i.e. fixation) of the biological tissues. Investigations into more durable tissue tanning were based on analyses of explanted tissue valves.[7–9] The most outstanding features encountered, using light microscopy, were an extensive foreign body reaction, collagen disruption and calcification. The so-called 'accepted' glutaraldehyde tissue tanning process used for decades needs to be revised so that tissues do not react with the host cells, thereby avoiding the foreign body reaction that triggers the chain of events leading to tissue failure and calcification.

Assessment of the No-React™ tissue-tanning process

The breakthrough afforded by the No-React™ tissue treatment is, in the authors' opinion, a major step towards improving the durability of heart valve tissues. This process uses only substances already approved by the FDA, many for parenteral use or well known in heart valve manufacture. Evidence that this new anticalcificant method is well accepted by the host and gives great strength to the treated tissue, while preserving its elastic properties near normal, has been obtained from (1) experiments *in vitro*, (2) implants in animals, (3) accelerated fatigue testing and (4) fibroblast culture.

Tensile testing *in vitro* of No-React™ processed tissues, in comparison with the glutaraldehyde tanning process

The tensile tests were conducted on a closed-loop feedback hydraulic material testing system (MTS, Minneapolis, Minnesota, USA), at a constant stroke rate of 0.67 mm/s. Each specimen was firmly held with minimal slack. The distance between the grip faces was 35 mm; the gauge length of each specimen was, therefore, 35 mm load/stroke. Data were recorded graphically and numerically on a microprocessor (Macintosh SE 30) with commercially available A/D conversion hardware and software. These data were then processed, on the basis of specimen cross-sectional area, to yield stress/strain relationships.

Animal experiments

Subcutaneous implants in Sprague–Dawley rats

Tissue segments (5 mm) were implanted under the abdominal skin in two groups of 19 rats, with glutaraldehyde-tanned tissue always on the right side of the abdomen and the No-React™-treated tissue on the left, in the same rat. One group of 19 rats had tissue implanted for 4 weeks and the other group had implants for 12 weeks. The implanted fragments were subsequently removed, rinsed, dried and weighed individually. They were then subjected to acid hydrolysis and the calcium content of the hydrolysates were assayed by fluorometric titration in a calcium analyser or, when necessary, by atomic absorption spectroscopy.

Orthotopic bioprosthetic implants in calves and sheep

Glutaraldehyde-treated bioprostheses were implanted in the calf in the mitral position (historical comparison), for periods of 3 months or longer in six animals. No-React™-treated bioprostheses were implanted in four weanling sheep (6 months old). The explanted bioprostheses were given code numbers and analysed radiologically and biochemically. Each leaflet was numbered and excised, dividing it in two parts, designated leaflet body and leaflet base.

Accelerated fatigue testing

Four standard bioprostheses were tested in the control or glutaraldehyde group and the same number of valves in the No-React™ group. Over 100 000 × 10^6 cycles were achieved, in accordance with Gabbay's[10,11] protocol for tissue valve fatigue testing. (Some of the fatigue tests and animal implant experiments were conducted in Professor S. Gabbay's laboratory, New Jersey Medical School, New Jersey, USA.)

Live fibroblast culture

This extremely sensitive test was used to determine the effect on live cells of contact with the tissue fragments. The control or glutaraldehyde group and the No-React™ group each consisted of 120 cultures. The culture medium used was RPMI-1640 plus non-essential amino acids (1% w/v) plus foetal bovine serum (15%, w/v) plus antifungal antibiotics (1% w/v) in flasks. Fibroblasts were removed from flasks treated with EDTA–Trypsin (Gibco). The cell line utilized was 4.1 fibroblast 1521 and 4.2 fibroblast 153 from normal human skin and lung, respectively. The culture period was 72 h, with preliminary assessments at 24 and 48 h.

Results of assessments

Tensile testing

The uniaxial tensile tests showed the No-React™ tissues to be five times stronger in tension, as measured by maximum stress to failure, than the glutaraldehyde-treated tissues (Figure 12.6).

Animal experiments

Implants in rats

The aldehyde-treated tissues had a mean (±SD) calcium content of 170 ± 2.8 μg/mg dry tissue, whereas no calcium was detected in the No-React™-treated tissues.

Mitral implants in calves and sheep

Of the six calves receiving glutaraldehyde-treated bioprostheses, three died of heart failure at 6 and 18 weeks. Two further animals died at 16 and 20 weeks, and one animal survived for 24 weeks (6 months). The explanted valves showed different degrees of tissue degeneration and calcification: the one recovered after 6 months showed some thickening and tissue degeneration, but those recovered earlier showed greater degrees of tissue degeneration, with tissue calcification that ranged from 70 to 192 μg/mg dry tissue.

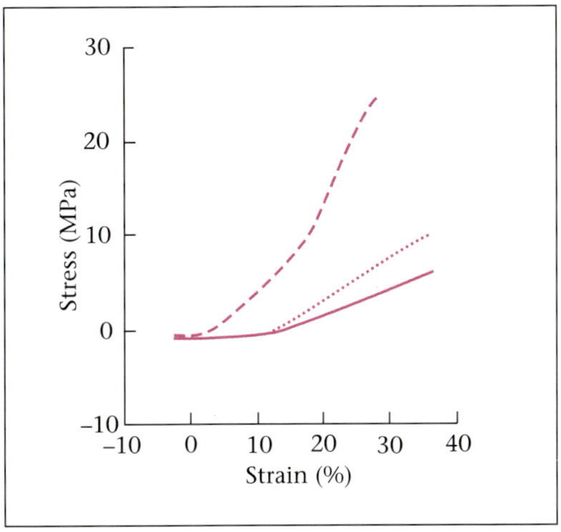

Figure 13.6 Tissue tensile testing: the tensile performance of 12 samples of No-React™-treated tissues (- - - -) was five times greater than that of glutaradehyde-fixed tissues (………, ——) two groups each of six samples of Ionesan–Shiley pericardial valves stored in 4.5% buffered formaldehyde.

In the No-React™ group, only one valve was explanted as all four sheep had an uneventful course. The No-React™ valve explanted electively at 8 months had a normal appearance and soft, pliable cusps, without any sign of tissue failure or calcification. No calcium was detected and the histological and electron microscopical appearance of the leaflets showed well-preserved structures. There was no sign of foreign body reaction. The remaining sheep will be killed on a yearly basis.

Accelerated fatigue testing

The accelerated fatigue tests showed that valves in the No-React™ group ranged from fairly well preserved to well preserved, compared with glutaraldehyde-treated bioprostheses.

Fibroblast cultures

The cytotoxic tests showed total cytocompatibility of the No-React™-treated tissue fragments. The live cells displayed normal function, division and maturation, providing a full and compact monolayer in less than 72 h. Conversely, the glutaraldehyde group showed complete cell death in less than 24 h (Figure 12.7).

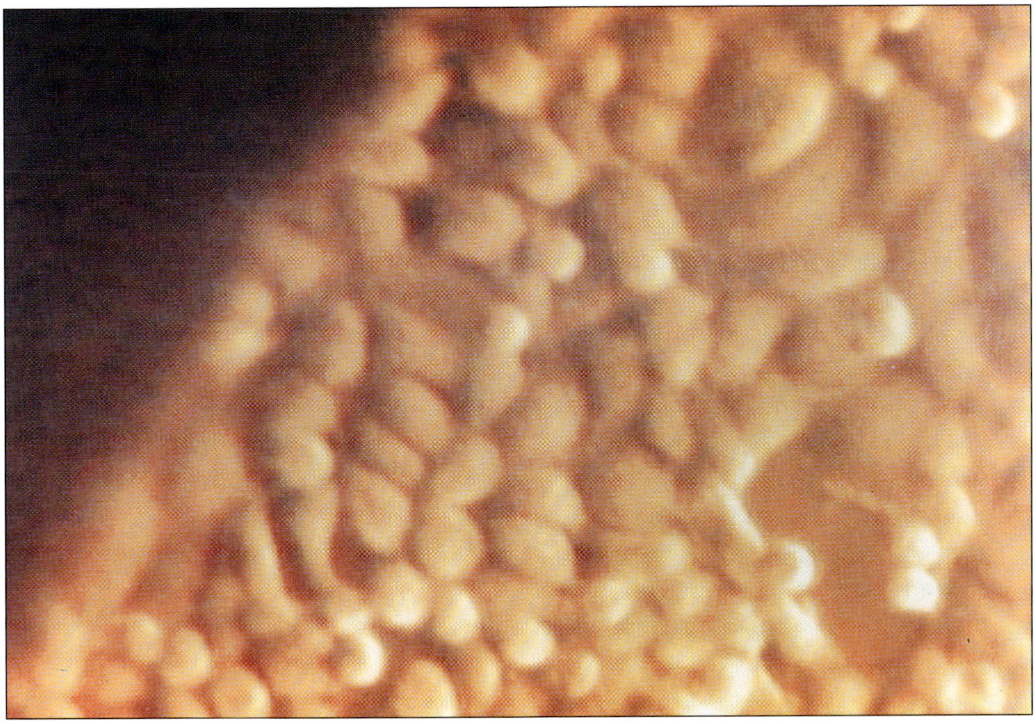

Figure 12.7. Live cell fibroblast culture containing No-React™ tissue fragment showed full cell growth at 72 h.

Conclusions

There is no doubt that current tissue heart valve substitutes can be improved. The extensive literature regarding the long-term outcome for bioprostheses has shown the need to improve design and tissue tanning.

The stentless concept is appropriate to improve the primary drawback of the current models and is in complete harmony with the anatomical design and function of the heart valve substitutes.

Tissue tanning based on prevention of the foreign body reaction appears to be the only means of enabling processed tissue to be accepted by the host and this was accomplished by the No-React™ tissue treatment. The experiments conducted to assess this process have shown that tissues strengthened in this way preserve a great deal of their softness and elasticity and are entirely cytocompatible. Thus, the good performance on fatigue testing and complete cytocompatibility of the treated tissue, and the excellent results obtained in over a year of animal implants (sheep), corroborated by the natural appearance of the explanted No-React™ bioprostheses at 6 months, clearly attests to the potential of this new tissue treatment to improve valve durability, which is of particular importance in children and young patients.

Although better quality of life and patient survival, and longer tissue valve durability, are anticipated,[12,13] nevertheless, as with any new tissue treatment and design, cases must be selected carefully for a well-conducted randomized study.

In conclusion, there is always room for improvement. The stentless concept and No-React™ tissue tanning have shown promise and thus call for randomized trials and close follow-up, since only time will prove their worth.

References

1. Hufnagel C A 1951 The surgical correction of aortic insufficiency. Bull Georgetown Univ Med Center 4: 128–130
2. Gibbon J H Jr 1954 Application of a mechanical heart and lung apparatus to cardiac surgery. Minn Med 37: 171–185
3. Hammermeister K E, Sethi G K, Henderson W G, Oprian C, Kim T, Rahimtoola S 1993 A comparison of outcomes in men 11 years after heart-valve replacement with a mechanical valve or bioprosthesis. N Engl J Med 328: 1289–1296
4. Belt E 1969 Leonardo the anatomist: Logan Clendering Lectures on the History and Philosophy of Medicine. Greenwood Press, New York
5. Robicseck F 1992 Leonardo Da Vinci's Anatomical Drawings. Grossi A, Donatelli F, Corno A, Brodman R (eds) Cardiology and cardiac surgery: current topics. Futura, Mount Kisco, New York, ch 1, pp 3–29
6. Ross D 1988 Aortic root replacement with a cardiac allograft: the infected aortic root. In: Yankah A C, Hertzer R, Miller D C *et al* (eds) Cardiac valve allograft 1962–1987. Springer, New York, p 167
7. Ferrans V J, Spray T L, Billingham M E, Roberts W C 1978 Structural changes in glutaraldehyde treated porcine heterografts used as substitute cardiac valves. Am J Cardiol 40: 1150–1184
8. Dahm M, Lyman W, Schwell A, Factor S, Frater W M 1990 Immunogenicity of glutaraldehyde-tanned bovine pericardium. J Thorac Cardiovasc Surg 99: 1082–1090
9. Schoen F J, Levy R J 1986 Pathophysiology of bioprosthetic heart valve calcification. In: Bodnar E, Yacoub M (eds) Biologic and bioprosthetic valves. Yorke Medical, London, pp 418–429
10. Gabbay S, Kadam P, Factor S, Cheung T K 1988 Do heart valve bioprostheses degenerate for metabolic or mechanical reasons? J Thorac Cardiovasc Surg 94(2): 208–215
11. Gabbay S, Bortolotti U, Josif M 1986 Mechanical factors influencing the durability of heart valve pericardial bioprostheses. Trans Am Soc Artif Intern Organs 32: 282–287
12. Vrandecic M P, Morea M, Radegran K 1992 Clinical use of a new stentless porcine aortic bioprostheses: a multicenter study. In: Grossi A, Donatelli F, Corno A, Brodman R (eds)

Cardiology and cardiac surgery: current topics. Futura, Mount Kisco, New York, pp 223–234

13. Vrandecic M P, Gontijo B F, Fantini F A, Paula E, Silva J A, Barbosa J, Cristina G, Barbosa M R and Morea M 1993 Heterologous mitral valve transplant. New alternative for mitral valve replacement: one year clinical follow-up. In: D'Alessandro L C (ed) Heart surgery. Casa Editrice Scientific Internazionale, Rome, vol 7: 57–68

Chapter 13

Stentless heart valves: biocompatibility issues associated with new antimineralization and fixation agents

D. J. Myers, J. Gross and G. Nakaya

Medtronic (Irvine, California, USA) has developed a glutaraldehyde-fixed stentless porcine aortic root bioprosthesis known as the Freestyle™ valve.[1] This valve is treated with an antimineralization agent, α-amino oleic acid (AOA), which has been shown to be effective in mitigating mineralization of treated tissue in both the rat subcutaneous[2] and sheep circulatory[3–5] models.

Recently, the fixation method ('root pressure fixation') used to prepare Freestyle™ porcine aortic roots has been employed to crosslink stentless valves experimentally with diepoxide (ethylene glycol diglycidyl ether) rather than with glutaraldehyde. In addition to making the tissue more compliant,[6] epoxide fixation has been shown to improve resistance to mineralization compared with aldehyde-crosslinked tissue.[7–9]. Having been successfully subjected to the battery of biocompatibility tests required by the United States Food and Drug Administration[10] and the International Standards Organization,[11] Freestyle™ valves are now on clinical trial both in the United States and overseas. Biocompatibility test protocols for epoxide-fixed valves are currently being developed on the basis of experience gained in testing the aldehyde-fixed and AOA-treated stentless devices.

The following report reviews initial results that unexpectedly indicated that aldehyde-fixed and AOA-treated valves were cytotoxic and that the epoxide-fixed valves demonstrated relatively little resistance to mineralization upon animal implantation. Reasons why the initial assays yielded these unexpected results are provided.

Biocompatibility: AOA-treated stentless valves

Preclinical biological evaluation of new replacement heart valves, regardless of whether they have been treated with antimineralization agents, must include testing for biocompatibility to determine whether they are safe for chronic (permanent) human implant (see Table 13.1). Specifically, in this study, AOA pre-treatment had to yield tissue that is non-toxic/non-pyrogenic, non-sensitizing, non-irritating, and haemocompatible (non-thrombogenic and non-haemolysing), as well as preventing or mitigating mineralization. Supplementary evaluations could also have included tests for genotoxicity, carcinogenicity and teratogenicity (reproductive/developmental toxicity). It was recognized that biological testing,

Table 13.1. United States medical device regulatory agency and International Standards Organization guidance for initial and supplementary biocompatibility tests†

(a) Guidance for initial evaluation tests

Device category		Biological effects§							
Body contact	Contact duration††	1	2	3	4	5	6	7	8
Blood	A	*	*	*	*			*	*
	B	*	*	*	*	*	*	*	*
	C	*	*	*	*	*	*	*	*

†NB: each device shall be considered on its own merits;
††A, limited (<24 h); B, prolonged (24 h to 30 days); C, permanent (>30 days); §1, cytotoxicity; 2, sensitization; 3, irritation or intracutaneous reactivity; 4, systemic toxicity; 5, subchronic toxicity; 6, genotoxicity; 7, implantation; 8, haemocompatibility.

(b) Guidance for supplementary evaluation tests

Device category		Biological effect		
Body contact	Contact duration††	Chronic toxicity	Carcinogenicity	Reproductive/ developmental Biodegradation
Blood	A			
	B			
	C	*	*	

†, †† As in (a)

which often relies upon animal models, could not be shown conclusively to elicit the same response in humans. Furthermore, an adverse reaction in animals, or in a test *ex vivo* or *in vitro*, for one device application might not be a problem with the same pretreatment in another application. Finally, actual testing could be supplemented with relevant experience which, with demonstrable evidence of safe use, might obviate the need for additional evaluations. Each device and pretreatment was therefore considered on its own merits; additional tests could have been indicated, as well as certain assays waived.

Biocompatibility testing of AOA neat

The antimineralization agent itself was biologically tested first, as a neat compound, before testing of AOA-treated valves. Table 13.2 lists the tests employed, the laboratories that performed the tests, and the test results.

The neat compound was found to be completely biocompatible. Specimens of AOA-treated porcine valve cusps were also found to be biocompatible, using the cage implant system in rabbits developed by Dr James M. Anderson at Case Western Reserve University.[12] This assay quantitatively monitors the animal's inflammatory response to subcutaneously implanted test materials enclosed within a stainless steel wire cage; empty cages act as controls. Cell/implant interactions are evaluated by periodic measurements of various types of leucocytes in the cage exudates.

Biocompatibility testing of clinical-quality valves

AOA-treated Freestyle™ valves were tested for biocompatibility following satisfactory results with the neat agent. Table 13.3 indicates the initial results of these assays. It can be seen that the treated tissue failed the test for cytotoxicity (agar diffusion using mouse fibroblasts), and both intracutaneous and systemic injection of saline extracts in rabbits and rats (USP Class V Test), but passed all the

Table 13.2 Biocompatibility evaluation of neat α-amino oleic acid (AOA)

Test method	Laboratory	Result
Cytotoxicity: agar diffusion mouse L929 fibroblasts	PRL*	Non-cytotoxic
Static haemolysis: rabbit RBC	PRL	Non-haemolytic
USP Class V:		
Intracutaneous injection: rabbit	PRL	
Cottonseed oil		Non-irritating
0.9% saline		Non-irritating
Systemic injection: mouse	PRL	
Cottonseed oil		Non-irritating
0.9% saline		Non-irritating
Mutagenicity, Ames test: *Salmonella typhimurium*	Nelson (Utah)	Non-mutagenic
Sensitization: guinea pig	U.S. TESTING (New Jersey)	
Cottonseed oil		Non-sensitizing
0.9% saline		Non-sensitizing
Pyrogenicity: rabbit	U.S. TESTING	Non-pyrogenic
Cage implant: rabbit (AOA-treated cusps)	Case Western Reserve University	Biocompatible

*PRL, Medtronic Physiological Research Laboratories, Minneapolis, Minnesota, USA

Table 13.3 Biocompatibility evaluation of AOA-treated Freestyle™ valves

Test method	Laboratory	Result
Cytotoxicity: agar diffusion mouse L292 fibroblasts	PRL*	Cytotoxic
Static haemolysis: rabbit RBC	PRL	Non-haemolytic
USP Class V:		
Intracutaneous injection: rabbit	PRL	
Cottonseed oil		Non-irritating
0.9% saline		Irritating
Systemic injection: mouse	PRL	
Cottonseed oil		Non-irritating
0.9% saline		Irritating
Mutagenicity, Ames test *Salmonella typhimurium*	Toxikon (Mass.)	Non-mutagenic
Sensitization: guinea pig	Toxikon	
Cottonseed oil		Non-sensitizing
0.9% saline		Non-sensitizing
Pyrogenicity: rabbit	Toxikon	Non-pyrogenic
Cage implant: rabbit (AOA-treated cusps)	Case Western Reserve University	Biocompatible (post week 2)

*As in Table 13.2

other assays. Using untreated glutaraldehyde-fixed valves as controls, it was subsequently shown that even this tissue caused a cytotoxic tissue culture response and yielded irritant saline solution extracts (Table 13.4). It was concluded from these results that residual glutaraldehyde leached from the clinically rinsed valves, not the new antimineralization process, was the most likely cause of the unexpected toxic/irritant response.

Table 13.4 Biocompatibility evaluation of untreated and AOA-treated Freestyle™ valves

	Results	
Test method	AOA-treated	Untreated
Cytotoxicity: agar diffusion mouse L929 fibroblasts	Cytotoxic†	Cytotoxic†
USP Class V:		
Intracutaneous injection: rabbit		
Cottonseed oil	Non-irritant	Non-irritant
0.9% saline	Irritant††	Irritant††
Systemic injection: mouse		
Cottonseed oil	Non-irritant	Non-irritant
0.9% saline	Irritant*	Irritant*

†Over 50% cells lysed, crenated, non-adherent; ††significant erythema and oedema vs negative control; *significant reaction, any animal/time, vs negative control.

Glutaraldehyde elution studies

Elution studies were performed on AOA-treated and control (untreated) valves to determine the extraction profile of aldehyde from the tissue. Clinical quality 27 mm Freestyle™ valves were separately rinsed in sterile 0.9% saline (2 min soak in each of three 500 ml aliquots of solution) and then individually further extracted in 50 ml aliquots of saline for increasing periods of time (2–120 h). Samples of each rinse solution and rinsed valve extract were then analysed by gas chromatography for glutaraldehyde concentration (0.5 ppm limit of detection). The results of this study showed that 3.4 mg glutaraldehyde per rinsed valve could have been released into the saline extracts prepared for the standard biocompatibility tests. For each test, under the extraction conditions utilized, the calculated glutaraldehyde concentration was approximately 85 ppm, a level considerably higher than the threshold of only a few ppm known to elicit a toxic response.[13]

Health risk assessment

A health risk assessment was conducted to determine the potential risk to patients due to residual glutaraldehyde in the rinsed valve. Table 13.5 summarizes the findings of this study, which, for the largest implant (27 mm), indicate that the peak level of glutaraldehyde in the blood would be 0.069 ppm. This concentration of glutaraldehyde was considered to be 100 times lower than that which has been reported to be clinically toxic.

Repeat cytotoxicity/USP V testing with revised protocol

The cytotoxicity and USP V irritant tests were repeated on clinically rinsed 27 mm valves using 10-fold and 100-fold dilutions of the standard saline extracts previously employed. These more clinically relevant dilutions all gave negative responses in both tissue culture and rabbit or rat irritancy assays. As the diluted

Table 13.5 Results of health risk assessment and biocompatibility tests

1. A *glutaraldehyde leach rate constant* was calculated to be 0.73 h^{-1}, with a leaching half-life of 1 h

2. Assuming an entire 27 mm valve was implanted (worst case), the *peak concentration of glutaraldehyde in the blood* was calculated to be 0.069 mg/l and would occur 4 h after implantation. This level was considered well below (100 ×) that which has been reported to be acutely toxic

3. All *acute cytotoxicity and USP V testing* performed with both 10- and 100-fold dilutions (compared with the original extracts) gave negative responses. These tests utilized solutions of glutaraldehyde up to 300 times greater than the maximum levels expected in the body fluids of patients implanted with the largest valve

extracts still contained up to 300 times the maximum blood level of glutaraldehyde expected in a patient, rinsed stentless valves were considered to be safe for clinical implantation.

Rationale for revision of biocompatibility test protocol

It was established that residual fixative in the clinically rinsed stentless valves was the cause of their failure to pass the initial cytotoxicity and irritant bioassays. This was anticipated on the basis of 'classic' (more concentrated) extraction conditions employed previously on stented valves; these valves had never failed to pass the same biocompatibility tests. Owing to the presence of a greater amount of aortic wall, stentless valves have a larger mass of glutaraldehyde-fixed material than do stented valves. The major component of this larger mass is aortic wall tissue, a very dense material that is slow to release extractable glutaraldehyde compared with the thinner cusps and the non-absorptive/non-biological components of the stented valves such as polyester cloth, polypropylene stent, and suture.

Biocompatibility: diepoxide-fixed stentless bioprostheses

Background

The physical and mechanical properties have been investigated of stentless porcine aortic bioprostheses crosslinked with the diepoxide compound ethylene glycol diglycidyl ether (Denacol EX-810, manufactured by Nagase Chemicals, Ltd, Osaka, Japan). Crosslinking tissue with epoxy compounds has been reported to (a) improve the biochemical properties of vascular grafts,[14] (b) provide canine aortic valve conduits as RV–PA implants that are more thromboresistant, more hydrophilic and more resistant to calcification than glutaraldehyde-fixed valves[15] and (c) yield porcine aortic valves similar in shrink temperature to glutaraldehyde-fixed tissue, but higher in moisture content and more resistant to pannus formation and calcification in studies of young sheep.[16]

Epoxide chemistry

EX-810 is one of a group of difunctional and trifunctional epoxide compounds that have been tested as tissue crosslinkers by the authors and by others.[17–21] The Denacol epoxide compounds are not supplied as pure di- or triepoxide but as mixtures of one or more of these agents.[22] A typical gel-permeation chromatogram of the material marketed as Denacol EX-810, and utilized to crosslink the porcine aortic heart valves in this study, is presented as Figure 13.1. The active agents in this solution are primarily diglycidyl ethers of various chain length; the major peak represents ethylene glycol diglycidyl ether (EDGE). Typical gas

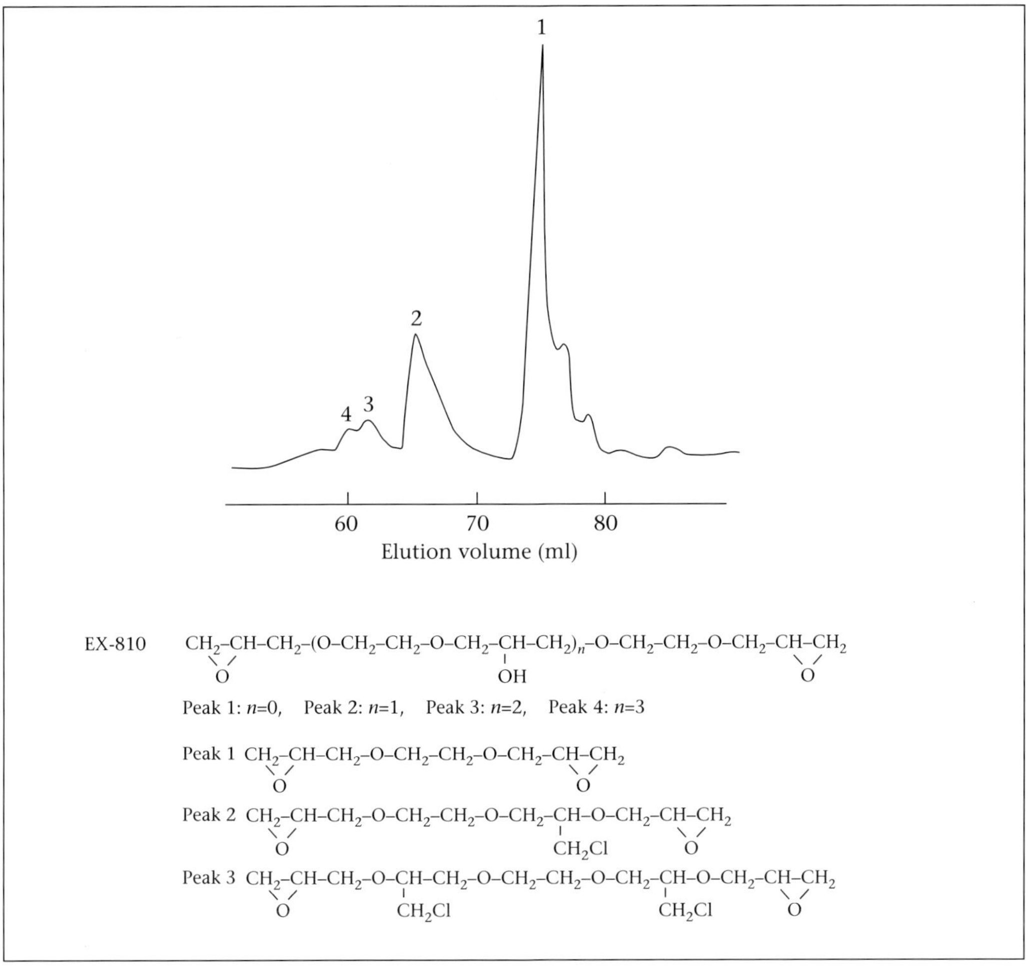

Figure 13.1. Composition of Denacol EX-810 (mixture of epoxides) as shown by gel-permeation chromatographic analysis.

chromatograms of two lots of Denacol EX-810 used in our study are illustrated in Figure 13.2; the area of the major peak in each assay indicated that EDGE makes up 62–65% of the mixture.

Fixation kinetics

Fixation kinetics studies were peformed to determine the optimum conditions of crosslinking porcine aortic and mitral heart valves with EDGE. A phosphate-buffered (pH 7.4) aqueous solution of EDGE at 3% concentration was tested initially.

Aortic roots and the mitral valve apparatus were carefully dissected from fresh porcine heart valves, rinsed in cold phosphate buffer, and then separately immersed in 100 ml aliquots of the diepoxide solution. The cups containing the valves were left to stand in the dark at 25°C for 18 days.

Periodic measurements were made of solution pH and shrink temperature (degree of crosslinking) of valve cusps. As the pH of the solution increased fairly rapidly (about 0.1 pH unit/day), the fixative in each cup was replaced approximately every 5 days in order to maintain the pH below 8. Figure 13.3 demonstrates the increase in shrink temperature of aortic cusps with increasing duration of exposure to the fixative. Stable shrink temperatures of about 77–78°C were attained after 14 days of treatment.

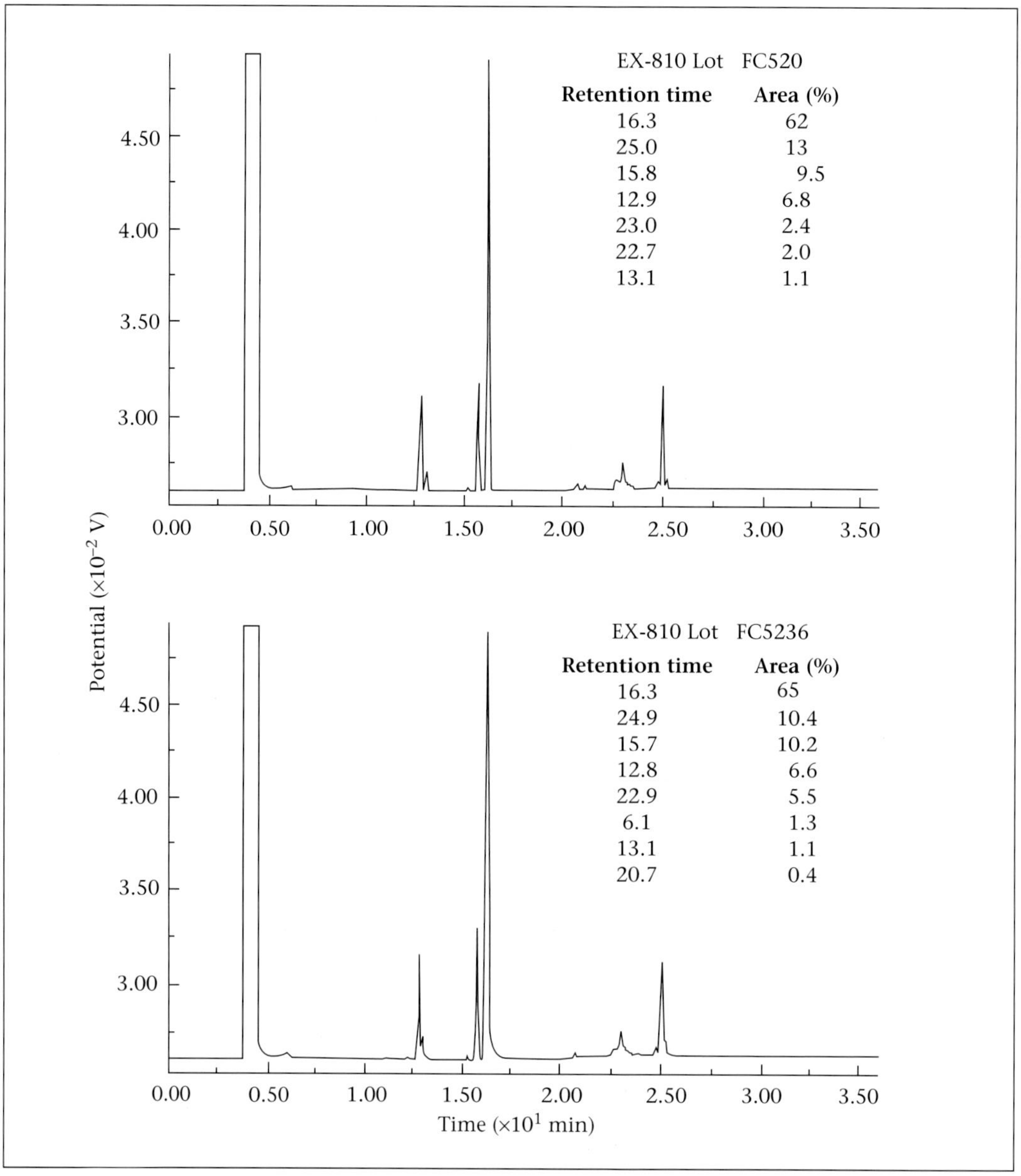

Figure 13.2. Gas chromatograms of two lots of Denacol EX-180. Area of the major peak (ethylene glycol diglycidyl ether) represents 62–65% of the epoxides in each lot.

Stabilizing pH of fixation solution and increasing rate of fixation

Experiments were then conducted to increase epoxide crosslinking efficiency. Alcohol was added to the fixation solution (20% isopropyl alcohol v/v) in an attempt to stabilize the pH over longer periods of time. Alcohol was effective in reducing the rate of change in pH while shrink temperature increased steadily to stabilize at about 80°C (Figure 13.4); the fixative solution was changed only once during this experiment. With a relatively stable pH, increasing concentrations of diepoxide (3, 4 and 5%) in alcoholic phosphate buffer were tried: increasing the concentration of diepoxide effectively accelerated the rate of fixation and increased the terminal shrink temperature of aortic valve cusps (Figure 13.5). Finally, an increase in the initial alkalinity of the buffer system was investigated in an attempt to accelerate the rate of fixation moderately and to increase the final shrink temperature of the tissue. Phosphate was thus replaced with 0.04 M borate

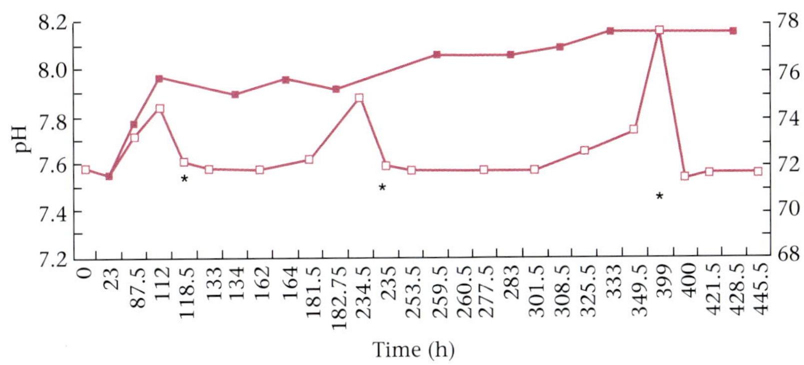

Figure 14.3. Fixation of porcine aortic cusp with aqueous 3% diepoxide solution: change in ph (□) and shrink temperature (T shrink; ■) with treatment time. Asterisks indicate change to fresh diepoxide solution.

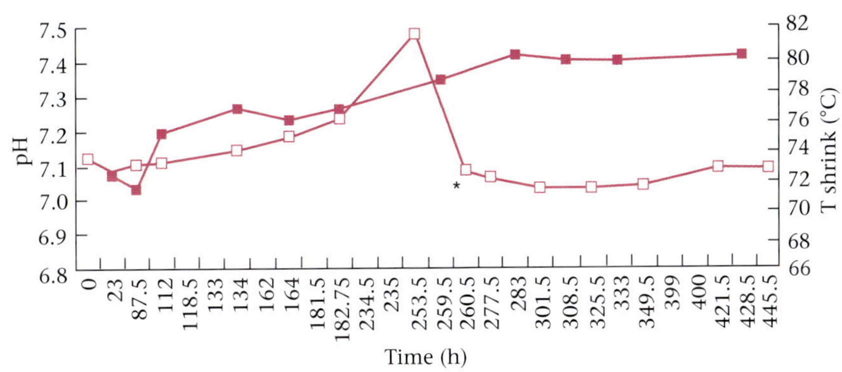

Figure 14.4. Fixation of porcine aortic cusp with alcoholic (isopropyl alcohol, 20% v.v) 3% diepoxide solution: change in pH (□) and shrink temperature (T shrink; ■) with treatment time. Asterisk indicates change to fresh diepoxide solution.

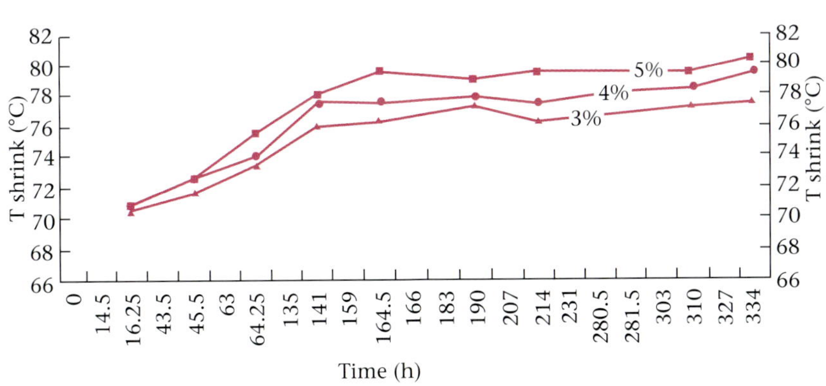

Figure 14.5. Fixation of porcine aortic cusp with alcoholic (isopropyl alcohol, 20% v/v) 3% (▲), 4% (●) and 5% (■) diepoxide solutions: change in shrink temperature (T shrink) with treatment time.

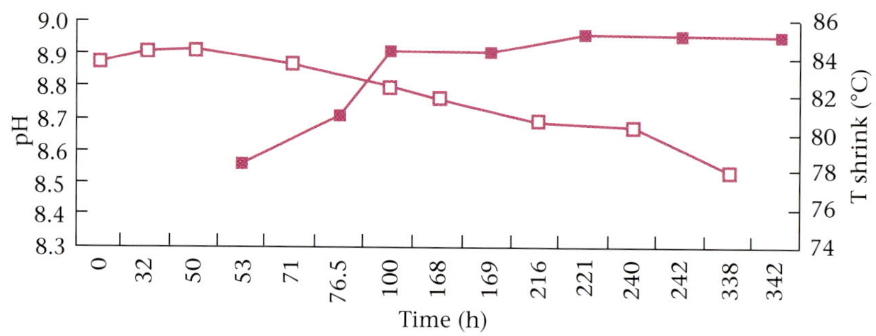

Figure 14.6. Fixation of porcine aortic cusp with 5% diepoxide containing 20% alcoholic (isopropyl alcohol) borate buffer: change in pH (□) and shrink temperature (T shrink; ■) with treatment time.

buffer in order to start the fixation reaction at pH 8.3 rather than pH 7.4. Figure 13.6 shows that porcine aortic cusps fixed in 5% diepoxide fixative containing 20% alcoholic borate buffer attained a stable shrink temperature in excess of 84°C within 10 days. The shrink temperature of diepoxide-fixed porcine aortic valve cusps was thus shown to equal that attained with exposure of cusps to 0.2% glutaraldehyde.

Characterization of diepoxide-fixed porcine aortic valve cusps and wall tissue

Porcine aortic roots fixed with the 5% diepoxide solution described above were tested to determine denaturation temperature, resistance to enzymatic digestion and resistance to mineralization. These tests were conducted for Medtronic by Biomedical Design, Inc. of Atlanta, Georgia, USA. The biaxial mechanical properties of the epoxide-fixed valve leaflets were also investigated and these tests were performed by Systematic Solutions Limited, Auckland, New Zealand. Glutaraldehyde-fixed (phosphate-buffered 0.2% glutaraldehyde) and fresh porcine aortic roots were used as controls.

Valve preparation

Fresh porcine aortic roots were fixed for epoxide or glutaraldehyde crosslinking employing the same 'root pressure fixation' process used to glutaraldehyde-fix the Freestyle™ valve.[23] In this method, the inflow and outflow segments of the aortic root are maintained at equal internal pressure. The leaflets, not having been exposed to a differential hydraulic pressure, are 'zero-pressure-fixed' thus maintaining the native structural/functional morphology compared with even low-pressure-fixed valves. After 10 days of exposure to the diepoxide or glutaraldehyde fixative, the fixed valves were then sterilized by 24 h exposure to borate-buffered 1% glutaraldehyde containing 20% isopropyl alcohol. The valves were then aseptically transferred to shipping jars containing borate-buffered 0.2% glutaraldehyde solution. The temperature of these containers was raised to 37°C for 24 h to enhance the biocidal activity of the packaging solution.

Test procedures

Thermal denaturation testing

Thermal denaturation (shrink temperature) testing was used to compare the degree of collagen crosslinking attained with diepoxide and glutaraldehyde fixation. An abrupt change in length of the tissue specimen occurs (due to the transition of long asymmetric collagen molecules to random coils) as the material is slowly warmed in an aqueous environment.[24] Porcine aortic valve cusps fully tanned with glutaraldehyde exhibit a shrink temperature of about 86°C, whereas fresh cusp tissue shrinks at a temperature of only 65°C. Testing was conducted on eight each of glutaraldehyde-fixed, epoxide-fixed and fresh porcine aortic roots. The cusps were dissected from the valves and rinsed in saline for 3 h before testing. A rinsed cusp from each valve was then cut into an appropriate test specimen and the specimens were individually tested by immersion in distilled water at 55°C. The water temperature was increased at 5°C/min until the specimens started to shrink. There was no significant difference between the shrink temperature of glutaraldehyde-fixed (86.2 ± 0.7°C) and epoxide-fixed (85.2 ± 0.2°C) cusps (Figure 13.7). It should be noted that preliminary testing of epoxide-fixed cusps, not subsequently exposed to glutaraldehyde, did not demonstrate any significant difference in shrink temperature from epoxide-fixed cusps later exposed to glutaraldehyde (i.e. the epoxide-fixed valves were apparently crosslinked to such an extent that subsequent exposure to glutaraldehyde did not further enhance their thermal stability).

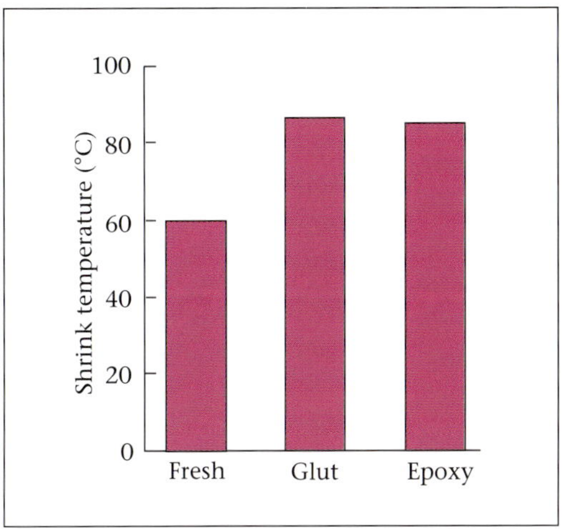

Figure 14.7. Shrink temperature of porcine aortic valve cusps crosslinked with phosphate-buffered 0.2% glutaraldehyde (Glut) or 5% Denacol EX-810 (Epoxy) in 0.04 M borate buffer containing isopropyl alcohol 20% (v/v), with a fresh cusp (Fresh) as control.

Resistance to enzymatic digestion

Samples of fixed and control tissues were tested for resistance to digestion by (1) protease and (2) collagenase. These tests were performed as an additional means of assessing the degree of biological stability afforded by diepoxide versus glutaraldehyde fixation.

Protease digestion test method

Protease (Sigma P5147, 4 units/mg) digestion of cusp and wall specimens was performed in HEPES buffer at 50°C with added glycine to neutralize residual aldehyde. Specimens were incubated in the enzyme for 22 h with shaking, plus, after drying and weighing, an additional 36 h of incubation in fresh protease solution. The percentage of digested tissue was calculated based on the weight of cusp and aortic wall specimens before and after exposure to the enzyme.

Results Figures 13.8 and 13.9 summarize the results of the protease digestion tests for cusp and wall respectively. There was no significant difference in resistance to protease digestion between the glutaraldehyde-fixed and epoxide-fixed cusp or wall.

Collagenase digestion test method

Collagenase (Sigma C-0773, 115 units/ml) digestion of minced cusp and wall specimens was performed in HEPES buffer at 37°C for 72 h with shaking. Aliquots of digestate were removed and tested for liberated amino acid using ninhydrin; a standard curve was prepared using L-norleucine. After an additional incubation period of 48 h at 37°C the ninhydrin test was repeated.

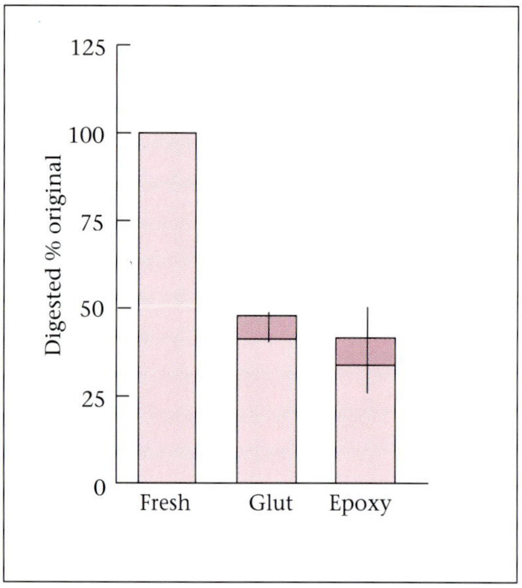

Figure 14.8. Protease digestion, for 22 (□) or 58 (■) h, or porcine aortic valve cusps crosslinked with glutaraldehyde (Glut) or diepoxide (Epoxy) compared with fresh cusp as control. Vertical rules are SEM. There is no significant difference in resistance to digestion between gluraraldehyde- and diepoxide fixed leaflets.

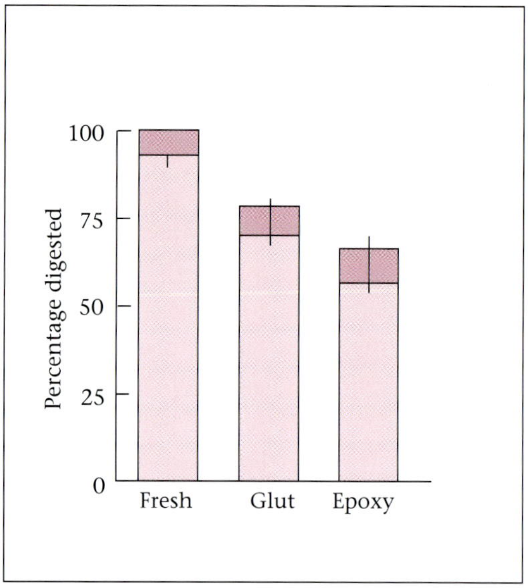

Figure 14.9. Protease digestion, for 22 (□) or 56 (■) h, of porcine aortic wall crosslinked with glutaraldehyde (Glut) or diepoxide (Epoxy) compared with fresh aortic wall as control. Vertical rules are SEM. There is no significant difference in resistance to digestion between glutaraldehyde- and diepoxide-fixed aortic wall.

Results Figures 13.10 and 13.11 summarize the results of the collagenase digestion tests for cusp and wall, respectively. There was no significant difference in the amount of released amino acids (nanomoles L-norleucine equivalent) per milligram dry tissue for glutaraldehyde versus epoxide-fixed leaflets. The epoxide-fixed aortic wall appeared to exhibit a slightly higher (although not statistically significant) resistance to collagenase digestion than glutaraldehyde-fixed aortic wall.

Biaxial mechanical properties

The elastic response of leaflets from root pressure diepoxide-fixed valves was assessed using the biaxial testing technique.[25] Crosslinking of tissue by any method acts to reduce the possible deformational degrees of freedom of the leaflet, making it stiffer, less extensible and therefore less flexible. The stiffer the tissue, the greater the bending stresses during leaflet flexure, and this can cause repetitive creasing/kinking leading to premature leaflet failure.[26] The elasticity of epoxide-fixed leaflets was tested to see if they were less stiff than the controls and might thus possibly open and close more efficiently. Buffered isotonic 0.2% glutaraldehyde-fixed valve leaflets were used as controls.

Sample preparation

Twenty-four leaflets from each of eight epoxide and eight glutaraldehyde-fixed porcine valves were tested. Each leaflet was removed from the aortic wall and immersed in 0.9% saline. Samples of cusp (10 mm^2) were cut from the centre of each leaflet while it was fully relaxed. Figure 13.12 illustrates the alignment of the radial and circumferential directions of the test specimens. Each specimen was then mounted on the biaxial testing apparatus with four No. 6 suture needles along each edge of the square sample. Small grains of carbide grit were sprinkled on the central area of the specimen as video targets for tracking the motion of the

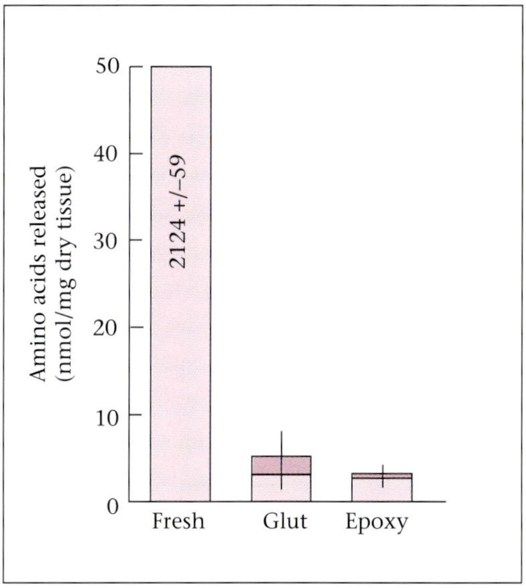

Figure 14.10. Collagenase digestion, as shown by amino acids released (expressed as nanomoles L-norleucine equivalent), for 72 (□) or 120 (■) h, of porcine aortic valve cusps crosslinked with glutaraldehyde (Glut) or diepoxide (Epoxy) compared with fresh cusp as control. Vertical rules are SEM. There is no significant difference in resistance to digestion between glutaraldehyde- and epoxide-fixed leaflets.

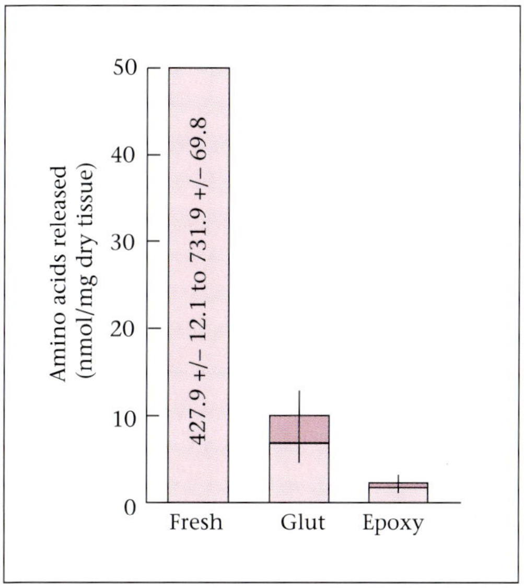

Figure 14.11. Collagenase digestion, for 72 (□) or 120 (■) h, of porcine aortic wall crosslinked with glutaraldehyde (Glut) or diepoxide (Epoxy) compared with fresh aortic wall as control. Vertical rules are SEM. Diepoxide-fixed aortic wall is slightly more resistant to collagenase digestion than the glutaraldehyde-fixed specimens.

tissue surface during loading. Following preconditioning (the sample was fully loaded and unloaded three times), the four hydraulic actuators on each side of the specimen were programmed to apply the same degree of tension simultaneously throughout the whole force range (0–80 Nm^{-1}) in 700 load steps over 90 s. For each load step, the video image of four preselected carbide grains was digitally captured and their geometric coordinates used to calculate the first and second principal strains close to the centre of the sample. Each sample was tested at least five times and median traction/strain curve was chosen; successive tests generally agreed to better than 3% strain.

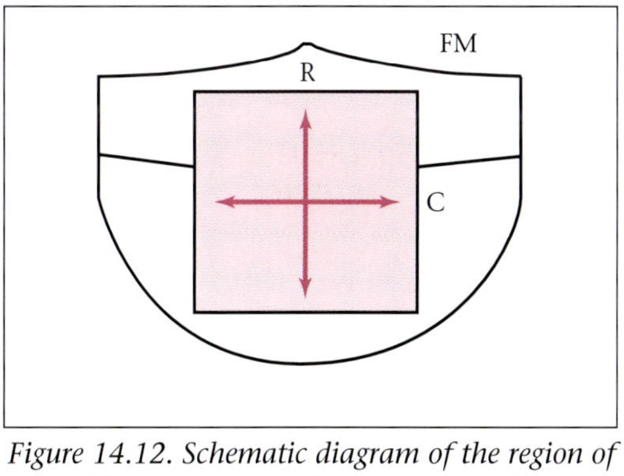

Figure 14.12. Schematic diagram of the region of excised leaflet from which a square sample was cut for biaxial testing. The edges of the sample were aligned in the radial (R) and the circumferential (C) directions, as defined by the free margin (FM).

Results

Figures 13.13 and 13.14 illustrate traction/strain curves for, respectively, the diepoxide and glutaraldehyde-fixed porcine leaflets. The biaxial elastic response for the two sets of fixed cusps was very similar. The leaflets of valves fixed with either agent showed significant circumferential extension consistent with previous studies of leaflets fixed with no differential pressure applied across them. In the radial direction, the small differences in extensibility are significant for tractions less than 6 Nm^{-1} and again for tractions greater than 40 Nm^{-1}.

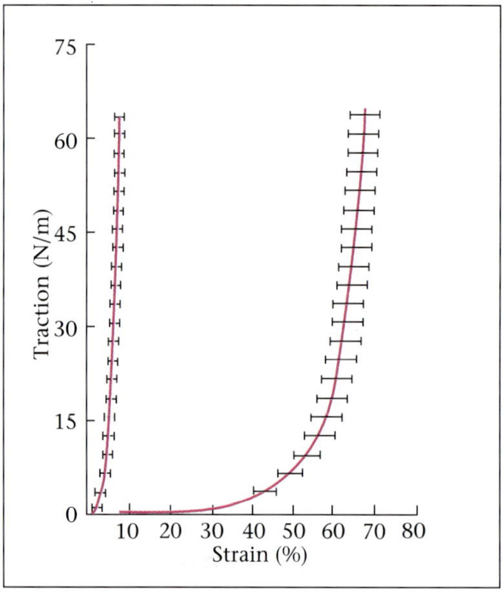

Figure 14.13. Mean (n = 24) biaxial traction/strain curves (left, circumferential; right, radial) for diepoxide-fixed porcine aortic valve leaflets. Horizontal bars are SD in strain at each value of traction. The leaflets show significant circumferential extension consistent with the root pressure-fixation process, i.e. where there is no differential pressure applied across the cusps of the valve during crosslinking.

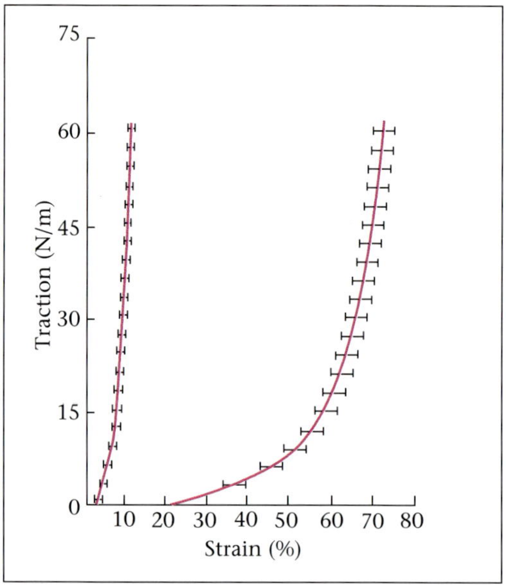

Figure 14.14. Mean (n = 24) biaxial traction/strain curves (left, circumferential; right, radial) for glutaraldehyde-fixed porcine aortic valve leaflets. Horizontal rules are SD in strain at each value of traction. The leaflets show significant circumferential extension consistent with the root pressure-fixation process, i.e. where there is no differential pressure applied across the cusps during crosslinking.

Comparison of the radial traction curves for the two sets of valve cusps demonstrates that the diepoxide-fixed tissue appears to be more flexible and exhibits a more clearly defined transition from low-modulus to high-modulus elastic behaviour (Figure 13.15). The epoxide-fixed tissue achieves greater strain under small loads than does the glutaraldehyde-treated tissue (i.e. the epoxide-fixed leaflets are softer, more extensible, and can bend with less effort). The shape of the curve for epoxide-fixed leaflets is closer to that of fresh, unfixed cusp than that for glutaraldehyde-fixed tissue. In the light of the comparability in the degree of crosslinking between the two tissues, the difference in flexibility may be accounted for by a 'flexible joint' in epoxide linkages compared with the 'stiff joint' of glutaraldehyde linkages.[6]

Resistance to mineralization

Specimens of glutaraldehyde and diepoxide-fixed aortic cusp and aortic wall tissue were implanted subcutaneously in 3-week-old rats to assess the resistance of these tissues to calcification. This method of testing has been demonstrated to mimic the pathology of mineralization seen in dystrophic calcification of clinically retrieved bioprostheses.[27] Typically, glutaraldehyde-fixed aortic cusps can exhibit levels of calcium in excess of 150 mg/g dry weight of tissue after 4–6 weeks of implantation in this model. In the present study, glutaraldehyde-fixed porcine aortic wall was also shown to accumulate calcium (over 50 mg/g) after 4-week implantations. Pretreatment of glutaraldehyde-fixed porcine aortic cusps with antimineralization agents such as surfactants or other agents has been reported to be capable of reducing cuspal mineralization by up to two orders of magnitude (i.e. 1–2 mg calcium/g dry tissue) compared with untreated controls.[28]

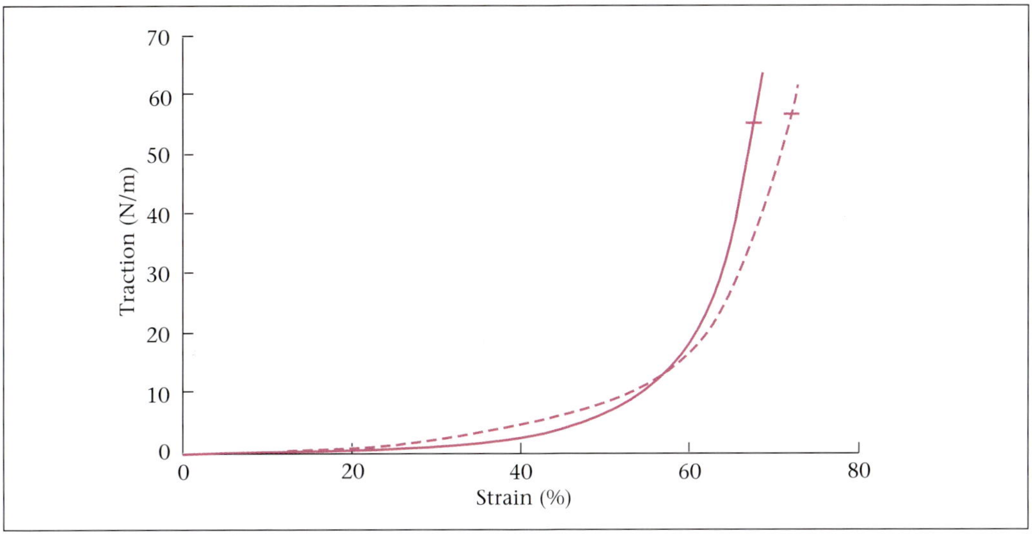

Figure 14.15. Radial traction/strain curves for diepoxide-fixed (——) and glutaraldehyde-fixed (- - - -) porcine aortic valve leaflets. Diepoxide-fixed tissue exhibits a more clearly defined transition from low-modulus to high-modulus elastic behaviour.

Implantation procedure and calcium analysis

Cusp and wall specimens dissected from saline-rinsed glutaraldehyde- and epoxide-fixed porcine aortic valves were randomly implanted subdermally in 3-week-old Sprague–Dawley male rats. Four specimens, approximately 0.5 × 1 cm each, were implanted in each animal for 4 weeks. Following retrieval and removal of surrounding tissue, a 1 mm band was cut from the edges of each wall specimen. The cusps and trimmed walls were then extensively washed in deionized water, lyophilized, weighed, and hydrolysed in 6 N redistilled ultrapure HCl at 85°C. Diluted samples of the hydrolysates were then submitted to calcium quantification using Inductively Coupled Plasma (ICP) analysis. This method is one of two commonly employed chemical assays to quantitate the concentration of metal ions in solution. In this case, the method was used to quantitate the amount of calcium in explanted specimens of tissue. As described, tissue specimens were washed, dried, weighed, and dissolved in acid. The acid solutions were appropriately diluted and sampled by the instrument; calcium concentrations of standard solutions containing 0–10 μg/ml pure calcium salts were used to calibrate the instrument. The ICP instrument employed is manufactured by Perkin-Elmer Company (Model Number 6000) and was used with a Perkin-Elmer Data Station Model Number 7500. The results, adjusted for dilution and sample weight, were statistically analysed using Student's *t*-test for independent variables.

Initial results

Figure 13.16 shows the results of calcium assays by ICP on explanted diepoxy- and glutaraldehyde-fixed cusp and wall tissue. There was no statistical difference in the level of calcification between the two methods of fixation for either the cusps (*t*-test, $p>0.42$) or the walls (*t*-test, $p>0.75$). These results were not anticipated. Previous publications by others had indicted that epoxide-fixed biomaterials implanted subcutaneously in rats or rabbits showed significantly lower calcium levels than glutaraldehyde-fixed controls.[6–19,15] Recently, polyepoxy-fixed porcine aortic valves, implanted for 5 months in weanling sheep as stented mitral valves reportedly showed only 0.33 mg calcium/g dry cusp tissue;[29] no mention was made of whether the aortic wall remnant of these stented implants was calcified.

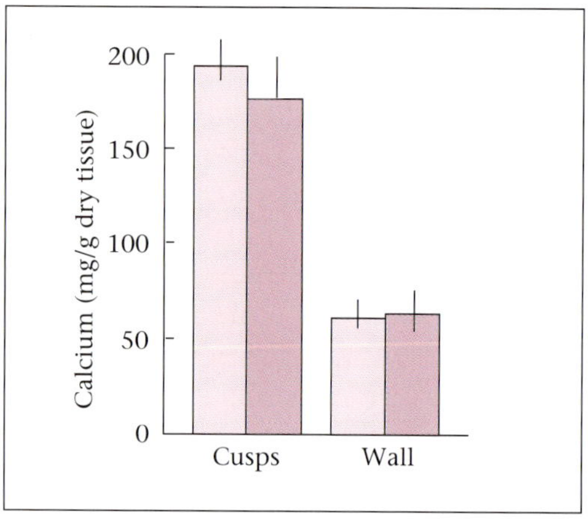

Figure 14.16. Calcium content of explanted diepoxy- (□) and glutaraldehyde-fixed (■) porcine aortic valve cusp and wall tissue. Diepoxide-fixed valves were sterilized in 0.2% glutaraldehyde solution. Vertical rules are SEM. There is no statistically significant difference in the level of calcification (at 4 weeks of implantation) between the two methods of crosslinking, for either cusp (t-test, $p > 0.42$) or wall (t-test, $p > 0.75$).

It was suspected that the method utilized to sterilize the epoxy-fixed valves terminally (24 h soak in 0.2% glutaraldehyde at 39°C) may have incited the mineralization. Consequently, subdermal rat implantations were repeated utilizing epoxide and glutaraldehyde-fixed valves prepared identically, except that the epoxide-fixed valves were not exposed to glutaraldehyde.

Additional procedure and results

Subdermal implantation of epoxide and glutaraldehyde-fixed cusp and aortic wall specimens was performed in 3-week-old rats, as described above. Sufficient samples of tissue were implanted to allow assessments of mineralization at 4 days, 8 days, 4 weeks and 8 weeks after implantation.

The level of calcification in epoxide-fixed cusp samples at 4 weeks, quantified using ICP assay, was significantly less ($p<0.0001$) than that measured in the glutaraldehyde-fixed controls (Figure 13.17). There was also a significantly lower level of calcium in the epoxide-fixed aortic wall implants compared with the controls ($p<0.03$). This tends to confirm the authors' suspicion that exposure to glutaraldehyde (despite previous epoxide crosslinking and thorough rinsing before implant) may have contributed to mineralization of diepoxy-fixed tissues. An assessment of the results following 8 weeks of implantation is currently under way.

Discussion

Subdermal and systemic vascular implantation of aldehyde-fixed biomaterials in several animal models has demonstrated that the crosslinking agent itself evidently plays a critical, but as yet undefined, role in dystrophic

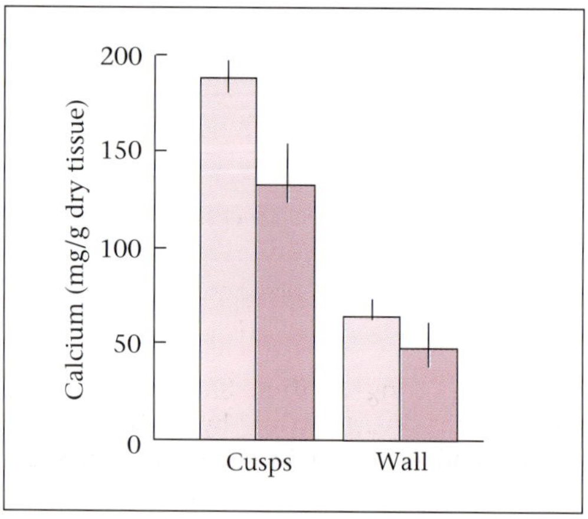

Figure 14.18. Calcium content of explanted diepoxy- (□) and glutaraldehyde-fixed (■) porcine aortic valve cusp and wall tissue. The diepoxide-fixed tissue was not exposed to glutaraldehyde at any time during processing. Vertical rules are SEM. There is a significant difference in the level of calcification at 4 weeks of implantation for both cusp ($p < 0.0001$) and wall ($p < 0.03$) tissue fixed with diepoxide.

mineralization.[30,31] Soaking glutaraldehyde-fixed tissue in L-glutamic acid has been shown to reduce mineralization significantly by blocking free aldehyde release.[32–34] Other antimineralization pretreatments of glutaraldehyde-fixed bioprostheses (such as AOA and T6) have proved to be even more effective in mitigating the tendency of tissue to calcify in experimental animals, but this efficacy has yet to be demonstrated clinically.[35]

Clearly, it would be advantageous for a stabilized biomaterial to resist mineralization without having to undergo further processing before implantation. Crosslinking biomaterials with polyepoxide compounds has thus recently gained considerable attention, since natural tissues fixed with such agents not only have a more attractive appearance, but are demonstrably less stiff, more hydrophilic, and calcify less than glutaraldehyde-fixed controls.[15,20]

These investigations have demonstrated that porcine aortic valves fixed with ethylene glycol diglycidyl ether have cuspal denaturation temperatures and resistance to enzymatic digestion equal to or greater than that exhibited by glutaraldehyde-fixed porcine aortic valves. The biaxial properties of the diepoxide-fixed cusps show slightly improved elastic behaviour at low traction levels compared with aldehyde-fixed controls. So far, in these studies, diepoxide-fixed aortic valves have been shown to be only marginally resistant to mineralization, and only when the tissue has not been subsequently exposed to glutaraldehyde. This finding tends to further confirm indications that the mere presence of free aldehydes in crosslinked tissue can incite mineralization in the rat subdermal model.

Additional studies are planned, not only to characterize further the biocompatibility and mechanical properties of diepoxide-fixed tissues but also to determine whether enhanced resistance to mineralization may be afforded by judiciously but significantly lowering the shrink temperature.

References

1. Kitzman D W, Kon N D, Stewart K P, Nomeir A M, Klopfenstein H S, Haumschild D J, Cordell A R 1993 Echo-doppler comparison of a new non-stented porcine aortic valve prosthesis with the aortic homograft. J Am Soc Echocardiol 6 (3): Pt 2
2. Girardot M N, Girardot J-M, Schoen F J 1991 Alpha amino-oleic acid, a new compound, prevents calcification of bioprosthetic heart valves. Trans Soc Biomater p 114
3. Girardot J-M, Gott J P, Guyton R A, Schoen F J 1991 A novel technology inhibits calcification of bioprosthetic heart valves: a sheep pilot study. Trans Soc Biomater p 60
4. Gott J P, Pan-Chih, Dorsey L M A, Jay J L, Jett G K, Schoen F J, Girardot J M, Guyton R A 1992 Calcification of porcine valves: a successful new method of antimineralization. Ann Thorac Surg 53: 207–216
5. Hall J D, Whitlark J, Horsley S, Dorsey L, Pan-Chih, Girardot J M, Girardot M N, Schoen F J, Gott J P, Guyton R A 1993 Antimineralization of bioprostheses: an improved amino oleic acid technique. ASAIO Meeting, New Orleans, p 93
6. Xi Tingfei, Fuying Liu, Xi Baoshu 1992 Effect of pretreatment with epoxy compounds on the mechanical properties of bovine pericardial bioprosthetic materials. J Biomater Appl 7: 61–75
7. Imamura E, Sawatani O, Koyanagi H, Yasuharu Noishiki, Miyata T 1989 Epoxy compounds as a new cross-linking agent for porcine aortic leaflets: subcutaneous implant studies in rats. J Cardiac Surg 4: 50–57
8. Sawatani O 1991 Usefulness of epoxy compounds as a new crosslinking agent for xenograft bioprostheses — comparison with glutaraldehyde treated porcine valve. Jpn Artif Organs 20 (5): 1452–1459
9. Xi Tingfei, Ma Jiazhen, Tian Wenhua, Lei Xuehui, Long Shuhui, Shi Baoshu 1992 Prevention of tissue calcification on bioprosthetic heart valve by using epoxy compounds: A study of calcification tests *in vitro* and *in vivo*. J Biomed Mater Res 26: 1241–1251

10. Final Draft Replacement Heart Valve Guidance, Version 4.0, December 1993 United States Food and Drug Administration, Rockvile, MD
11. European Committee for Standardization Document CEN/TC 285 WG3 TF 1: Cardiac Valves
12. Spilezewski K L, Anderson J M, Schaap R N, Salamon D D 1988 *In vivo* biocompatibility of catheter materials. Biomaterials 9: 253–256
13. Nimni M E, Cheung D T, Strates B, Kodama M, Sheikh K 1988 Biotechnology. In Nimni M E (ed) Collagen, 1st edn. CRC Press, Boca Raton, pp 1–38
14. Lee J M, Thyagarajan K, Pereira C A, McIntyre J, Tu R 1991 Crosslinking of a prototype bovine artery xenograft: Comparison of the effects of glutaraldehyde and four poly (glycidyl ether) compounds. Artif Organs 15: 303
15. Okoshi T, Noishiki Y, Tomizawa Y, Morishima M, Taira T, Kawai T, Itoh H, Miyata T, Koyanagi H 1990 A new bioprosthetic cardiac valve with reduced calcification. ASAIO Trans 36: M411–M414
16. Sung HW, Shen S-H, Tu R, Lin D, Hata C, Noishiki Y, Tomizawa Y, Qui jano R C 1993 Comparision of the crosslinking characteristics of porcine heart valves fixed with glutaraldehyde or epoxy compounds. ASAIO J 39: M532–M536
17. Lohre J, Baclig L, Sagartz J, Guida S, Tu R 1991 Evaluation of two diepoxy ether compounds for biocompatible potential. Artif Organs 15 (4): 124
18. Sung H W, Tu R, Shen S H, Lu C-L, Lin D, Hata C, McIntyre J P, Wang E, Qui jano R C 1993 Crosslinking characteristics of a polyepoxy compound preserved porcine bioprosthesis. Abst 19th Mtg Soc Biomater April 28 – May 2, 1993, 25
19. Murayama Y, Satoh S, Oka T, Imanishi J, Noishiki Y 1988 Reduction of the antigenicity and immunogenicity of xenografts by a new crosslinking reagent. Trans Am Soc Artif Intern Organs 34: 546
20. Tu R, Lu C-L, Thyagarajan K, Wang E, Nguyen H, Shen S, Hata C, Qui jano R C 1993 Kinetic study of collagen fixation with polyepoxy fixatives. J Biomed Mater Res 27: 3–9
21. Lohre J M, Baclig L, Sagartz J, Guida S, Thyagarajan K, Tu R 1992 Evaluation of two epoxy ether compounds for biocompatible potential. Artif Organs 6: 630
22. Nagase Chemicals 1993 Denacol Epoxy Compounds. Product catalogue, Nagase Chemical Ltd, Shinmachi 1-chome, Nishi-ku, Osaka, 550–91 Japan, p 14
23. Barratt-Boyes B G, Christie G W, Raudkivi P J 1992 The stentless bioprosthesis: surgical challenges and implications for long-term durability. Eur J Cardio-Thorac Surg 6 (Suppl 1): S39–S43
24. Woodroof E A 1979 The chemistry and biology of aldehyde treated tissue heart valve xenografts. In: Ionescu M I (ed) Tissue heart valves. Butterworth, London, pp 347–362
25. Christie G W 1992 Anatomy of aortic heart valve leaflets. The influence of glutaraldehyde fixation on function. Eur J Cardio-Thorac Surg 6 (Suppl 1): S25–S33
26. Broom N D, Thomson F J 1979 Influence of fixation conditions on the performance of glutaraldehyde-treated porcine aortic valves: towards a more scientific basis. Thorax 34: 166–176
27. Fishbein M C, Levy R J, Ferrans V J, Dearden L C, Nashef A, Goodman A P, Carpentier A 1982 Calcification of cardiac valve bioprosthesis: Biochemical, histologic, and ultrastructural observations in a subcutaneous implantation model system. J Thorac Cardiovasc Surg 83: 602–609
28. Jones M, Eidbo E E, Hilbert S L, Ferrans V J Clark R E 1989 Anticalcification treatments of bioprosthetic heart valves: *in vivo* studies in sheep. J Card Surg 4: 69–73
29. Hata C, Sung H W, Shen S, Witzel T, McIntyre J, Thyagarajan K, Rowland M, Noishiki Y. Explant analysis of a porcine heart valve bioprosthesis cross-linked with polyepoxy compound. Am Soc Artif Int Organs Abstr 40th Ann Mtg April, San Francisco
30. Gong G, Ling Z, Seifter E, Factor S M, Frater R W M 1991 Aldehyde tanning: the villain in bioprosthetic calcification. Eur J Cardio-Thorac Surg 5 (6): 288–293
31. Liao K, Gong G, Hoffman D, Seifter E, Factor S M, Macaluso F, Frater R W M 1993 Spontaneous host endothelial growth on bioprosthetic valves and its relation to calcification. Eur J Cardio-Thorac Surg 7 (11): 591–596
32. Liao K, Seifter E, Yellin E, Hoffman D, Frater R W M 1991 Improved biocompatibility by post fixation treatment of aldehyde fixed porcine aortic valves. Artif Org 15(4) Abstr 342
33. Grimm M, Eybl E, Grabenwöger M, Griesmacher A, Losert U, Bock P, Muller M M, Wolner E 1991 Biocompatibility of aldehyde-fixed bovine pericardium. J Thorac Cardiovasc Surg 102: 195–201

34. Grimm M, Eybl E, Grabenwöger M, Spreitzer H, Jager W, Grimm G, Bock P, Muller M M, Wolner E 1992 Glutaraldehyde affects biocompatibility of bioprosthetic heart valves. Surgery 111 (1): 74–78
35. Levy R J 1994 Glutaraldehyde and the calcification mechanism of bioprosthetic heart valves. J Heart Valve Dis 3 (1): 101–104

Chapter 14

Acylazide treatment: an alternative to glutaraldehyde processing

H. Petite, Ph. Menasché, A. Piwnica and A. Huc

Collagen, which represents about 40% of the proteins of plants and animals, endows connective tissues with most of their properties, such as mechanical resistance, haemostatic capability and contribution to cell development. These three characteristics, associated with the good biocompatibility and biodegradability of collagen, have led numerous researchers to develop collagen-based biomaterials.[1] Some applications need high mechanical resistance and also very low biodegradability of the biomaterials. Such is the case, for instance, with heart valves or vascular substitutes. For these applications, it is possible to use the whole natural tissue from which the cell residues have been removed. The great advantage of this technique is that the helical and fibrillar structures of collagen and, consequently, its native mechanical properties, are retained. However, if this natural tissue is implanted, it is digested progressively by collagenases and by proteases so that it loses its mechanical properties and disappears. It is therefore necessary to tan collagen to prevent enzymes from recognizing the protein. In the tanning process, the side-chain groups are blocked and new crosslinks between peptide chains are achieved. The tanning process can also be used to improve the mechanical properties of biomaterials made of extracted collagen and to reduce their biodegradability, but in this case the mechanical performance will be weaker than that of natural connective tissues, because the native fibrillar structure is destroyed by the extraction treatment, even when the helical structure has been saved.

Hitherto, the most widely used technique for crosslinking collagen-based biomaterials has been glutaraldehyde tanning. This process introduces into the protein non-natural crosslinks (Figure 14.1) and free residues of aldehydes in the form of polymers that are difficult to remove from the biomaterial. The presence of glutaraldehyde inside the collagen brings about loss of the biological properties of the protein, mainly because of the cytotoxic characteristics of glutaraldehyde. Moreover, glutaraldehyde appears to enhance calcification of the implanted tissue. This drawback of glutaraldehyde processing is of particular relevance to bioprosthetic heart valves, the long-term durability of which is known to be severely limited by leaflet mineralization.[2] Even if mechanical,

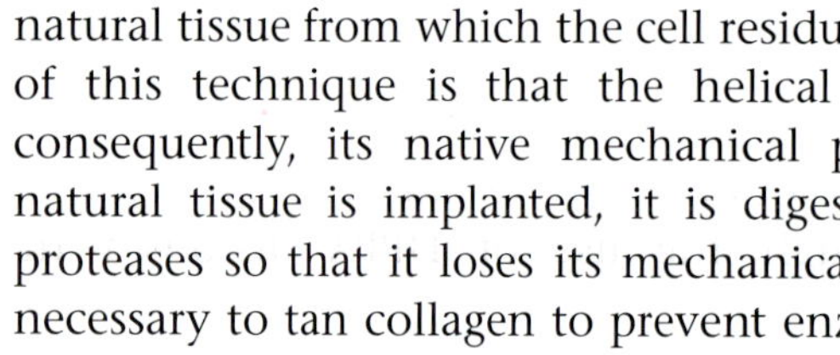

Figure 14.1.

stress-related factors significantly contribute to this mineralization process, biochemical, glutaraldehyde-related factors also play an important role.

Many researchers are therefore seeking new processes for tanning collagen that would be as effective as glutaraldehyde but would better preserve the biocompatibility of the material. Most of the crosslinking techniques studied leave residues in the collagen and thus modify the biological properties of the protein. One, however, is of interest because it gives rise to natural crosslinks between peptide chains without leaving any residue. In this method, carbodiomides are used. Unfortunately, the crosslinking obtained is rather weak compared with that from glutaraldehyde.[3] To address this issue the authors have developed a new tanning treatment of collagen in order to obtain a high level of crosslinking with the least modification of the biological properties of collagen. The technique consists of forming azide groups on carboxylic side-chain groups of the peptide chains and of obtaining natural bonds by reaction of these new groups with amino acid chain groups.

The various properties of a collagen material treated by this technique were determined in order to evaluate its performance as a biomaterial. The endpoints included thermal stability of collagen, mechanical properties, digestibility by enzymes, behaviour of endothelial cells on the material and calcification after implantation *in vivo*. The same experiments were performed on the same collagen material treated by glutaraldehyde (GTA) and the properties of the two materials were compared.

Processing techniques

Procedure of crosslinking using diphenylphosphorylazide (DPPA)

The first step[4] consists of creating azide groups on the side-branched chain carboxylic groups of the collagen through the use of DPPA dissolved in dimethylformamide (DMF). In a second step the amino side-chain groups react in a buffered medium with azide groups to give peptide bonds (Figure 14.2).

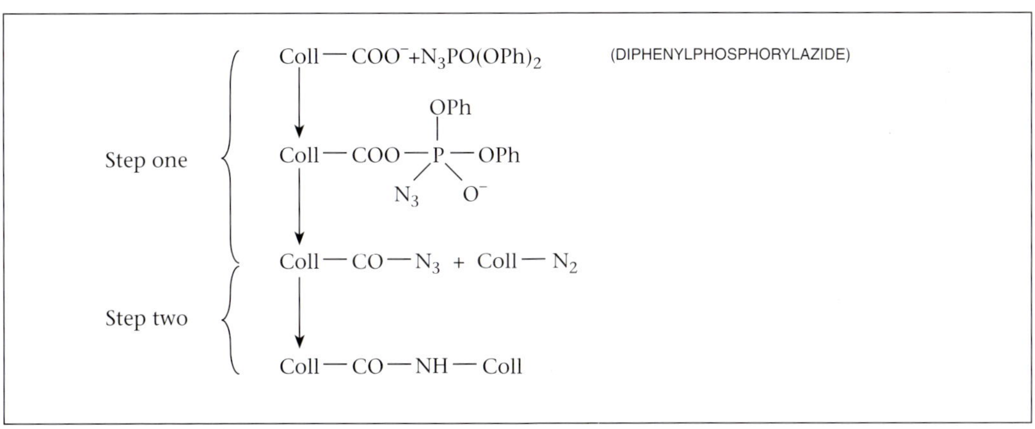

Figure 14.2.

This new procedure was applied to calf pericardial tissue obtained from pericardial sacs taken from young calves, stripped of all visible fat and stored at 4°C in 70% ethanol until treatment. Pericardial samples (10 mg dry weight) were immersed in 5 ml of a solution of DPPA in DMF at 4°C for 24 h (step 1), rinsed in a borate buffer (0.04 M disodium tetraborate, 0.04 M boric acid; pH 8.9) and then incubated overnight at 4°C.

The greatest crosslinking is obtained for a concentration of 0.5% DPPA in DMF (referred to below as DPPA strong).

GTA procedure

Pericardium was treated with a fresh solution of GTA (0.6, 0.5 or 0.05%) for 4 days at 4°C as described by Petite *et al.*[5]

Results

Thermal stability of collagen

The greater the collagen crosslinking, the higher the temperature of denaturation in water of this protein. Thus, determination of the thermal stability of the material indicates the efficacy of the crosslinking procedure. To this end, thermal changes were measured with a differential scanning calorimeter (SETARAM, type 111) and thermal changes were recorded at a constant rate of 2°C/mm from 20 to 100°C.

The results are shown in Table 14.1. It can be seen that the stability obtained with the DPPA procedure is almost as great as that due to GTA treatment.

Table 14.1 Thermal stability and mechanical properties of unimplanted pericardium (average of 20 samples)

Group	Thermal stability (°C)	Break at elongation (%)	Tensile strength (N/mm²)
GTA 0.5%	85.5 ± 0.6	38.6 ± 10.4	12.8 ± 4.8
FA 4% and GTA 0.5%	85.5 ± 0.6	38.6 ± 10.4	12.8 ± 4.8
GTA 0.5% and AA solution	81.4 ± 1.1	50.2 ± 11.0	16.9 ± 4.0
DPPA strong	81.9 ± 1.1	48.5 ± 9.2	16.4 ± 5.9
Fresh tissue	61.0 ± 1.0	21.9 ± 6.1	9.1 ± 5.4

GTA: glutaraldehyde; AA: amino acid; FA: formaldehyde; DPPA: diphenylphosphorylazide

Mechanical properties

These determinations involved an extensiometer (INSTRON, type 6022). The samples were tested after being dried.

The breaking point after elongation and the tensile strength were determined. The results, shown in Table 14.1, indicate that the DPPA process is as efficient as GTA for increasing the mechanical properties of the fresh pericardium.

Collagenase digestion

Samples of pericardium (10 mg) were incubated in 5 ml bacterial collagenase solution (3300 U/ml) in buffer pH 7.4 (10 mM Tris, 25 mM calcium dichloride). After 24 h the mixture was centrifuged for 10 min at 3000g. The collagen content of the supernatant was determined by hydroxyproline assay. The percentages of undigested collagen from the biomaterials (fresh, GTA 0.5%, GTA 0.05% and DPPA strong) are given in Table 14.2. The results show that DPPA strong protects the tissue against enzyme digestion as effectively as GTA.

Cytocompatibility study

For this study, the behaviour was examined of endothelial cells of chicken embryos in culture in the presence of the biomaterial.[6] Fragments of aortas of chicken embryos were cultured in contact with pericardia at 37°C for 8 days. After

Table 14.2

Pericardium	Percentage of collagen not digested after 24h in collagenase (average of 5 determinations)
GTA 0.6%	99 ± 1
GTA 0.05%	99 ± 1
DPPA strong	96 ± 2
Fresh	3 ± 1

GTA: glutaraldehyde; DPPA: diphenylphosphorylazide

this period samples were withdrawn (Figure 14.3): some were used to evaluate areas of migration and the density of the cells; other were examined by scanning electron microscopy.

As a reference, a plastic material (Thermanox, LUX Corporation) treated for tissue culture was used. The results, given in Table 14.3, show that for GTA 0.6% the density of cells is high and the migration area small compared with the reference. In contrast, with pericardium treated by DPPA strong, the density of cells is about the same as for the reference and the migration areas seem to be larger (Figure 14.4). These results mean that the cells on pericardia treated with GTA 0.6% or GTA 0.05% are organized in dense layers with weak migration in order to try to resist the toxic effects of glutaraldehyde.

For DPPA-treated pericardium, migration appears to be excellent in comparison with the other materials and the cultured cells are well spread out.

This study shows the excellent biocompatibility of DPPA-treated pericardium.

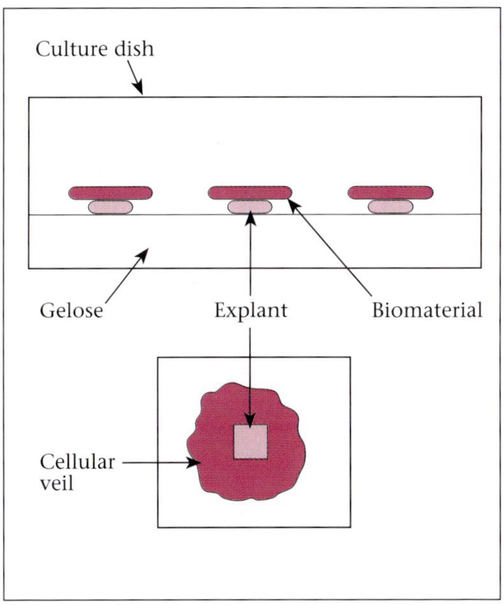

Figure 14.3.

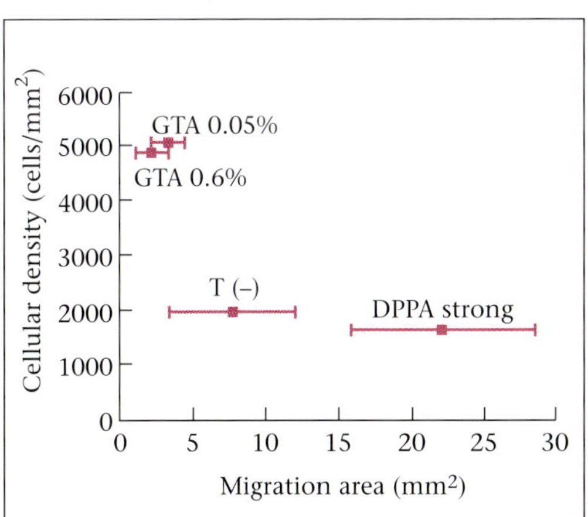

Figure 14.4

Calcification

To evaluate the calcification of the biomaterials *in vivo*, GTA 0.6% pericardium and DPPA pericardium were implanted subcutaneously in the abdominal region of Sprague–Dawley rats.[7] For this purpose, four abdominal pouches were surgically prepared. After the samples had been inserted, the pouches were sutured. At 2, 4 and 8 weeks, the pericardial samples were explanted. For quantitative analysis of calcium, the samples were dissected free of host tissue and extracted for 36 h at 78°C in 6 N HCl. The calcium content of the extracts was then determined by atomic absorption spectrophotometry.

The results are shown in Table 14.4. It appears that the calcium content of DPPA-treated pericardium is much less than that of GTA-treated tissue.

Table 14.3

	Number of samples	Migration areas (mm^2)	Cellular density average ($cells/mm^2$)
Aorta Thermanox (T–)	8	7.65 ± 4.38	1933
Aorta pericardium GTA 0.6%	7	2.09 ± 1.17	4866
Aorta pericardium GTA 0.05%	7	3.26 ± 1.18	5041
Aorta pericardium DPPA strong	5	22.2 ± 6.3	1604

Table 14.4 Calcium content of implanted pericardium samples (mg/g dry weight – average of 20 samples) Unimplanted pericardium: 0.5–1 mg/g dry weight

Group	2 weeks	4 weeks	8 weeks
GTA 0.5%	78	103	111
FA 4% and GTA 0.5%	45	55	139
GTA 0.5% and AA solution	36	50	127
DPPA strong	2	8	9

GTA: glutaraldehyde; AA: amino acid; FA: formaldehyde; DPPA: diphenylphosphorylazide

Conclusions

Determinations of the denaturation temperatures, mechanical properties and digestibility by collagenase indicate that crosslinking of pericardial tissue by the DPPA technique is as effective as that achieved with the use of the standard GTA procedure. As for biocompatibility, the cell cultures performed on the treated tissues suggest that DPPA treatment causes no disruption of cell development, in contrast to the GTA procedure. Moreover, it appears that this new crosslinking procedure reduces calcification of the tissue when implanted. For all these reasons, crosslinking by acylazide appears to be a very promising crosslinking treatment for collagen-based biomaterials. This conclusion is reinforced by the studies by Simmons and Kearney,[3] who compared several crosslinking procedures of collagen. Their results show that the most interesting treatment is acylazide.

The excellent properties of this new treatment have been demonstrated in practice by the presence on the European market of two collagen-based biomaterials: the first is a membrane for guided tissue regeneration (Paroguide, Coletica SA) for inducing bone reformation around the teeth; the second is a collagen thread (Colladerm, Coletica SA) for correcting the skin defects due to scarring. These products have been used in several thousand cases and have now proved their efficacy and their very good tolerance in human applications.

References

1. Huc A 1993 Les biomatériaux à base de collagène. Lyon Pharm 44: 309–319
2. Schoen F J, Tsao J N, Levy R 1986 Calcification of bovine pericardium used in cardiac valve bioprostheses. Am J Pathol 123: 134–145
3. Simmons D M, Kearney J N 1993 Evaluation of collagen crosslinking techniques for the stabilization of tissue matrices. Biotechnol Appl Biochem 17: 23–29

4. Petite H, Frei V, Huc A, Herbage D 1994 Use of diphenylphosphorylazide for crosslinking collagen-based biomaterials. J Biomed Mater Res 28: 159–165
5. Petite H, Rault I, Huc A, Menasche P, Herbage D 1990 Use of the acyl azide method for crosslinking collagen rich tissues such as pericardium. J Biomed Mater Res 24: 179–187
6. Sigot-Luizard M F, Lanfranchy M, Duval J L, Benslimane S, Sigot M, Guidoin R G, King M N 1986 The cytocompatibility of compound polyester–protein surfaces using an *in vitro* technic. In Vitro Cell Dev Biol 22: 234–240
7. Gatti A M, Galloni M, Monari E, Noera G, Pasquino E, Vallana F 1991 Calcification of chemically treated bovine pericardium. Int J Artif Organs 14: 647–654

III Clinical Experience

Chapter 15

Homograft valves in congenital heart disease: a 7-year experience

G. Bloch, A. Bel, B. Faris, A. Nguyen Van-Cao, T. du Roy de Chaumaray and P. Menasché

Homograft valves have been implanted in humans for three decades, since their first implantation by Ross[1] and Barratt-Boyes[2] in 1962. Despite their good early and mid-term results, their use has been limited by lack of availability in a large range of sizes, and by the variability of durability. This was due to the multitude of preservation techniques, some deleterious to the valve, and led to the development of standardized processing, in a wide range of sizes, of bioprostheses with long-term storage capabilities. Recent publications reporting excellent long-term results,[3] as well as improvements in storing techniques (e.g. cryopreservation) have boosted the use of homografts, especially in the young and particularly for congenital heart diseases.

Preparation and storage of valves

During the last 7 years, 136 cryopreserved homograft valves have been implanted in children, the youngest being 5 years old. These valves originated from heart explantations (at the time of the transplantation) in 70%, or from multiorgan donors when the heart was unsuitable for transplantation in 30%. Hearts were maintained in a sterile condition and brought to the tissue bank. Homograft valves are said to be viable because of the short period of warm (between aortic crossclamping to heart immersion in 4°C saline bath) and cold (until cryopreservation proceeding) ischemia, a mean of 7 h (range 2–18 h). Viability has been assessed in previous studies,[4] including morphological and biochemical tests, electron-microscopic examination and cell culture. Valves, with their matrix of collagen fibres and fibroblasts and their endothelial cell lining, retain their viability for periods extending to 30 h at 4°C. Above this limit, even at 4°C, as demonstrated by Yankah,[5] cell viability is compromised, with no more endothelial cells at 8 days, and only 20% of viable fibroblasts at 14 days. Moreover, it has not yet been determined whether these viable cells are functional and produce collagen and other structural proteins, allowing for natural turnover and maintenance of valvular function.

With the current cryopreservation technique (nutrient medium of RPMI, foetal calf serum, 10% dimethylsulfoxide as a cryoprotectant and no added antibiotic; temperature reduced by 2°C/min to –80°C and then storage in vapour phase liquid nitrogen at –196°C), viability is not altered.[6]

Viability could be an important factor in valve durability, as already noted by O'Brien.[7] This corroborates some authors' 18-year experience with autograft aortic replacements (i.e. viable valves). Matsuki[8] noted a very low rate of valve degeneration of 0.09% patient-year.

The authors insist on strict compatibility in ABO blood group but not in HLA typing, because of the limited availability of homografts. Even if endothelial cells still express surface antigens, immunosuppressive therapy is not mandatory in the authors' opinion.

Indications for operation

A total of 136 homograft valves have been implanted, for repair of the right outflow tract in 58 (either as a root replacement or as a monocusp patch), for tricuspid valve replacement in 19, mounted on a stent, or for repair of the left outflow tract in 59 with the freehand technique.

Operative indications on the right heart were as follows:

1. *Surgical enlargement of the outflow tract*, with the homograft used as a conduit in 26 (atrioventricular in 12; atriopulmonary in four for tricuspid atresia or univentricular heart; ventriculopulmonary in 10 for pulmonary atresia with intact ventricular septum, TGA, VSD and pulmonary stenosis);
2. *Revalvation of the pulmonary tract*, after pulmonary annulus transection to avoid postoperative pulmonary insufficiency, and use of the homograft wall as an enlargement patch for the pulmonary artery trunk. Twenty-four monocusp patches were used in 23 cases of Fallots tetralogy or DORV with pulmonary stenosis, and once in a univentricular heart to create a valve in a bicavopulmonary derivation.
3. *Pulmonary valve replacement.* This was mandatory in eight patients with persistent severe pulmonary insufficiency following corrective surgery in infancy of Fallot's tetralogy, or for pulmonary valve stenosis, which often is extremely calcified in adults.
4. *Tricuspid valve replacement.* Operative indications were tricuspid rheumatic lesions in 10, endocarditis (often with associated VSD) in five, failure of conservative surgery in three, and annulus dilatation in an Ebstein-like disease that was non surgically curable, in one.

Operative indications on the left outflow tract included aortic valvular lesions following rheumatic fever in 40, endocarditis on congenital lesions in five, aortic valve dystrophy in four, failure of aortic conservative surgery or balloon valvuloplasty in three, contraindication to oral anticoagulant treatment in two, Takayasu disease in one, and subaortic lesions such as VSD and aortic valve replacement failures in three or tunnel stenosis recurrence in one. The implantation technique was freehand with a double running suture line. Most of the homografts were 16–25 mm internal diameter. However, six patients required an aortic annulus enlargement, which was achieved quite easily with the anterior mitral leaflet adjoining the homograft: these involved five posterior enlargements and one aortoventricular plastic procedure.

Results

With a mean follow-up of 32 months (1–84 months) there have been no early deaths after homograft valve implantation in the right outflow tract or for tricuspid replacement. Two early deaths occurred after surgery on the left outflow tract, one from cardiac failure, the other from septic shock.

Four patients (2.94%) underwent reoperation: two of these were for aortic insufficiency, one from valvular prolapse, with aortic replacement by a mechanical valve at 6 months, the other from size mismatch at the time of implantation, with replacement by a mechanical valve at 13 months. Another patient, with homograft degeneration, subsequently received a mechanical valve at 17 months. The fourth patient developed homograft endocarditis and received another homograft valve at 48 months.

There was only one late death at 12 months, which was the sudden death of a patient with a hypertrophic heart. There were no thromboembolic complications. Clinical and echocardiographic monitoring showed satisfactory results in all patients, with little if any, regurgitation and no noticeable transvalvular gradient.

Conclusions

At mid-term follow-up, it appears that homograft valves are well suited for repair of the right and left outflow tracts, tricuspid valve replacement and endocarditic valvular lesions in children.

Even though growth of the homograft is possible, it can be disrupted by the fibrous healing process. However, in the left outflow tract, an internal diameter of 16 mm appears to be adequate for a lifetime, because of the very low transvalvular gradient. The freehand implantation technique used, even though more demanding than standard aortic replacement, is safe, easier and less awkward in the event of reoperation than inclusion techniques or root replacements. The principal obstacle to the widespread use of homograft valve remains their limited availability.

References

1. Ross D N 1962 Homograft replacement of the aortic valve. Lancet 2: 487
2. Barratt-Boyes B G 1964 Homograft aortic valve replacement in aortic incompetence and stenosis. Thorax 19: 131–150
3. Angell W W, Oury J H, Lamberti J J, Koziol J 1989 Durability of the viable aortic allograft. J Thorac Cardiovasc Surg 98: 48–56
4. Wolfinbarger L Jr, Hopkins R A 1989 Biology of heart valve cryopreservation. In: Hopkins R A (ed) Cardiac reconstructions with allograft valves. Springer-Verlag, New York, pp 21–36
5. Yankah A C, Hetzer R 1987 Procurement and viability of cardiac valve allografts. In: Miller D C, Ross D N, Somerville J, Yacoub M H (eds) Cardiac valve allograft 1962–1987. Springer-Verlag, New York, pp 23–26
6. Brockbank K G M, Bank H L 1987 Measurement of postcryopreservation viability. J Card Surg 1: 145–151
7. O'Brien M F, Stafford F G, Gardner M A H *et al* 1987 The viable cryopreserved allograft aortic valve. J Card Surg 1: 153–167
8. Matsuki O, Okita Y, Almeida R S *et al* 1988 Two decades' experience with aortic valve replacement with pulmonary autograft. J Thorac Cardiovasc Surg 95: 705–711

Chapter 16

Pulmonary autograft (Ross) operation

S. Westaby

Ross described pulmonary autograft replacement of the aortic valve in 1967, five years after his landmark subcoronary implantation of an aortic homograft.[1] These remarkable developments occurred at a time when myocardial protection and postoperative care were limited and when aortic homografts were restricted in availability. The complexity of this double valve procedure, together with a significant rate of reoperation during the learning curve, discouraged most surgeons from attempting the technique.[2–4] The enormous benefits of autogenous tissue, perfect valve design and indefinite durability without risk of thromboembolism were passed over for the simplicity of insertion of the newly marketed Starr and Carpentier valves. Fortunately, Ross never shared the view that the pig closely resembles the human and he had an inherent dislike for mechanical prostheses and the risks of anticoagulation. His persistence with the autograft operation has produced persuasive long-term results that indicate that the pulmonary autograft is the 'dream ticket' for aortic replacement in infancy, childhood and the teenage years.[5–6] With increasing availability of cryopreserved homografts for reconstruction of the right ventricular outflow tract, protagonists now advise the operation for any patient with a life expectancy greater than 20 years.[7] As long as normal leaflet coaption is established at the time of operation, the pulmonary autograft is competent, durable, remains alive and grows with the developing child or teenager. The biological problems of homograft degeneration are transferred to the low-pressure pulmonary arterial position where optimal valve function is less important and failure occurs slowly.[8] For the aortic position the pulmonary autograft achieves the criteria for the ideal valve substitute, including indefinite durability, central non-obstructive flow, zero risk of thromboembolism or haemolysis and no requirement for anticoagulation.[9] If such a commercial valve existed and proved to be the least expensive option, all surgeons would use it. Currently, the overall hospital mortality for aortic replacement with a stented mechanical or bioprosthetic device is less than 2%, and essentially zero for young patients. In experienced hands the pulmonary autograft procedure has equivalent risk but the great majority of surgeons are not prepared to accept additional mortality during the learning curve for what usually amounts to a double root replacement operation. There is now a registry for the Ross procedure in order to further evaluate long-term outcome. Excluding the Ross experience, less than 500 patients have benefited from this operation to date and less than 100 surgeons worldwide have performed the operation.[7] The pulmonary autograft operation deserves more widespread consideration and perhaps should be undertaken on a supraregional basis.

Aspects of surgical technique

The operation is applicable to most patients with aortic pathology and provides a durable, non-antigenic valve free from thromboembolic bleeding or degenerative complications. The operation is contraindicated if the aortic annulus or ascending aorta are excessively dilated, as in annulo-aortic ectasia and Marfan's syndrome. Size mismatch between the pulmonary and aortic annulus is a potential source of technical error and autograft regurgitation; however, the native, untreated pulmonary wall is so distensible that the elasticity compensates for margins of error in size mismatch. Ross considers that the pulmonary autograft can be employed as an intra-aortic implant without strict consideration of the diameter differences. Others advise against mismatch and consider an aortic annulus of diameter greater than 27 mm to contraindicate the Ross procedure.

It is usually possible to decide in favour of the Ross operation from preoperative investigations and before inspecting the pathological aortic valve. Careful preoperative echocardiography can predict the size of the native pulmonary and aortic annulus and aid selection of an appropriate pulmonary or aortic homograft to reconstitute the right ventricular outflow. The operation begins by dissecting the adventitia between the ascending aorta and pulmonary artery; this is assisted by a tape around the ascending aorta. Cardiopulmonary bypass is then commenced with bicaval cannulation with arterial return to the aortic arch. With the heart still beating, dissection of the aortic fat pad is carried out to identify the origin of the right coronary artery. Excision of the pulmonary root is performed in the same way for both intra-aortic implant or root replacement; the author achieves this with the aortic crossclamp in place and the heart arrested with anterograde St Thomas' solution. Systemic temperature is reduced to 28°C. The pulmonary artery is opened at the origin of the right pulmonary artery in a transverse plane and before transection the pulmonary valve is inspected (Figure 16.1). Occasionally, a bicuspid or quadricuspid valve will preclude its use. If the pulmonary valve is normal and the aortic annulus less than

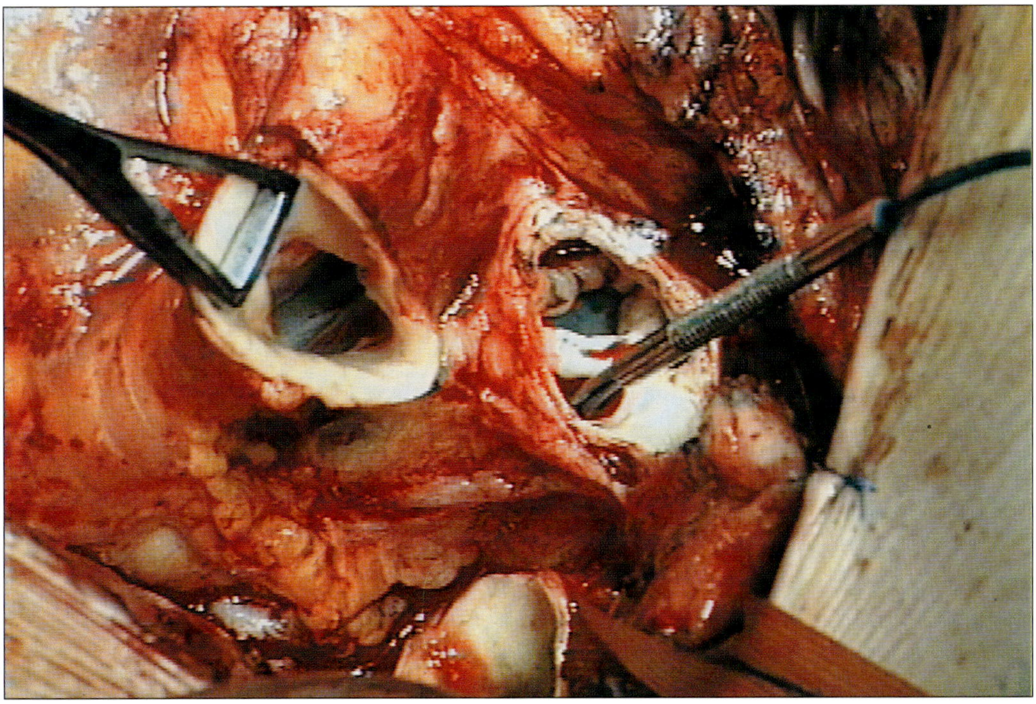

Figure 16.1.

27 mmHg, the pulmonary artery is divided and the adventitia along the posterior wall dissected, staying close to the pulmonary artery but without removing the adventitia. The dissection is kept close to the pulmonary artery, to avoid the left main stem and left anterior descending coronary. Once the right ventricular myocardium is identified posteriorly a right-angled dissector is passed through the pulmonary valve into the right ventricle and used to identify a point on the anterior wall 3–4 mm proximal to the pulmonary valve. A transverse ventriculotomy is then made, extending the incision posteriorly towards the surgeon and the previously dissected posterior plane (Figure 16.2). The plane of

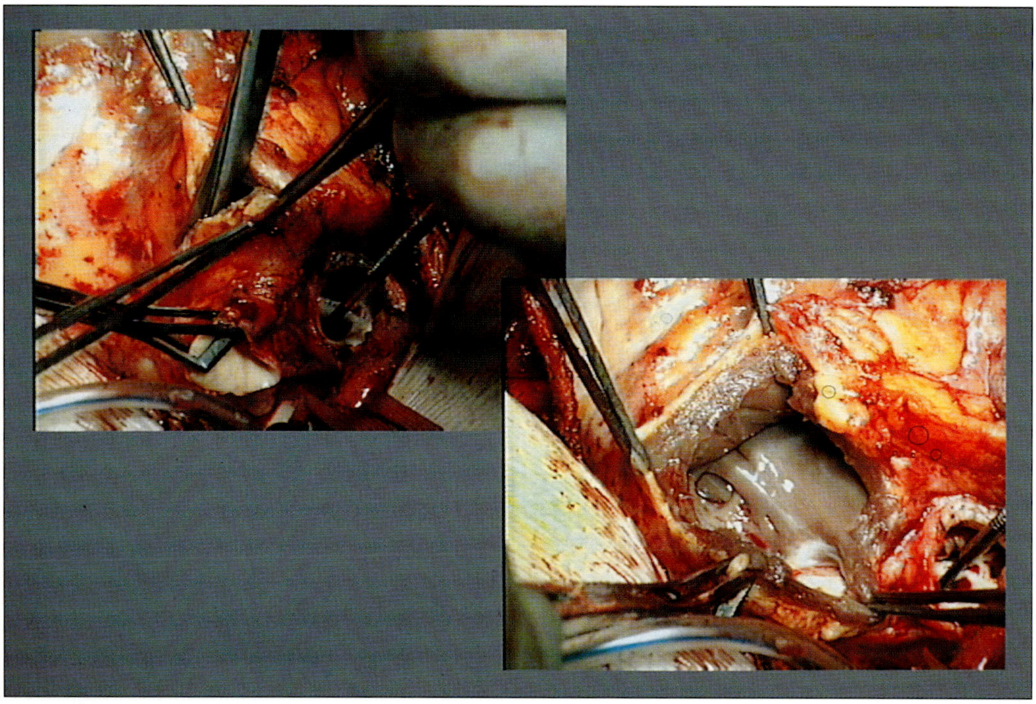

Figure 16.2.

dissection is maintained parallel with the operating table to avoid the left anterior descending coronary and its first septal branch. Only 2–3 mm myocardium are excised on the pulmonary root and, when the incision is carried away from the patient, the myocardium is shaved superficially off the septum. On removing the pulmonary root, fat and excess myocardium are removed to leave 2–3 mm myocardium proximal to the annulus in a single plane (Figure 16.3). The pulmonary wall is thin and adventitia should not be removed. The viable pulmonary root is then placed under blood in the pericardium while the aortic valve is excised.

The author prefers a transverse aortotomy 2–3 mm above the sinotubular junction and at least 5 mm above the origin of the right coronary artery. Appropriately placed stay sutures eliminate the need for retractors in the operative field during intra-aortic implantation. For infants and children with congenital aortic stenosis and often undergoing a second or third procedure, aortic root replacement is undertaken. The native coronary ostia and a short length of the main coronary artery are first mobilized from the aortic wall. The native valve is then excised and, if necessary, the root enlarged (Figure 16.4).

Implantation is performed using an inflow suture line of interrupted 4/0 polypropylene sutures. Three sutures are used to orientate the pulmonary valve by

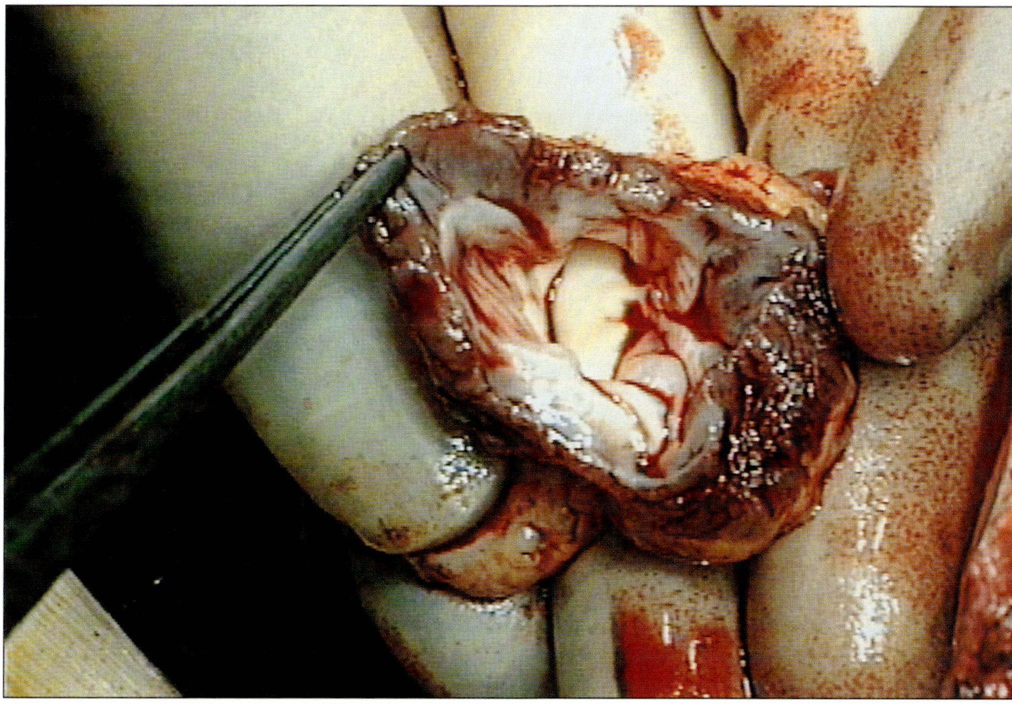

Figure 16.3.

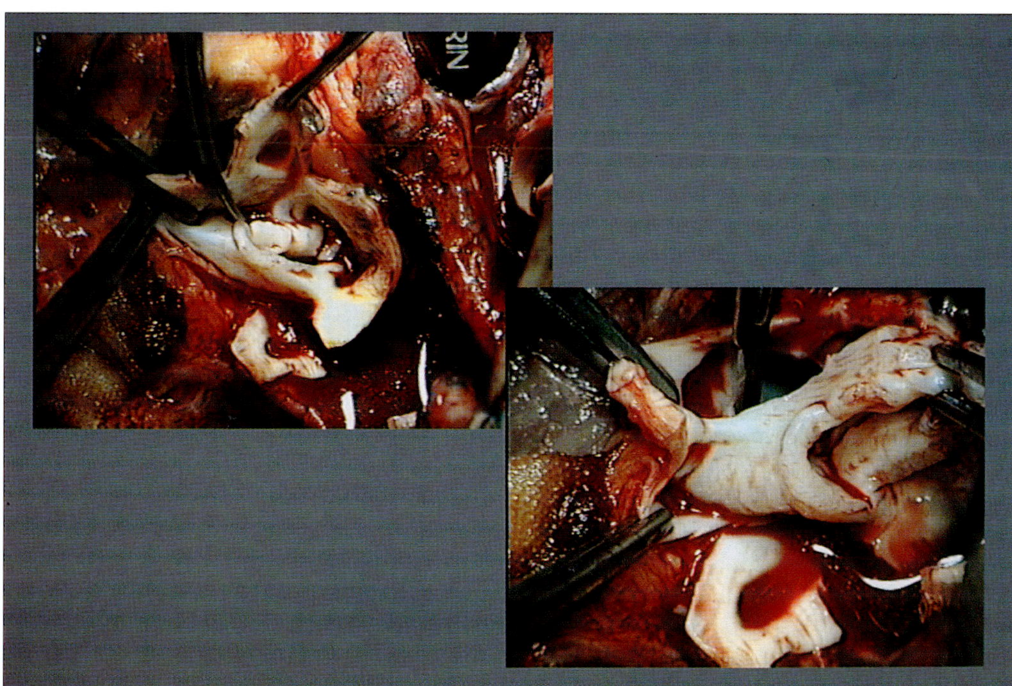

Figure 16.4.

placing one suture between the base of one of the pulmonary sinuses and a point on the aortic annulus below the ostium of the left coronary artery. A second suture is placed below the right coronary ostium and a third to trifurcate the native annulus. The remaining sutures are inserted in a horizontal plane below the annulus at the commissures. For aortic root replacement a cuff of autogenous pericardium is placed within the loop of the suture to assist haemostasis when the knots are tied. The sutures are gathered in three equal groups and the pulmonary valve run down into place. The pulmonary root is inverted to facilitate tying the knots. The pulmonary artery and valve are then restored to anatomical position and inspected to ensure integrity of the valve cusps. For subcoronary valve replacement the commissural attachments are obtained by horizontal mattress sutures of 4/0 Prolene placed 1 mm above the pulmonary valve commissure and attached 3–4 mm above the native aortic commissures. The pulmonary root cylinder is thin-walled and unforgiving of technical errors. Freehand subcoronary insertion must take into account semilunar valve functional anatomy and normalized aortic root geometry. A dilated aortic root must be corrected by aortoplasty to ensure a technically perfect implant. The only cause for autograft failure is surgical inaccuracy at this stage, which has led to the majority of users to perform aortic root replacement; this guarantees spatial orientation of the commissural posts. Ross completes the subcoronary implant by scalloping the pulmonary sinuses adjacent to the coronary ostia. A continuous suture of 4/0 Prolene then joins the pulmonary outflow to the internal aspect of the native aorta. This method is directly equivalent to implantation of an aortic homograft. The Elkins group in Oklahoma insert the pulmonary cylinder using a root inclusion method, which maintains the circumference of the pulmonary outflow.[10] Holes are made with a punch in the pulmonary arterial wall to accommodate the coronary ostia; these are then included in the repair by circumferential suture. Elkins considers this modification to prevent distortion of the commissures while the native aorta provides external support for the distensible pulmonary root and limits bleeding. For aortic root replacement, the native coronary ostia are implanted directly into the pulmonary root, using continuous 5 or 6/0 Prolene. We then perform the distal suture line between the pulmonary autograft and native aorta and identify the site for right coronary implantation by distending the pulmonary artery with cardioplegic infusion. It is important that the right coronary is not implanted too close to the proximal suture line when distension of the autograft can cause kinking. Care must be taken to avoid the commissure between the right and non-coronary sinus of the pulmonary valve; this can be avoided by inserting a marker suture prior to closure of the distal suture line. For root replacement it is important that the pulmonary valve annulus is implanted at or below the native valve annulus in order to ensure external support by left ventricular muscle. Root replacement in a supra-annular position leads to dilatation of the autograft annulus and progressive aortic regurgitation. Although use of the autograft as a free root replacement is associated with better immediate valve competence, bleeding is a potential complication. Pacifico and colleagues use a circumferential wrap of bovine pericardium to support the distensible root and prevent bleeding.[11] This modification has reduced their incidence of complications and provided a hospital mortality of less than 2%.

The right ventricular outflow is reconstituted using a pulmonary homograft of similar size to the native aortic annulus (Figure 16.5). The proximal suture line to the right ventricular outflow is initiated in the midpoint of the posterior aspect of the right ventricular outflow tract (Figure 16.6). Using a continuous suture line of 4/0 Prolene, the surgeon sutures towards himself, taking full-thickness 'bites' of myocardium (Figure 16.7). When suturing away from this midpoint, more superficial bites are used in order to avoid the first septal perforating branch of the left anterior descending coronary. During this manoeuvre systemic rewarming is undertaken and the proximal anastomosis completed by full-thickness sutures along the anterior aspect of the right ventriculotomy. During completion of this anastomosis the aortic crossclamp is released and the area of dissection for

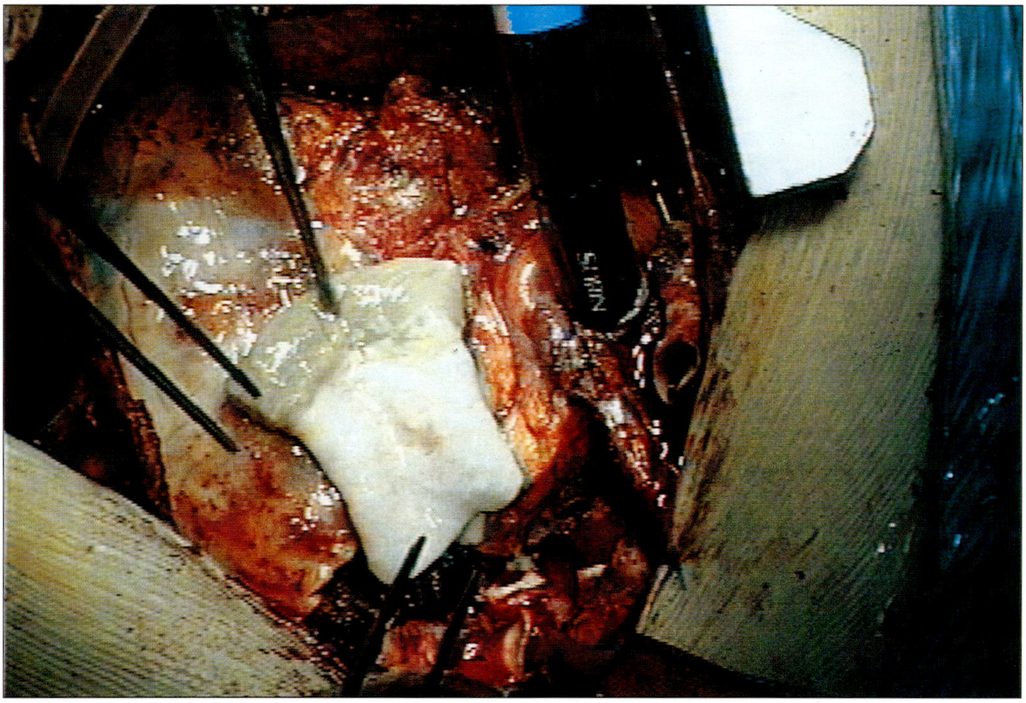

Figure 16.5.

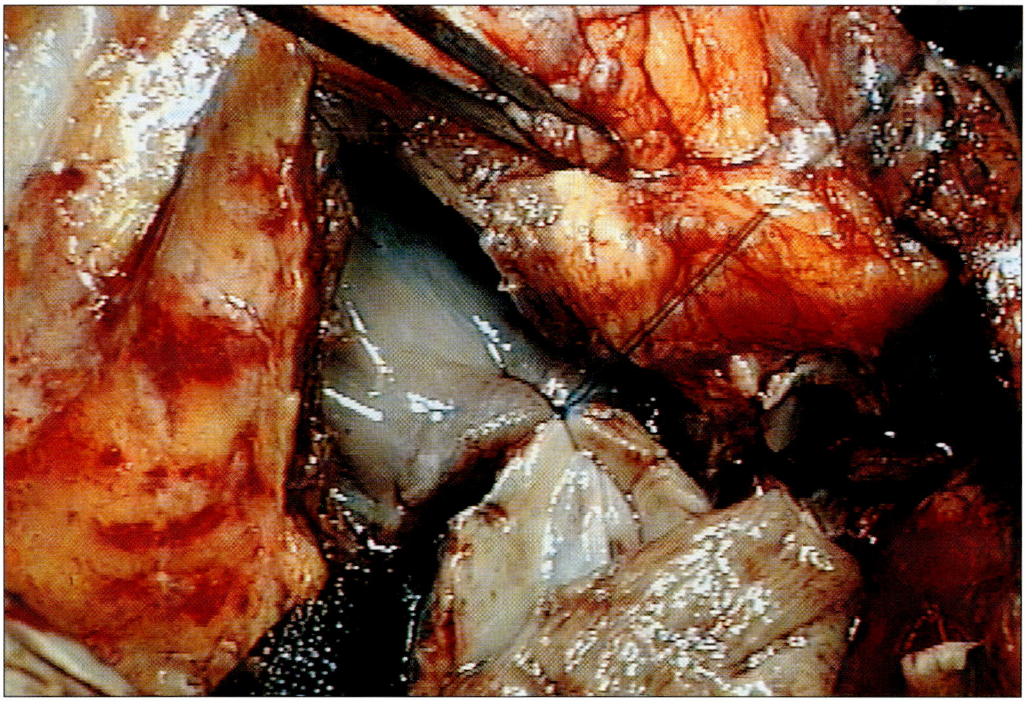

Figure 16.6.

removal of the pulmonary autograft checked for bleeding points. When haemostasis is secure the distal suture line is sewn, taking care not to purse-string this anastomosis. Anastomotic narrowing of this distal suture line has been an important late complication of the Ross procedure.

The completed 'Switch' operation is shown in Figure 16.8. After full rewarming and thorough de-airing, cardiopulmonary bypass is discontinued. It is useful to verify competence of the pulmonary autograft with intra-operative

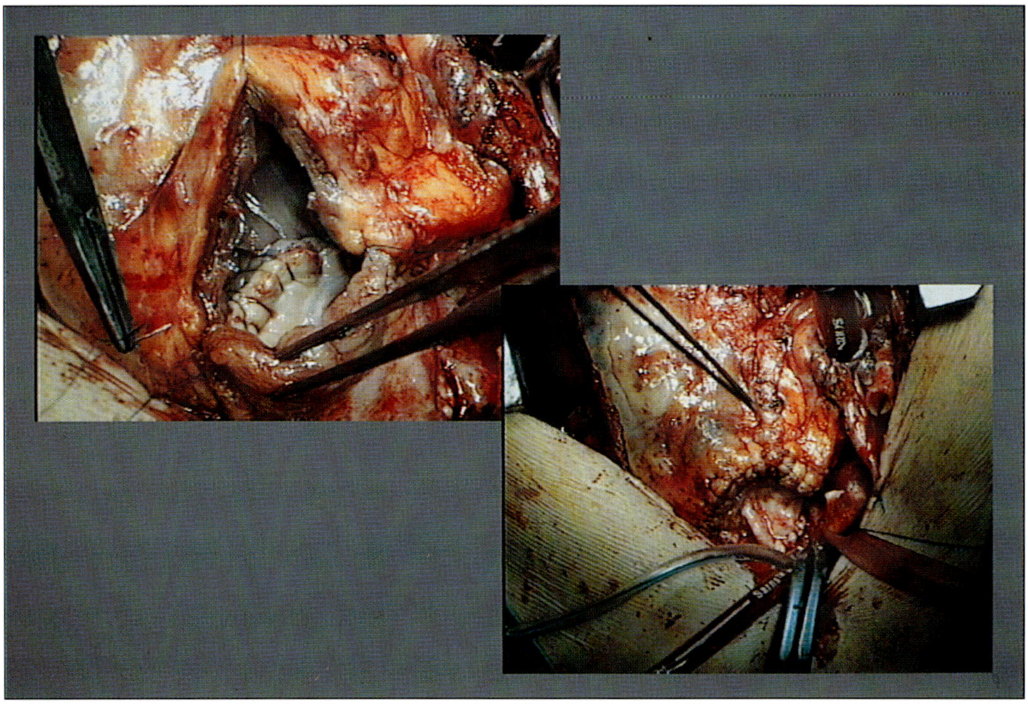

Figure 16.7.

transoesophageal echocardiography. Mild aortic insufficiency, identified by colour flow Doppler, is acceptable and tends to disappear with time. Aortic insufficiency (2–3+) usually results from a technical error with distortion of the pulmonary valve with leaflet prolapse. This usually results from failure to suspend all three commissures to a similar level during aortic closure and is addressed by taking down the distal suture line and resuspending the commissures.

Haemodynamics, durability and clinical outcome

The Ross procedure has now been used for patients aged one day to 70 years. All series have an early reoperative rate (up to 5%) for aortic regurgitation. This is due to imperfect implantation or autograft patient mismatch in a large annulus or funnel-shaped ascending aorta. Transvalvular gradients are negligible and clearly superior to those with stented bioprostheses or mechanical valves. The incidence of autograft incompetence is less with the root replacement and with external pericardial support. In homograft surgery, the valve is selected so that the internal diameter is 2–3 mm smaller than the internal diameter of the native aortic annulus. Ross considers that the pulmonary autograft can be employed as an intra-aortic implant without strict consideration of diameter differences. The native, untreated pulmonary arterial wall is so distensible that the elasticity compensates for margins of error in size mismatch. The quality and strength of the pulmonary arterial wall are greater in children and young adults, together with those who have undergone previous cardiac surgical procedures. These subsets may be better suited to the use of an unwrapped autograft for root replacement.

Matsuki and others reporting Ross's patients (241 in 21 years), cite bacterial endocarditis and technical failure as the main reasons for reoperation. Prolapse of the non-coronary cusp and detachment of sutures at the upper and lower margins of the graft were the predominant cause of incompetence. Primary tissue degeneration is an exceptionally rare event (0.09%/patient/year) and calcification

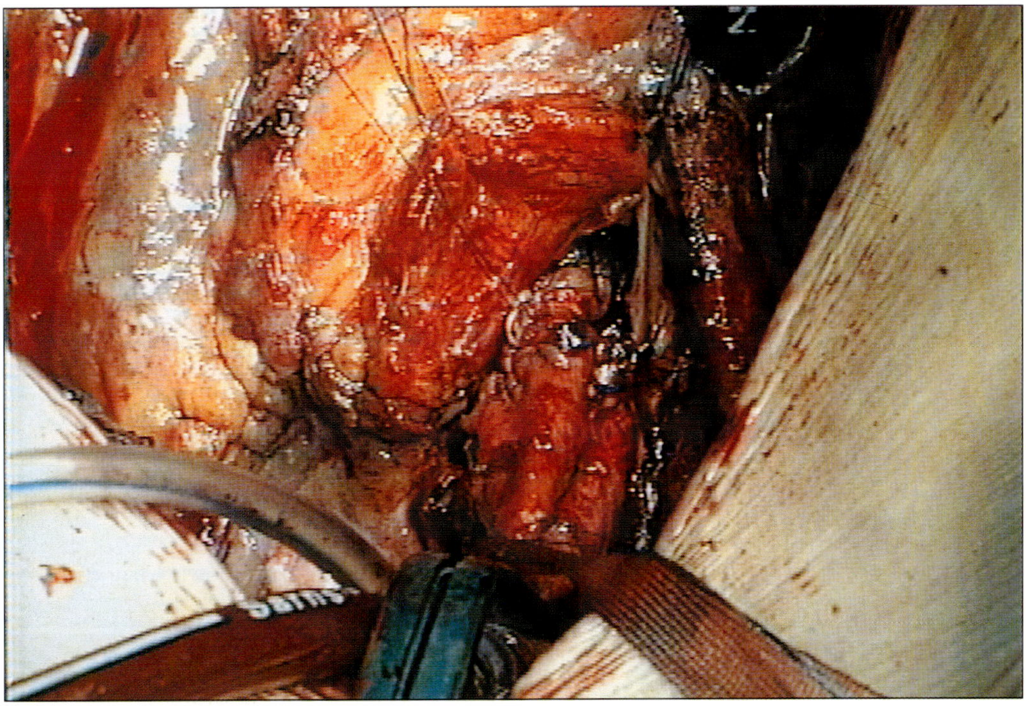

Figure 16.8.

can only be identified histologically. At reoperation the quality of the autogenous pulmonary valve is such that prolapsing leaflets can sometimes be resuspended, avoiding valve re-replacement. Ross's early experience produced an actuarial freedom from reoperation of 50% at 20 years. This figure must take into account the influence of a learning curve when surgical materials and techniques were relatively primitive. About 20% of patients required reoperation for right vetricular outflow reconstruction after fascia lata or pericardial valved conduits were employed. Aortic and pulmonary homografts have demonstrated excellent durability in the right ventricular outflow tract following the Ross operation. Calcification in right-sided homografts has been largely confined to the graft wall, the valve itself showing a low incidence of calcification and degeneration. Somerville documented the fate of aortic homografts used for right ventricular outflow reconstruction after aortic and mitral valve replacement with pulmonary autografts.[8] The incidence of homograft wall calcification was only 27% for these patients, in contrast to 92% for patients undergoing right-sided homograft insertion for cyanotic congenital heart disease. This is partly accounted for by the younger age of patients with the congenital condition and the presence of pulmonary hypertension in this group. The wall of the pulmonary homograft is thinner and has even less potential for calcification.

There is a zero incidence of thromboembolism with the pulmonary autograft, so that anticoagulation is not required at any stage of treatment. Freedom from embolism and anticoagulant-related bleeding provides an excellent quality of life and, together with the indefinite durability of the valve, provides the strongest argument for its use in children and young adults. There is now indisputable evidence that the valve grows with the young patient. Elkins has recorded a progressive increase in size of the autograft in children, proportional to the rate of growth and unlikely to be due to the propensity to stretch; the size increase does not result in aortic regurgitation. Paradoxically, autografts with mild regurgitation postoperatively may become fully competent in time. Long-term event-free survival is clearly superior to any other category of aortic valve replacement, including the allegedly viable, freehand-sewn, antibiotic-preserved homograft or the homovital homograft from a transplanted heart. At reoperation for right ventricular reconstruction, the autograft remains thin, pliable and resembles the normal aortic valve.

Recently it has been suggested that the pulmonary autograft, with its full viability and immune compatibility, should provide the best solution to aortic root infection, particularly in childhood. Despite the complexity of the autograft procedure in patients with infection and a bleeding tendency, early experience has been encouraging. Oswalt has recently reported autograft aortic root replacement in patients with endocarditis and periannular destruction.[12] The authors consider the autograft to be technically well suited to the treatment of endocarditis, since the native pulmonary arterial tissue is less friable than that of the cryopreserved homograft and moulds well to the irregular and unforgiving surfaces of the infected root. The autograft is inserted in a subannular position and sewn to the proximal borders of débrided root abscesses. The authors prefer a continuous suture, rather than tying a large number of individual sutures deep in the outflow tract.

References

1. Ross D N 1969 Observations on homograft aortic valves. Guys Hosp Reports 118: 5–11.
2. Matsuki O, Okita Y, Almeida M D, McGoldrick J P, Hooper T L, Robles A, Ross D N. 1988 Two decades experience with aortic valve replacement with pulmonary autograft. J Thorac Cardiovasc Surg 95: 705–711
3. Geens M, Gonzalez-Lavin L, Daurbarn C, Ross D N 1971 The surgical anatomy of the pulmonary artery root in relation to the pulmonary valve autograft and surgery of the right ventricular outflow tract. J Thorac Cardiovasc Surg 62: 262–267
4. Stelzer P, Jurer D J, Elkins R C 1989 Aortic root replacement with pulmonary autograft. Circulation 80(Suppl. III): 209–213
5. Schoof P H, Goome-Dijkhuis A H, Bogers A J J C, *et al.* 1994 Aortic root replacement with pulmonary autograft in children. J Thorac Cardiovasc Surg 107: 367–373
6. Elkins R C, Knott-Craig C J, Ward K E, McCue L, Lane M M 1994 Pulmonary autograft in children: realised growth potential. Ann Thorac Surg 57. In press
7. Oury J H, Eddy A C, Cleveland J C 1994 The Ross Procedure: A progress report. J Heart Valve Dis 3: 361–364
8. Saravaki O A, Somerville J, Jefferson K E 1980 Calcification of aortic homografts used for reconstruction of the right ventricular outflow tract. J Thorac Cardiovasc Surg 80: 909–920
9. Kouchoukos N T, Davila-Roman V G, Spray T L, Murphy S F, Perrillo J B 1994 Replacement of the aortic root with a pulmonary autograft in children and young adults with aortic valve disease. N Engl J Med 330: 1–6
10. Elkins R C, Santangelo K L, Stelzer P, Randolph J D, Knott-Craig C J 1992 Pulmonary autograft replacement of the aortic valve: An evolution of technique. J Cardiac Surg 7: 108–116
11. Pacifico A D, Kirklin J K, McGiffin D C, Matter C J, Narda D C, Deitheim A G 1994 The Ross operation – Early echocardiographic comparison of different operative techniques. J Heart Valve Dis 3: 365–370
12. Oswalt J D, Dewar S J 1993 Aortic infective endocarditis managed by the Ross operation. J Heart Valve Dis 2: 380–384

Chapter 17

Aortic valve replacement with a stentless porcine aortic valve

C. M. Feindel, T. E. David, J. Bos, Z. Sun, H. E. Scully and H. Rakowski

Durability continues to be one of the most important shortcomings of a bioprosthesis for valve replacement. Tissue valves fail because of a combination of breakdown of the glutaraldehyde-fixed collagen fibres at high stress points and gradual destruction by the body's immune system. Since mechanical stresses are particularly important in tissue valves supported by semirigid stents, the authors believe that elimination of the stents may improve the long-term durability of biological valves used in the aortic position, and therefore developed a stentless porcine aortic valve for human use.[1,2] There are no artificial components in this valve other than a single fine layer of Dacron cloth on the outside of the aortic wall to facilitate implantation and explantation. The valve is low-pressure fixed and no anticalcification treatment is used. The valve is made by St Jude Medical (St Paul, Minnesota, USA) and its trademark is the Toronto SPV™. This chapter reports the authors' experience with this stentless porcine valve, described previously.[1,2]

Patients and methods

From July 1991 to April 1994, 120 patients underwent aortic valve replacement (AVR) with a stentless porcine aortic valve (SPV). Table 17.1 shows the preoperative clinical data of these patients. Left ventricular ejection fraction was estimated by echocardiography or contrast ventriculography. Table 17.2 shows the operative data. The SPV was secured in the patient's aortic root with two layers of sutures in a manner similar to that during AVR with a freehand-sewn aortic valve homograft. The selection of the size of the SPV was based on the diameter of the sinotubular junction at the level of the commissures rather than the size of the valve annulus. However, in 11 patients it was necessary to reduce the diameter of the sinotubular junction because it exceeded the diameter of the aortic annulus.

All patients have been monitored by Doppler echocardiographic study before discharge from hospital, 3–6 months later and annually thereafter. Eight patients were discharged from hospital on warfarin sodium because of atrial fibrillation, and 103 were discharged on aspirin.

Statistical analyses were performed with BMDP software (BMDP Statistical Software, Los Angeles, California, USA). Actuarial survival and freedom from complications were calculated using the life-table method. Results were expressed as percentages of the means and the percentages of the standard errors of the means. Echocardiographic parameters were expressed as the means and standard deviation of the means. The temporal changes in these parameters were compared by Student's *t*-tests.

*Table 17.1 Preoperative clinical data**

Number of patients		120
Age (years):	mean	61.9
	range	35–80
Sex:	male	87 (72)
	female	33 (27)
ECG:	sinus rhythm	109 (91)
	atrial fibrillation	9 (8)
	heart block	2 (2)
New York Heart Association:		
	Class I	8 (7)
	Class II	54 (45)
	Class III	45 (37)
	Class IV	13 (11)
Aortic valve lesion:		
	Stenosis	93 (77)
	Insufficiency	13 (11)
	Mixed	14 (12)
Left ventricular ejection fraction (%):		
	>54	54 (45)
	40 – 54	42 (35)
	21 – 39	22 (18)
	<21	2 (2)
Coronary artery disease:		
	Single vessel	13 (11)
	Double vessel	13 (11)
	Triple vessel	10 (8)

*Except where stated, values are numbers of patients, with percentage in parentheses; the sum of percentages may not equal 100% owing to rounding-up errors

*Table 17.2 Operative data**

Aortic valve pathology:			
	Degenerative calcific		62 (52)
	Congenital:	bicuspid	43 (36)
		other	4 (3)
	Rheumatic		3 (2)
	Annuloaortic ectasia		2 (2)
	Failed bioprosthesis		5 (4)
	Healed endocarditis		1 (1)
Size of the stentless porcine aortic valve (mm):			
	22		1 (1)
	23		9 (7)
	25		31 (26)
	27		40 (33)
	29		39 (32)
Coronary artery bypass:			
	One		14 (12)
	Two		8 (7)
	Three		9 (7)
	Four or more		5 (4)
Other procedures:			
	Ablation of WPW† pathways		1 (1)
	Repair of abdominal aortic aneurysm		1 (1)
	Carotid artery endarterectomy		1 (1)
Aortic crossclamping time (min)			81 ± 17
Cardiopulmonary bypass time (min)			98 ± 22

*As in Table 17.1; †WPW, Wolff–Parkinson–White

Results

Operative mortality and morbidity

There was only one operative death. This patient had an uneventful AVR and was discharged from hospital 1 week after surgery. She was readmitted on the fourth postoperative week with the diagnosis of acute infective endocarditis due to *Staphylococcus aureus*. At reoperation, she was found to have a large aortic root abscess involving the tricuspid valve, which required extensive aortic outflow tract reconstruction. She died 2 days after reoperation.

Eight patients experienced one or more postoperative complications. Two patients suffered perioperative myocardial infarction. Three patients developed respiratory failure and required prolonged ventilatory support. One patient suffered a stroke after spontaneous cardioversion of atrial fibrillation but recovered completely. Three patients developed fever and positive blood cultures (*S. aureus, Enterococcus faecalis* and *Listeria monocytogenes*, respectively) without clinical or echocardiographic evidence of endocarditis but were treated with intravenous antibiotics for 6 weeks.

Late mortality

Patients have been followed from 1 to 34 months. There have been three late deaths. A 79-year-old man died of skin cancer 9 months after AVR and repeat coronary artery bypass and a 58-year-old man committed suicide; the SPV was functioning normally in these patients 3 months after AVR. A 68-year-old man who had AVR and coronary artery bypass surgery had a cardiac arrest 2 months after surgery and was successfully resuscitated but never regained consciousness and died.

Haemorrhagic and thromboembolic complications

At the time of the last follow-up, 116 patients were alive and had their original SPV; five patients were taking warfarin sodium, 71 were taking aspirin and 40 were not taking oral anticoagulant or antiplatelet agents. There have been no haemorrhagic complications. Two patients have had transient cerebral ischaemic attacks. Clinical investigation established that one of these patients had extracranial cerebrovascular disease and CT scan showed cerebral infarction in this patient. The other patient had normal carotid arteries and normal CT scan of the brain.

Infective endocarditis

The only endocarditis occurred early after surgery and was described under 'Operative mortality'.

Reoperation

There have been two reoperations — one for infective endocarditis, described under 'Operative mortality' and one for left main stenosis in a 79-year-old man. The aortic valve was left undisturbed in this elderly patient because it was functioning normally.

Clinical assessment

Of 116 patients alive and with their original SPV, 102 (88%) are in New York Heart Association functional Class I, and 14 (12%) are in Class II. Auscultation of the aortic valve is normal in 92 patients; an ejection murmur can be heard in 24 patients, and no patient has a diastolic aortic murmur.

Haemodynamic assessment by Doppler echocardiography

Aortic insufficiency was detected in 15 patients: it was graded as trivial in six and as mild in nine patients. Table 17.3 shows the effective areas as well as the mean and peak systolic gradients obtained by Doppler echocardiography soon after surgery, after 3–6 months, and after 11–14 months, in patients with valves of 23 mm or larger. The effective areas increased significantly during the first 3–6 months after implantation ($p<0.001$). In addition, the mean and peak systolic gradients across the SPV decreased significantly during the first 3–6 months following implantation ($p<0.001$). These findings are probably explained by remodelling of the aortic root after implantation of the SPV and by the rapid regression of the left ventricular hypertrophy in patients with aortic stenosis.

*Table 17.3 Echocardiographic data**

Follow-up period (months)	SPV[†] size (mm)			
	23	25	27	29
	Aortic valve area (cm^2)			
< 1	1.6 ± 0.5	1.4 ± 0.3	1.7 ± 0.4	1.9 ± 0.5
3–6	1.8 ± 0.4	1.9 ± 0.7	2.3 ± 0.8	2.5 ± 0.8
11–14	1.8 ± 0.5	1.9 ± 0.6	2.2 ± 0.7	2.5 ± 0.7
	Mean systolic gradient (mmHg)			
<1	6.7 ± 3.1	7.1 ± 4.0	5.1 ± 1.9	4.4 ± 2.3
3–6	3.0 ± 0.9	3.9 ± 2.1	.8 ± 1.2	2.8 ± 1.8
11–14	2.9 ± 0.5	3.4 ± 1.8	2.9 ± 1.4	2.1 ± 1.3
	Peak systolic gradient (mmHg)			
<1	15.8 ± 7.4	16.9 ± 9.4	12.0 ± 4.5	10.5 ± 5.4
3–6	7.2 ± 2.1	9.2 ± 5.1	6.6 ± 2.8	6.6 ± 4.2
11 – 14	6.8 ± 1.2	8.1 ± 4.2	6.8 ± 3.3	5.0 ± 3.2

*Values are means ± SD; [†] SPV, stentless porcine valve

Discussion

AVR with a stentless biological valve was first reported by Ross in 1962[3] and Barratt-Boyes in 1964.[4] These surgeons implanted an aortic valve homograft in the subcoronary position using a technique described by Duran and Gunning.[5] Binet and associates described five patients who had AVR with stentless heterologous aortic valves in 1965.[6] O'Brien and colleagues accumulated a formidable experience with formaldehyde-preserved porcine stentless valves.[7] With the development of tissue fixation with glutaraldehyde[8] and with the availability of commercially stented porcine valves, the interest in stentless porcine valves dwindled until recently. There is now a renewed interest in stentless porcine valves and all major heart valve manufacturers now have their version of a stentless porcine valve available for clinical use in a number of countries.

Aortic insufficiency (AI) after implantation of stentless biological valves is common.[9,10] Although the AI is usually mild, it may result in premature valve failure due to either increased mechanical stress or infective endocarditis. When the authors first started implanting the stentless valve, they used the diameter of the aortic annulus to select the size of the valve, as was their practice during AVR with aortic valve homograft. This approach is satisfactory as long as the patient's aortic root is normal. However, most patients with aortic valve disease have an abnormal aortic root, and implantation of a normal aortic valve in an abnormal aortic root will result in AI. Other factors must be considered, therefore, during AVR with a stentless biological valve such as the Toronto SPV™. In humans with normal aortic roots the diameter of the aortic annulus is approximately 10–15% larger than the diameter of the sinotubular junction.[11,12] In pigs the relationship between these two diameters is similar to that in humans.[12–14] Because the spatial relationships of the three commissures of the aortic valve are crucial for proper leaflet coaptation, the diameter of the sinotubular junction is more important than the diameter of the aortic annulus during implantation of a stentless biological valve. Although the diameter of the aortic annulus must also be considered in the selection of size of a stentless valve, it is not difficult to fit an SPV one size larger than the size of the aortic annulus. Greater discrepancies between these two diameters require tailoring of the sinotubular junction after implantation of the SPV. This, the authors have accomplished, merely by plicating the sinotubular junction at the level of the commissures of the SPV; the diameter of the sinotubular junction has been reduced to the diameter of the SPV.

An anatomical feature that may affect the haemodynamics of the Toronto SPV™ is its orientation within the aortic root of the patient. In humans, the non-coronary cusp with its sinus is frequently larger than the right and left cusps and sinuses;[12–14] in pigs, the right cusp is usually the largest of the three.[12] The right cusp of a porcine aortic valve is the one attached to the muscular septum and is easily recognizable. During implantation of the Toronto SPV™ the authors orientate the cusp attached to the muscle shell toward the non-coronary aortic sinus of the patient. Since the non-coronary aortic sinus is frequently enlarged and sagging onto the roof of the left atrium in patients with longstanding aortic valve disease, suturing the inflow of the SPV in a supra-annular position along the non-coronary sinus allows for implantation of a larger valve than the diameter of the aortic annulus and its also corrects the unevenness of the level of the aortic annulus among the three aortic sinuses.

In experimental animals, stentless porcine aortic valves are less likely to calcify than are stented ones.[15] Since 1991, the authors have implanted the stentless valve only in older patients. The Toronto SPV™ is being implanted in Canada and in the United States under an Investigation Device Exemption (IDE) and it cannot be used in patients younger than 35 years of age.

The issue of durability of stentless porcine aortic valves remains unresolved.[8,16] Further follow-up will be required to determine whether they are more durable than stented porcine valves. The experience with aortic valve homografts suggests that the durability of this valve is shortened when it is mounted in a stent before implantation.[17] By extrapolation, one would expect the SPV to be more durable than stented porcine valves.

References

1. David T E, Ropchan G C, Butany J W 1988 Aortic valve replacement with stentless porcine bioprostheses. J Card Surg 3: 501–505
2. David T E, Pollick C, Bos J 1990 Aortic valve replacement with stentless porcine aortic bioprosthesis. J Thorac Cardiovasc Surg 99: 113–118
3. Ross D N 1962 Homograft replacement of the aortic valve. Lancet 2: 487
4. Barratt-Boyes B G 1964 Homograft aortic valve replacement in aortic valve incompetence and stenosis. Thorax 19: 131–150
5. Duran C G, Gunning A J 1962 A method for placing a total homologous aortic valve in the subcoronary position. Lancet 2: 488

6. Binet J P, Duran C G, Carpentier A, Langlois J 1965 Heterologous aortic valve transplantation. Lancet 2: 1275
7. O'Brien M F, Neilson G H, Galea E G *et al* 1970 Heterograft valves. An analysis of clinical results of valve replacement. Circulation 16, 17 (Suppl 2): 16–19
8. Carpentier A, Lemaigre G, Robert L *et al* 1969 Biological factors affecting long-term results of valvular heterografts. J Thorac Cardiovasc Surg 58: 467–483
9. Casabona R, De Paulis R, Zattera G F, di Summa M, Bottone W, Stacchino C, Vrandecic MO, Morea M 1992 Stentless porcine and pericardial valve in aortic position. Ann Thorac Surg 54: 681–685
10. Kirklin J K, Smith D, Novick W *et al* 1993 Long-term function of cryopreserved aortic homografts. A ten-year study. J Thorac Cardiovasc Surg 106: 154–166
11. Kunzelman K S, Grande K J, David T E, Cochran R P, Verrier E D 1994 Aortic root and valve relationship: Impact on surgical repair. J Thorac Cardiovasc Surg 107: 162–170
12. Grande K J, Kunzelman K S, Cochran R P, David T E, Verrier E D 1993 Porcine aortic leaflets arrangement may contribute to clinical xenograft failure. ASAIO J 39: 918–922
13. Sands M P, Rittenhouse E A, Mohri H, Merendino K A 1969 An anatomical comparison of human, pig, calf and sheep aortic valves. Ann Thorac Surg 8: 407–414
14. Silver M A, Roberts W C 1985 Detailed anatomy of the normally functioning aortic valve in hearts of normal and increased weight. Am J Cardiol 55: 454–461
15. Hazekamp M G, Goffin Y A, Huysmans H A 1993 The value of the stentless biovalve prosthesis. An experimental study. Eur J Cardio-Thorac Surg 7: 514–519
16. Hofig M, Nellessen U, Mahmoodi M *et al* 1992 Performance of a stentless xenograft aortic bioprosthesis up to four years after implantation. J Thorac Cardiovasc Surg 103: 1068–1073
17. Angell W W, Oury J H, Lamberti J J, Koziol J 1989 Durability of the viable aortic allograft. J Thorac Cardiovasc Surg 98: 48–56

Appendix

The following formulae were used for calculation of gradients and valve areas by Doppler echocardiography:

$$\text{Peak Systolic Gradient (mmHg)} = 4\ (Va^2 - Vo^2)$$

where Va is the peak velocity in the aortic valve, and Vo is the peak velocity in the left ventricular outflow.

$$\text{Mean Systolic Gradient (mmHg)} = 4\ \left[(0.65Va)^2 - (0.65Vo)^2\right]$$

where Va is the peak velocity in the aortic valve, and Vo is the peak velocity in the left ventricular outflow.

$$\text{Aortic Valve Area (cm.sq.)} = \frac{0.8 \times V_1 \times A_{LVOF} \times 100}{100.8 \times \sqrt{(V_2^2 - V_1^2)}}$$

where V_1 is the mean velocity in the left ventricular outflow, V_2 is the mean velocity in the aortic valve and and A_{LVOF} is the area of the left ventricular outflow tract.

Chapter 18

Early experience with the Toronto stentless porcine valve in aortic valve replacement

K. Wong, S. Brecker, A. Khaghani, M. H. Yacoub and R. J. Pepper

The main features of current biological valves relate to their design, which in turn affects the haemodynamic profile of the valve and may also influence its long-term durability. Stented bioprostheses have the disadvantages of high transvalvular gradients in small valve sizes, and limited durability. The design of stentless porcine valves allows a larger valve orifice, which would be a significant factor in smaller valve sizes.[1] In addition, stentless aortic valves allow flexibility at the commissures that are thought to be an important cause of early degeneration in stented bioprostheses.[2] The aim of this study was to evaluate the design and the haemodynamic profile of the Toronto stentless porcine valve (TSPV), which the authors have been using over a 23-month period. This valve is a low-pressure glutaraldehyde-fixed porcine aortic valve with the aortic sinuses and the subannular tissue trimmed to 1.5–2.0 mm from the base of the leaflets. The outer wall of the valve is covered by Dacron.

Patients and methods

Between January 1992 and November 1993, 59 patients underwent aortic valve replacement with a TSPV. The choice of valve was either indicated clinically (age, young women) or the patient had requested a bioprosthetic valve. There were 59 patients with a mean age of 66.7 years (range 35.1 – 84.6 years), comprising 38 men and 21 women. The aetiology of aortic valve disease in these patients was stenosis (52.5%), regurgitation (13.6%), mixed disease (32.2%) and native valve endocarditis (1.7%). Six patients had previously undergone cardiopulmonary bypass surgery and 26 patients had concomitant procedures.

Clinical data were collected at the operative and postoperative stages and at follow-up (3–6 and 11–14 months). Follow-up assessment of all patients included the documentation of any complications, functional class and the presence of an early diastolic murmur. Continuous wave and colour Doppler echocardiography was performed before the patient's hospital discharge when the patient was ambulant and at the follow-up periods. Mean and peak gradients were derived from the velocity across the valve. Aortic regurgitation was graded as mild, moderate or severe.

Results

Twenty-four (40.7%) patients received coronary artery bypass grafts in addition to valve replacements. Two other patients underwent other procedures (removal of left atrial clot, and mitral valve replacement). Implantation of the TSPV was no more demanding than freehand-sewn homografts; this is reflected in the crossclamp times, which were marginally less than when inserting a homograft. The mean crossclamp times were 62 min (range 43–98 min) for aortic valve replacement alone and 87.6 min (range 60–162 min) when concomitant procedures were performed.

Early mortality was 3.4% (*n*=2). The first patient died following a pulmonary embolus while another patient died from postcardiotomy heart failure. Major complications occurred in two patients: one patient had to be reopened because of pericardial tamponade following surgery; the other patient developed prosthetic valve endocarditis 6 weeks after her operation, which required reoperation where a homograft valve was used. There were no incidences of valve-related thromboembolism on follow-up. No patient received anticoagulant therapy routinely in the early postoperative period.

The mean follow-up period was 9.6 months (range 3–11 months). All patients were in NYHA class I or II. The mean peak transvalvular gradients in relation to the valve sizes are shown in Figure 18.1. These gradients are low even in smaller valve sizes and were similar, if not marginally less, in most cases when measured at follow-up. Central aortic regurgitation was noted in five patients on echocardiography (mild in four and moderate in one) although this was not clinically significant in any patient.

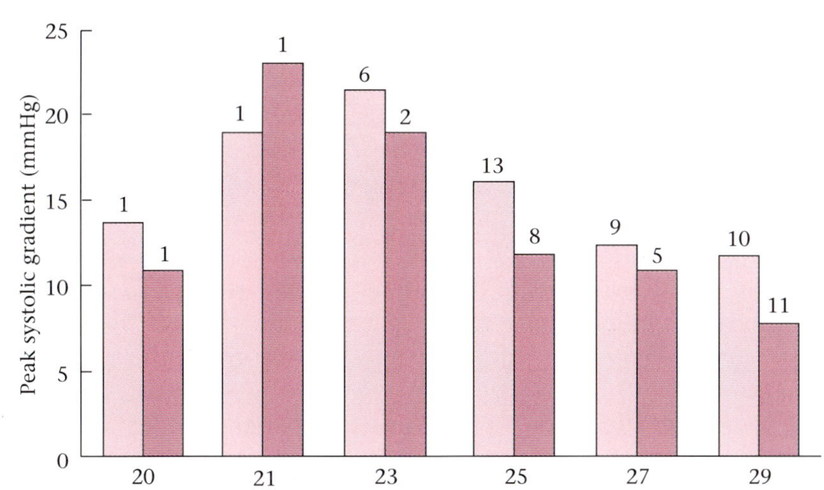

Figure 19.1. Peak systolic gradient (mean) at early postoperative stage (day 6; □) and at 3–6 month follow-up (■); numerals at the top of the histograms denote number of patients who had echocardiograms.

Discussion

Homografts offer distinct advantages in aortic valve replacement, including good haemodynamics, low valve-related complications and longer durability than bioprosthetic valves.[3] Their advantages in the setting of endocarditis and aortic valve reoperations have also been demonstrated.[4] Their main disadvantage is their limited availability and stentless porcine valves may bridge this gap, in this respect.

The insertion of these valves is not dissimilar to that of a homograft but may be technically more demanding for surgeons who have limited experience in this field. The TSPV is, in fact, much easier to insert as the valve is fixed in glutaraldehyde and backed by Dacron, making it easier to orientate. It has also been trimmed around the sinuses; this is, however, a major disadvantage as it reduces the option of performing a root replacement. Furthermore, the outer wall of Dacron would make its insertion in patients with endocarditis a potential risk for reinfection.

Central aortic regurgitation was noted in five patients on echocardiography. This has also been documented by other surgeons using stentless valves[1,5] and in homograft valve insertion. In all the authors' cases regurgitation has not been clinically significant, perhaps in part because of the sensitivity of colour Doppler echocardiography in detecting regurgitation. There is no doubt, however, that correct sizing and siting the commissures is critical in preventing significant regurgitation. David and colleagues[1] have suggested sizing the valve at the sinotubular junction. They found that the diameter of the annulus was often smaller than the sinotubular junction in patients with aortic stenosis and this would mean that the commissures were pulled apart, resulting in poor coaptation. The authors have sized valves by estimating the size of the annulus by eye ('eyeballing') and have not noted any higher (in fact, a much lower) incidence of regurgitation compared with other series.[1,5]

The mean crossclamp times and the mortality in our series is quite acceptable. The valves do take longer to insert than a bioprosthetic or mechanical valve, which only requires one suture line, but our results suggest that this does not influence the clinical outcome of these patients. In addition, there has been a very low incidence of valve-related complications.

The design of the valve (without the stents) and the technique of insertion allows for a larger effective valve orifice, which is important in small valve sizes. The transvalvular gradients obtained are very encouraging and, interestingly, have appeared to fall in most cases at follow-up. It is, perhaps, still too early to comment on these changes, as it is on the long-term durability of these valves. However, the theoretical advantage of having flexible stents, in reducing tensile stress at the commissures, should allow a more durable valve.

In conclusion, the early results reported here suggest that the TSPV provides excellent valve haemodynamics, can be inserted safely and has a low incidence of valve-related complications. It is also more readily available than homografts and although its long-term durability is unknown, these results encourage its further clinical application.

References

1. David T, Bos J, Rakowski H 1992 Aortic valve replacement with the Toronto SPV bioprosthesis. J Heart Valve Dis 1(2): 244–248
2. Christie G W, Barratt-Boyes B 1991 On stress reduction in bioprosthetic heart valve leaflets by the use of a flexible stent. J Card Surg 6(4): 476–481
3. Ross D N, Yacoub M H 1969 Homograft replacement of the aortic valve. A critical review. Prog Cardiovasc Dis 9: 275–279
4. Albertucci M, Wong K, Petrou M, Mitchell A, Somerville J, Theodoropoulos S, Yacoub M 1994 The use of unstented homograft valves for aortic valve reoperations. Review of a 23 year experience. J Thorac Cardiovasc Surg 107 (1): 152–161
5. Casabona R, De Paulis R, Zattera G, di Summa M, Bottone W, Stacchino C, Vrandecic M, Morea M 1994 Stentless porcine and pericardial valve in aortic position. Ann Thorac Surg 54: 681–685

Chapter 19

Stentless and stented valves in the aortic position: an early postoperative assessment

Xu Y. Jin, D. G. Gibson, M. H. Yacoub and J. R. Pepper

Assessment of left ventricular (LV) function and the outcome analysis after aortic valve replacement are usually based on the use of a mechanical or stented bioprosthetic valve. However, unlike the free-sewn homograft, these prostheses all cause a significant residual pressure drop and thus the resistance of the left ventricle to ejection might still be higher than normal, and, therefore, associated with incomplete regression of ventricular hypertrophy. The recently developed Toronto stentless bioprosthesis has been shown to cause less resistance to LV ejection. However, there are as yet no data available to compare the performance of this new valve with that of either the homograft or the conventional stented valve. More importantly, the potential benefit due to lower ejection resistance of the valve substitute, to early improvement in LV function after valve replacement, has not yet been elucidated. The aim of the study described here was, therefore, to compare the performance of the aortic homograft, Toronto stentless valve, and stented bioprosthesis, and to assess their possible implications in relation to functional changes of the left ventricle early after implantation.

Subjects

A total of 36 patients undergoing aortic valve replacement due to stenosis (28) or regurgitation (eight) were investigated: 17 patients received a homograft (HV), 12 were given a Toronto stentless valve (TV), and seven had a stented valve (SV). The clinical data were matched, apart from the younger population in the homograft group and the longer ischaemic time in stentless valve groups (Table 19.1).

Table 19.1 Clinical data of 36 patients undergoing aortic valve replacement†

	Valve		
Variable††	HV	TV	SV
Age (years)	52 ± 13**#	65 ± 9	71 ± 7
Sex (F/M)	4/13	5/7	3/4
AR/AS	5/12	2/10	1/6
LV mass (g/m^2)	190 ± 70	200 ± 65	215 ± 95
Co-CABG case	5	5	4
Valve size (mm/m^2)	12.5 ± 1.1	13.5 ± 2.0	13.0 ± 1.2
AoXCL time (min)	112 ± 18**	89 ± 13*	76 ± 10

†Values are means ± SD; ††AR/AS, aortic regurgitation/aortic stenosis; LV mass, left ventricular mass; Co-CABG case, concomitant coronary artery bypass graft; AoXCL time, aortic crossclamp time; §HV, homograft valve; TV, Toronto stentless valve; SV, stented valve; #p<0.01 vs TV; ** p<0.01 vs SV; *H vs TV = $p < 0.05$; TV vs SV = $p < 0.01$

Methods

A perioperative transoesophageal echocardiogram combined with high-fidelity LV pressure recording, together with thermodilution cardiac output measurement, were performed just before cardiopulmonary bypass and 20 h after valve replacement. From the simultaneous LV minor axis dimension, wall thickness and cavity pressure measurements, the peak value of circumferential wall stress (LVS) and peak rates of change in LV pressure (dP/dt) normalized to developed pressure (DP) were derived, using an offline digitizing computer system. LV stroke work index (LVSWI) was calculated from LV mean developed ejection pressure and stroke volume index (LVSVI). The valve pressure drop was calculated from the difference between peak LV systolic pressure and peak systolic blood pressure. The valve energy lost was calculated from the difference between LVSWI and the stroke work index in systemic circulation, and normalized to LVSWI.

Results

Twenty hours after valve replacement, LV peak systolic pressure (LVP) and LVS in SV were both significantly higher than in HV and TV. The percentage of mechanical energy lost at the valve was also significantly higher in SV than that in HV and TV (Table 19.2).

Table 19.2 Comparison of valve performance[†]

	Valve[§]		
Variable[††]	HV	TV	SV
Peak LVP (mmHg)	103 ± 14	105 ± 13	119 ± 14*[#]
Peak LVS (g/cm^2)	137 ± 37	127 ± 27	165 ± 40*[#]
Valve pressure drop (mmHg)	−9 ± 17	−10 ± 14	9 ± 12*[#]
Valve energy lost (percentage of LVSWI)	3 ± 18	4 ± 18	14 ± 16*[#]

[†]As in Table 19.1; [††]LVP, left ventricular pressure; LVS, left ventricular circumferential wall stress; LVSWI, left ventricular stroke work index; [§]as in Table 19.1; *$p<0.02$ vs HV; [#]$p<0.01$ vs TV.

At the same time, the percentage changes in systolic function (+dP/dt/DP), diastolic function (–dP/dt/DP) and LVSVI all showed more improvement in HV and TV compared with SV, and both HV and TV had significantly less decrease in LVSWI than that in SV (Table 19.3).

Table 19.3 Comparison of postoperative changes in LV function[†]

	Valve[§]		
Variable[††]	HV	TV	SV
+dP/dt/DP (%)	23 ± 5*	21 ± 6*	7 ± 5
–dP/dt/DP (%)	41 ± 8*	50 ± 9*	25 ± 5
LV EDP (%)	–18 ± 6*	–21 ± 7*	27 ± 10
LV SVI (%)	22 ± 6*	32 ± 9*	10 ± 8
LV SWI (%)	–13 ± 5*	–6 ± 5*	–20 ± 7

[†]As in Table 19.1; [††]+dP/dt/DP, percentage change in systolic function; –dP/dt/DP, percentage change in diastolic function; LV EDP, left ventricular end diastolic pressure; LV SVI, left ventricular stroke volume index; LV SWI, left ventricular stroke work index; [§]as in Table 19.1; *$p<0.01$ vs SV

Comment

Some recent reports have indicated that there is a lower pressure gradient across a stentless bioprosthesis than across a stented one in the aortic position. The data reported here are in agreement with these observations; furthermore, the mechanical energy lost at the valve, and the peak systolic pressure and wall stress of the left ventricle, were both much lower in the Toronto valve and homograft than in the stented valve. These findings clearly demonstrate a significantly lower resistance to ventricular ejection when a stentless valve is used. None of the parameters of Toronto valve performance differed from those found for the aortic homograft; thus, the performance of the Toronto valve was similar to that of a homograft at the early postoperative stage.

Complete relief of increased ejection resistance has a significant impact on the early improvement of LV function. The +dP/dt/DP represents the maximum shortening velocity of the myocardium which increased to a significantly greater extent after use of a stentless valve than after use of a stented valve. This can be explained, at least in part, by the effect of alteration in force–velocity relation, i.e. the greater the decrease in contracting force (peak wall stress), the greater the increase in contracting velocity (+dP/dt/DP). The changes in ventricular relaxation and end diastolic pressure were also more beneficial with the stentless valve, indicating that LV diastolic function could improve within first 24 h after operation. Thus, the significant increase in LV stroke volume and a stable LV SWI in patients receiving stentless valves, seems firmly based on the underlying changes in LV systolic and diastolic function.

In summary, the aortic homograft and Toronto valve, in causing less resistance to LV ejection, are associated with a greater early improvement in LV function than is the stented valve. The early performance of the Toronto stentless porcine valve closely resembles that of the aortic homograft; however, its long-term performance needs further investigation.

Chapter 20

Medtronic Freestyle™ bioprostheses: clinical study

H. A. Huysmans

The first clinical implant of a Freestyle™ bioprosthesis was performed on 6 August 1992; by 5 January 1994, 19 centres were participating in the clinical study — 13 centres in North America, five in Europe and one in New Zealand. The total number of implants at that date was 309. Results can be presented only on the basis of the forms received, approved and entered in the database so far. Those results come from 15 study centres with a total of 203 implants.

Patients

The implants were performed between 6 August 1992 and 4 November 1993. The number of implants per study centre varied from 1 to 45, with a mean of 10.7 per centre. There were 111 male patients and 92 females, with a mean age of 66.36 years (32–91 years). Only 4.5% were below the age of 50 years; 70.4% of patients were over 65 years of age. The mean age per study centre varied from 58.70 to 76.23 years, indicating a different policy for implantation of the Freestyle™ bioprosthesis.

Participating surgeons were free to select their patients. Although 82.8% of patients were in New York Heart Association class III or IV preoperatively, one study centre had 60% of its patients in class I or II, probably indicating a difference in policy for patient selection.

Methods of implantation

The Freestyle™ bioprosthesis was implanted in one of the following three ways:

1. As a valve replacement, whereby two or three sinuses of the prosthesis were completely excised.
2. Mini-root replacement, whereby only openings in the two coronary sinuses were made with preservation of a tissue ring above the sinotubular junction. The prosthesis was placed within the patient's aorta.
3. The third method was root replacement, whereby the patient's aortic root was excised and replaced by the bioprosthesis, and the two coronary ostia were reimplanted in the prosthesis.

The valve replacement technique was used in 65% of cases, the mini-root technique in 16.3% and root replacement in 18.7%. Of the 15 centres, five used only the root or mini-root technique and five used the valve replacement technique exclusively.

Sizes of bioprostheses used are shown in Table 20.1. It seems that most of the small prostheses were used in the valve technique.

*Table 20.1 Sizes of Medtronic Freestyle™ bioprostheses used**

Implantation technique	Size of bioprosthesis (mm)†				
	19	21	23	25	27
Valve	6	26	40	29	31
Mini-root	–	2	8	16	7
Root	2	3	7	10	16
Total	8	31	55	55	54

*n=203; †percentage of small sizes (19–21): valve 24%, mini-root 6%, root 13%

Of the 203 patients, 91 had concomitant procedures, predominantly coronary artery bypass grafting and ascending aorta repair. A total of 94 other procedures were performed in these 91 patients: they comprised 43.2% of the valve replacement cases, 36.4% of the mini-root cases and 57.9% of the root cases.

Crossclamp times needed for insertion of the stentless bioprostheses are given in Table 20.2.

*Table 20.2 Crossclamp time for bioprosthesis insertion**

Implantation technique	–Co† (n=112)	+Co† (n=91)
Valve	104 (70–171)	119 (56–166)
Mini-root	120 (80–164)	149 (98–235)
Root	103 (67–131)	119 (95–185)
Total	107 (67–171)	123 (56–235)

*Values are means (range in parentheses) in minutes; † – Co, + Co: without, or with, concomitant procedure

Results

Early postoperative mortality within 30 days amounted to 3.03% for the valve procedure, 3.03% for the mini-root procedure and 10.53% for the root replacement procedure, with a total of nine deaths (4.43%) in 203 cases. Seven of the deaths were from cardiac causes; none of the deaths was valve related. Early postoperative events, within 30 days, are shown in Table 20.3.

*Table 20.3 Early postoperative events (<30 days)**

Event	Implantation technique			
	Valve	Mini-root	Root	Total
Paravalvular leak	–	1	–	1
Unacceptable haemodynamics	1	1	–	2
Reoperation (explant bioprosthesis)	1	2	–	3
Thromboembolic				
CVA	3	–	1	4
TIA	–	–	1	1

*n=203

The follow-up was complete in 100% of cases. The duration is, of course, still very short, varying from 0 to 1.04 years, with a mean of 0.24 years. Discharge data were available in 190 patients, 3–6 months follow-up in 124 patients and 1-year follow-up in only eight patients.

There were four late postoperative deaths — three after using the valve technique, one following mini-root technique and none after root replacement. Only one late death was valve related and this was attributable to unacceptable valve haemodynamics. Late postoperative events are listed in Table 20.4.

*Table 20.4 Late postoperative events**

Event	Implantation technique			
	Valve	Mini-root	Root	Total
Paravalvular leak	1	–	–	1
Unacceptable haemodynamics	2	–	–	2
Reoperation	–	–	–	–
Thromboembolic				
CVA	1	–	–	1
TIA	1	3	–	4
Endocarditis	–	–	–	–

*n=132

The underlying pathology of the valves replaced is given in Table 20.5. The aortic lesion was stenosis in 34.5%, insufficiency in 12.3% and a mixed lesion in 53.2%. Tables 20.6 and 20.7 show the abnormalities of the aortic annulus and the degree of débridement necessary in each of the three techniques used. The most extensive calcification was present in the cases where the valve technique was used.

*Table 20.5 Pathology of valves replaced**

Aetiology	Implantation technique†				
	Valve	Mini-root	Root	Total	n
Senile degeneration	68.2	42.4	55.3	61.6	125
Rheumatic	6.8	3.0	7.9	6.4	13
Congenital	17.4	51.5	18.4	23.2	47
Myxomatous degeneration	3.0	0	21.1	5.9	12
Endocarditis	0.8	0	5.3	1.5	3
Failed prosthesis	5.3	3.0	5.3	4.9	10
Other	2.3	0	7.9	3.0	6

*n=203; †values are percentages

*Table 20.6 Condition of the annulus**

Condition	Implantation technique†				
	Valve	Mini-root	Root	Total	n
Normal	3.8	6.1	13.2	5.9	12
Calcified	90.2	87.9	68.4	85.7	174
Fibrotic	33.3	15.2	18.4	27.6	56
Dilated	3.0	0	13.2	4.4	9
Friable	0.8	0	2.6	1.0	2
Abcessed	0	0	2.6	0.5	1

*n = 203; †values are percentages

*Table 20.7 Degree of annular débridement necessary**

Débridement	Implantation technique†				
	Valve	Mini-root	Root	Total	n
None	3.8	3.0	18.4	6.4	13
Minimal	12.1	21.2	26.3	16.3	33
Moderate	50.8	60.6	21.1	46.8	95
Extensive	32.6	15.2	34.2	30.0	61

*n = 202; †values are percentages

All patients were clinically improved after surgery, with over 98% in New York Heart Association class I or II. Only 12.5% of patients received no anticoagulants postoperatively; 12.5% received warfarin and 75% platelet inhibitors. Postoperative echo-Doppler studies show a certain degree of transvalvular gradient. Table 20.8 shows the results in the three techniques and with different sizes of prosthesis at the time of discharge and at 3–6-month follow-up. The gradient appears to be size related and, to a certain extent, technique related, with the lowest gradients when the root replacement technique was used. More detailed information on the size 25 mm prosthesis is given in Table 20.9.

Table 20.8 Postoperative echocardiography: transvalvular gradient (in mmHg) at discharge and 3–6-month follow-up

		Implantation technique							
Prosthesis size (mm)	Follow-up period*	Valve		Mini-root		Root		Total	
		n	Mean	*n*	Mean	*n*	Mean	*n*	Mean
19	D	5	18.80	–	–	2	11.50	7	16.71
	3–6	3	18.33	–	–	1	9.00	4	16.00
21	D	25	15.76	2	13.00	2	8.00	29	15.03
	3–6	16	9.12	2	11.00	1	4.00	19	8.94
23	D	39	13.62	8	20.87	5	6.60	52	14.49
	3–6	22	9.95	7	15.14	4	3.75	33	8.94
25	D	28	10.18	15	10.80	9	3.89	52	9.27
	3–6	19	7.26	10	7.70	5	3.20	34	6.79
27	D	31	8.23	6	6.50	16	4.75	53	6.98
	3–6	21	5.29	3	7.67	9	3.89	33	5.12

*D, discharge; 3–6, 3–6-month follow-up

Table 20.9 Postoperative echocardiography: discharge haemodynamic data for the size 25 mm prosthesis

	Implantation technique		
Haemodynamic data*	Valve (*n*=28)	Mini-root (*n*=15)	Root (*n*=9)
Mean gradient (mmHg)	10.18	10.80	3.89
SD	5.31	6.78	2.67
Range	2–24	2–25	1–8
Mean EOA (cm^2)	1.76	1.48	1.97
SD	0.44	0.45	0.47
Range	0.9–2.66	0.91–2.42	1.10–2.63
Mean cardiac output (l/min)	5.51	4.55	4.58
SD	1.40	1.51	1.05
Range	3.5–9.9	2.6–7.6	3.1–6.2
Mean BSA (m^2)	1.83	1.90	1.76
SD	0.20	0.21	0.16
Range	1.40–2.37	1.40–2.20	1.46–1.92

*EOA, effective orifice area; BSA, body surface area

Some degree of regurgitation could be demonstrated in a moderate number of patients, but was very trivial in most of those cases. At 3–6-month follow-up only two patients showed a moderate degree of regurgitation (Table 20.10).

Table 20.10 Postoperative echocardiography: regurgitation at discharge and 3–6-month follow-up

Follow-up period	Regurgitation	Implantation technique			
		Valve	Mini-root	Root	Total
Discharge	None	80 (64)*	22 (71)	30 (88.2)	132 (69.5)
	Trivial/mild	45	9	4	58
	Moderate	0	0	0	0
3–6-months	None	48 (60)	16 (72.7)	20 (100)	84 (68.9)
	Trivial/mild	30	6	0	36
	Moderate	2	0	0	2

*Values are numbers of patients, with percentages in parentheses

Conclusions

The first results of the clinical study show an acceptable outcome after implantation of the Medtronic Freestyle™ bioprosthesis. Further studies are necessary to decide whether the technique of implantation has a significant influence on the postoperative results and whether this is due to technical problems or to other circumstances. Further improvements of techniques might lead to even better results.

Chapter 21

'Cylinder within a cylinder' method for porcine aortic root implantation

S. Westaby

Valve replacement with stentless aortic xenografts has been compared with aortic homograft implantation. Preliminary reports show ischaemic times that often exceed 100 min and prohibit stentless valve implants with associated procedures such as coronary surgery or mitral repair. In practice, the stentless xenografts have completely different handling characteristics and an inflow covered with Dacron cloth. With a carefully planned operation these valves can be implanted with ischaemic times of less than 45 min and into an aortic root with widespread calcification. The most important aspect, as with aortic allograft surgery, is to maintain the spatial orientation of the valve commissures to prevent distortion and aortic regurgitation. The following operative technique was designed for the Medtronic Freestyle™ stentless xenograft, which is presented as an intact porcine aortic root with ligated coronary arteries much the same as an aortic homograft (Figure 21.1). However, the preservation methods stiffen the porcine cylinder and with cloth covering the muscle bar and inflow these features simplify and expedite implantation. This chapter describes the 'cylinder within a cylinder' method which has been performed in 60 consecutive patients requiring aortic valve replacement with a bioprosthesis (by one surgeon). There were no exclusion criteria in this series. Patient selection for insertion of a bioprosthesis was the need for aortic valve replacement in a patient older than 70 years of age unless a tissue valve was specifically requested by a younger patient or anticoagulant therapy was contra-indicated.

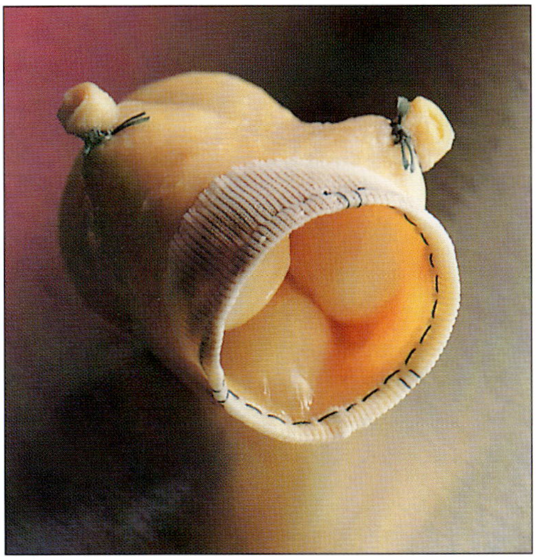

Figure 21.1.

Surgical methods

Cardiopulmonary bypass is established with an ascending aortic arterial cannula and a two-stage venous cannula in the right atrium unless mitral or tricuspid procedures are required. In this case dual caval cannulation is employed. Anterograde cold potassium cardioplegic infusion (St Thomas's solution) is used for myocardial protection and delivered either into the aortic root for pure aortic stenosis or into the coronary ostia in the presence of aortic regurgitation. A left ventricular vent is used to maintain a bloodless operative field and the perfusion temperature is maintained above 32°C. Distal anastomoses for saphenous vein grafts are applied before aortotomy and in the case of associated mitral procedure these are performed before implantation of the aortic xenograft.

With the aorta still distended by blood before crossclamp application, a plane of dissection is developed between the aorta and pulmonary artery to expose the anterior aspect of the sinotubular junction. The aorta is then opened with a transverse incision between 2 and 5 mm above the sinotubular junction but maintaining at least 5 mm clearance above the origin of the right coronary artery (Figure 21.2). The incision provides two-thirds transection of the aorta with easy access to both coronary ostia. Stay sutures are then applied to retract the distal

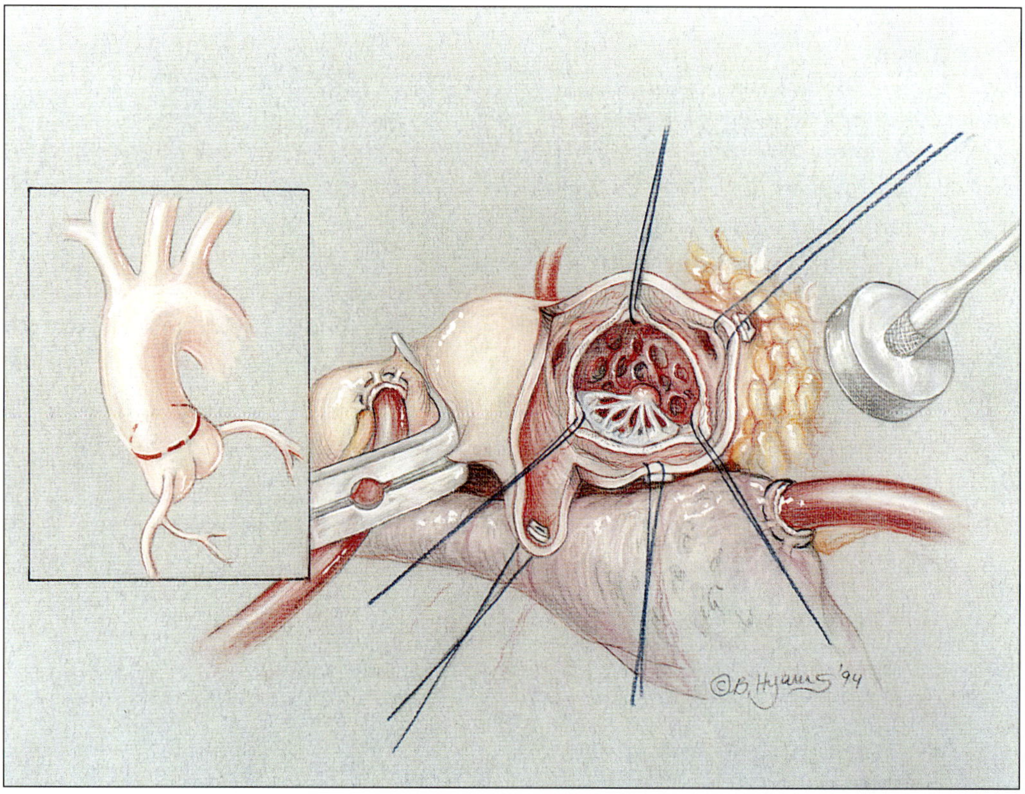

Figure 21.2.

aorta and to display the valve and right coronary ostium so that retractors are not required in the operative field. A Teflon pledgetted stay suture is placed directly above the right coronary and a second towards the surgeon in the middle of the non-coronary cusp. This holds open the cylinder of the aortic sinuses into which the cylinder of the porcine xenograft is inserted. The single assistant's hands are left free, since no further retraction is required. The diseased native valve is then excised, taking care not to excavate the site of the bundle of His. Thorough decalcification is performed to maximize the valve orifice. The coronary arteries are inspected so that the porcine coronaries can be carefully aligned. The porcine coronaries are situated closer together on the xenograft cylinder than the human coronaries and this is taken into account by not excising the porcine coronary buttons until the inflow aspect of the valve is securely seated.

Sizing of the xenograft is critical. It is particularly important not to distort the inflow aspect of the valve by oversizing, although a snug fit should be achieved between the porcine aortic cylinder and the human valve sinuses. The size of the patient's sinotubular junction is tailored to fit the outflow of the intact porcine xenograft cylinder. The circumference of the porcine outflow must be maintained to prevent distortion of the valve commissures. For the Freestyle™ valve the annulus is sized, not the sinotubular junction as for the Toronto valve. In the author's experience it has not been necessary to use a valve less than 21 mm in diameter. For a smaller aortic root, aortic root replacement would be employed, as opposed to the 'cylinder within a cylinder' method. When an appropriately sized valve has been chosen, the inflow sutures are applied to the aortic annulus in such a way as to maintain the porcine inflow in a single plane. Simple, single sutures of 2/0 Ethibond are applied to the commissures at a subcommissural level. Five or six simple interrupted sutures are then placed along the base of each cusp (Figure 21.3). The xenograft cylinder is held in an atraumatic VSD patch-holding

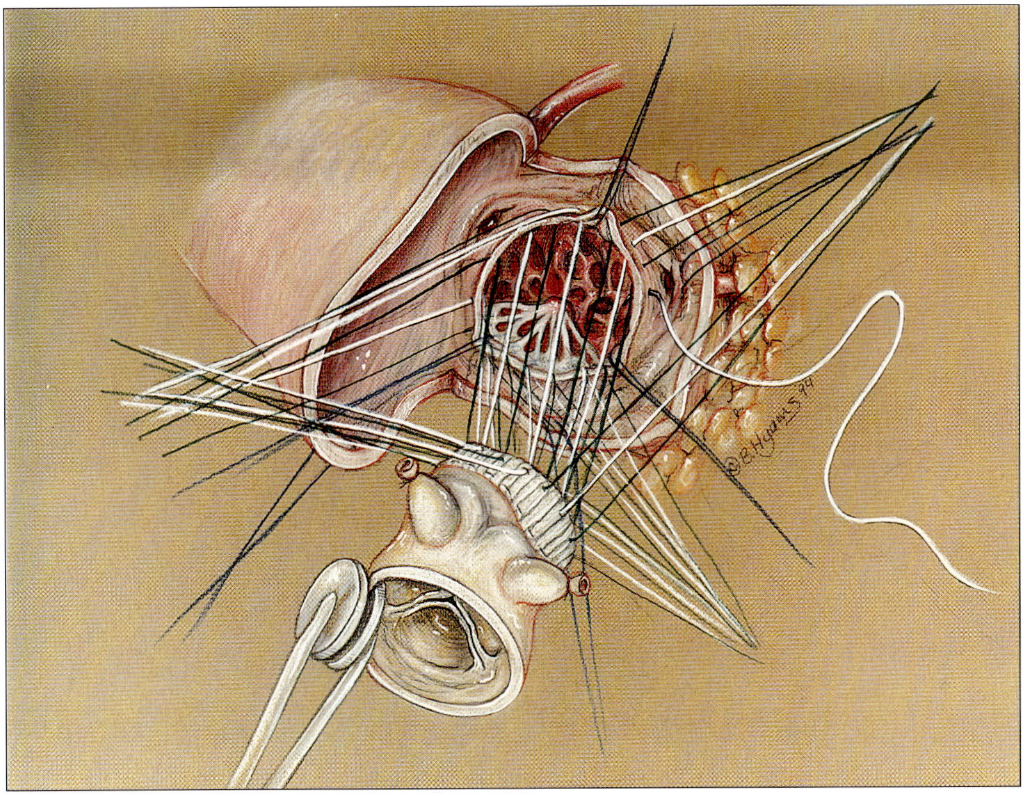

Figure 21.3.

forceps to facilitate handling. This is placed above the commissure between the pig coronaries on valve tissue which is later excised. The valve sutures are then inserted through the inflow cloth covering of the xenograft and a single throw placed on each suture for easy identification. Orientation of the valve is critically important so that the ligated porcine coronaries align with the human coronary ostia. The valve is lowered into place and the inflow sutures tied and cut (Figure 21.4). Attention is then turned to the native coronary ostia, which face the pig aortic sinuses. A U-shaped portion of the porcine aortic sinus is excised to remove the ligated porcine coronary directly adjacent to the human coronary ostium. This is performed for both left and right coronary arteries unless the patient's right coronary artery is completely occluded in its proximal 2–3 cm; in this case the porcine aortic sinus is left intact and only the porcine left coronary sinus is excavated (Figure 21.5). With the xenograft coronary sinuses appropriately

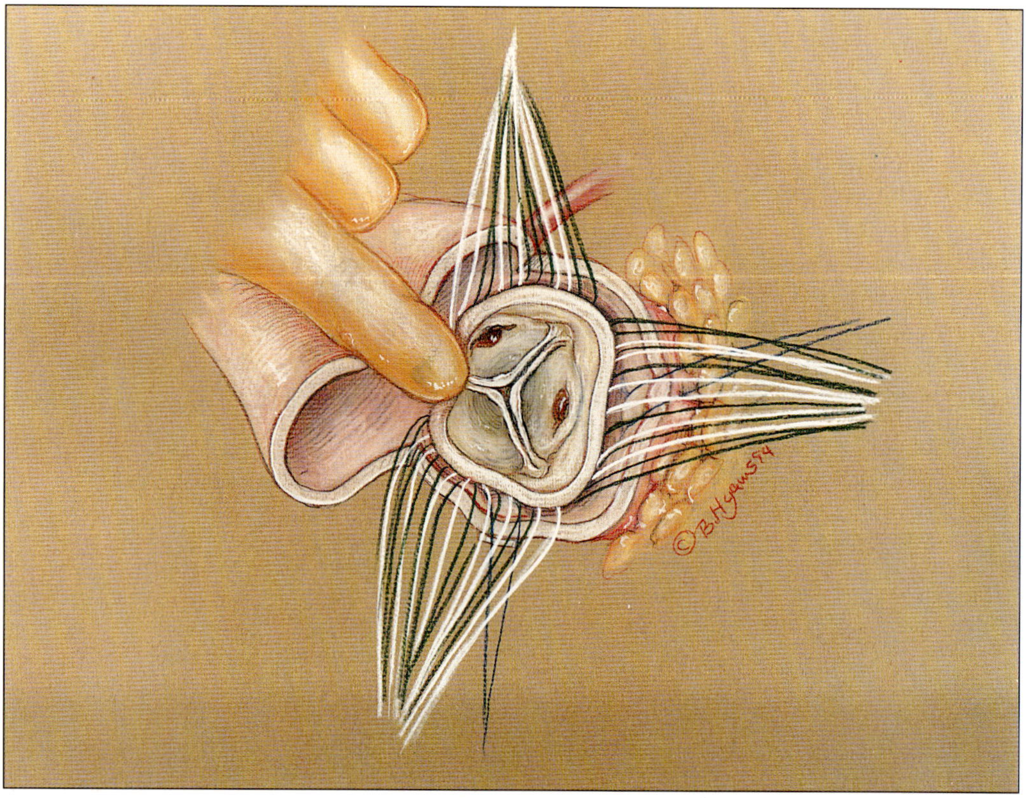

Figure 21.4.

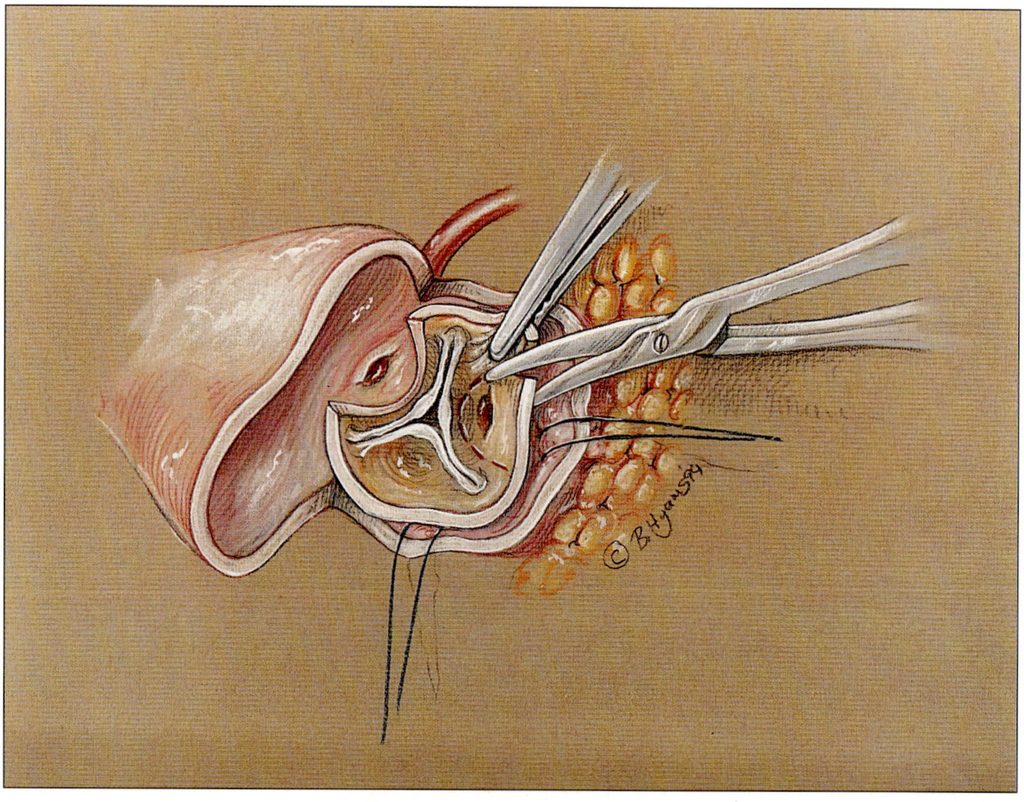

Figure 21.5.

excised, the outflow aspect of the xenograft is sutured into place with a single double-armed suture of 4/0 Prolene. The porcine aortic cylinder is trimmed to length appropriately so that the outflow suture line does not buckle the porcine cylinder. This begins opposite the surgeon by attaching the porcine commissure between left and right coronary cusps to the rim of the human aorta at the site of the equivalent human commissure. For the first few implants it proves helpful to attach the remaining two porcine commissures at an appropriate position on the human aortic root in order to guarantee maintenance of the circumference of the porcine outflow. The double-armed Prolene suture is first taken to the surgeon's right, attaching the outflow portion of the xenograft to the crest of the native aorta and then passing down under the patient's right coronary ostium toward the non-scalloped, non-coronary sinus. Access to the sub-coronary aorta is facilitated by traction on the stay suture above the right coronary. The suture line continues along the non-coronary sinus, joining the crest of the porcine valve to the crest of the hemi-transected native aorta. The left arm of the suture is taken beneath the patient's left coronary artery and around the crest of the aorta to meet the first limb of the suture (Figure 21.6). The sutures are tied together and then one limb of the Prolene is used to close the aortotomy as the suture approaches the mid-portion of the non-coronary sinus. Before completing aortic closure, freedom of access into the patient's coronary ostia is checked. Implantation is completed by a closing continuous suture of 4/0 Prolene, after which the heart is de-aired appropriately (Figure 21.7). The aortic crossclamp (ischaemic) times for isolated aortic replacement range from 35 min for a simple case with aortic regurgitation and no calcification to 59 min in severely calcified aortic sinuses. Cardiopulmonary bypass times range from 42 to 70 min. For combined

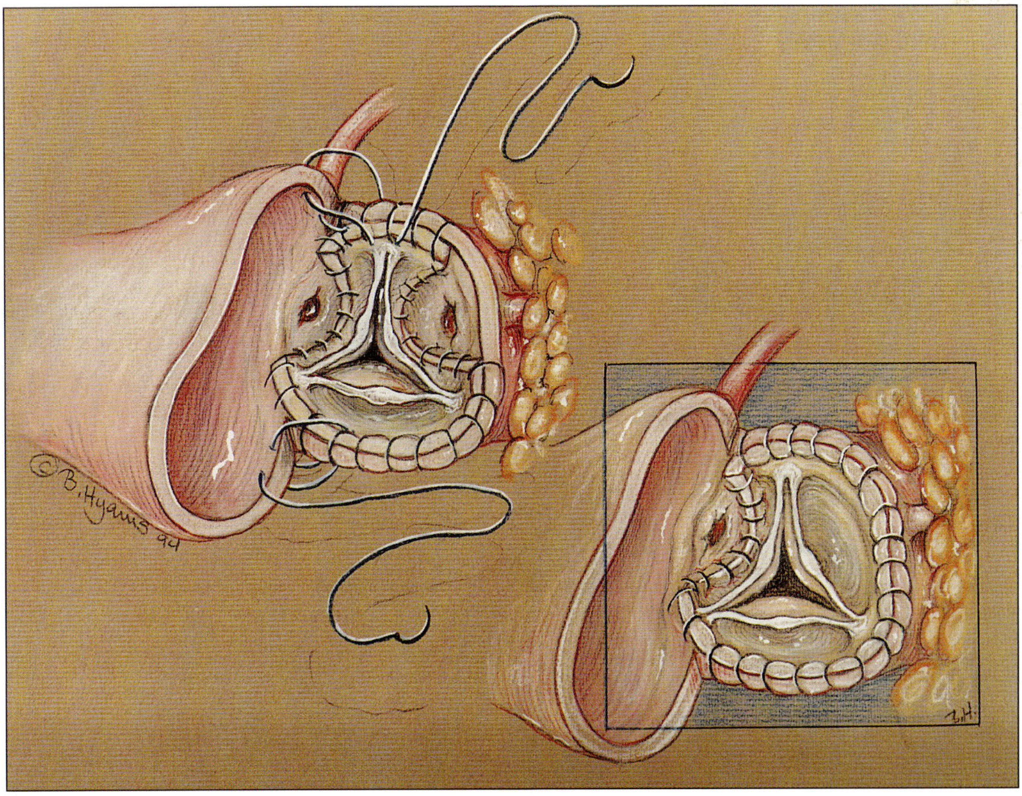

Figure 21.6.

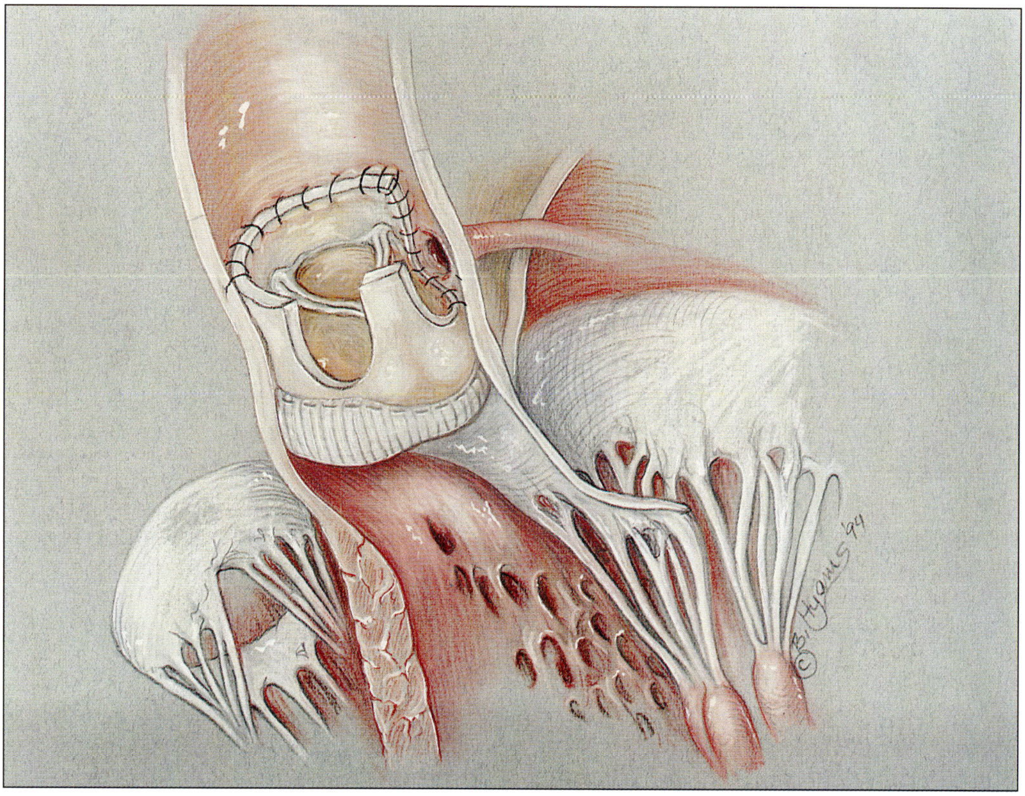

Figure 21.7.

procedures including coronary surgery and valve repair, ischaemic times range from 50 min to 108 min (triple valve operation).

Haemodynamics

Haemodynamics *in vitro* (pulse duplicator) and *in vivo* (Oxford patients) are shown in Tables 21.1–21.3. Transvalvular pressure drop has been found to be dependent on the surgical learning curve. The highest gradients occurred in female patients with a small aortic root (<23 mm); these patients have a relatively short distance between the valve annulus (adjacent to the His bundle) and the right coronary ostium. In order to accommodate the height of the inflow cloth over the muscle bar, a degree of distortion occurred and compromised the valve orifice. When this problem was recognized early in the experience the cloth over the muscle bar was rotated to the non-coronary sinus.

Early postoperative gradients fell significantly with time (6-month assessment). This may be due to several factors, including resolution of perivalvular haematoma and inflammation, distension or remodelling of the porcine root to fit the human sinuses and involution of septal hypertrophy with reduced gradient in the outflow tract. Gradients at 6 months were directly comparable to those with aortic homografts and those described for the Toronto SPV™. Clinical improvement was seen in all patients (Table 21.4).

Comment

The 'cylinder within a cylinder' method is simple and easily reproducible, and can be undertaken with ischaemic times only slightly longer than those for stented valves. There is a learning curve in this procedure, for both valve sizing and

Table 21.1 Haemodynamic data in vitro: *effective orifice area (cm²) of different sizes of Freestyle™ aortic bioprostheses*

Cardiac output (l/min)	Bioprosthesis size (mm)				
	19	21	23	25	27
4.5	1.83	2.07	2.59	3.19	3.34
7.5	1.98	2.42	3.03	3.85	4.32

Table 21.2 In vitro *effective orifice area (cm²): Freestyle™ vs Hancock bioprostheses*

	Bioprosthesis			
Cardiac output (l/min)	Freestyle™ (19 mm)	Hancock (19 mm)	Freestyle™ (23 mm)	Hancock (31 mm)
3.0	1.64	0.93	2.23	2.29
4.5	1.83	1.08	2.59	2.45
6.0	1.89	1.13	2.89	2.75
7.5	1.98	1.19	3.03	2.91

Table 21.3 Mean systolic gradients for the first 55 patients within 6 days of operation

Valve size (mm)	*n*	Mean systolic gradient (mmHg)	Range (mmHg)
21	13	13.8	7–23
23	20	9.6	2–16
25	17	6.9	1–15
27	5	5.0	2–9

Table 21.4 New York Heart Association classes of patients receiving Freestyle™ stentless xenografts

	NYHA class					
Assessment	V	VI	III	II	I	*n*
Preoperative	4	39	12	0	0	55
Postoperative	0	0	0	11	41	52*

*Three patients have follow-up periods of less than 3 months

surgical technique. The last 20 isolated aortic replacements were performed with ischaemic times of less than 45 min, even in the presence of calcified aortic sinuses. More important than duration of operation is the avoidance of complications. The transverse incision provides easy access to the sub-coronary aorta so that the potential for technical coronary obstruction is avoided. The spatial organization of this method makes it easy to avoid distortion of the commissures and the author has not encountered any significant aortic regurgitation using this method.

The Freestyle™ valve is easy to use by this method, since the stiffened glutaraldehyde-preserved aortic sinuses sit up within the human sinuses and facilitate suturing. Considerably fewer valve sutures are required than for an aortic homograft, because of the Dacron inflow cover. However, it is important to keep the Dacron cloth away from the coronary ostia, since this may stimulate tissue overgrowth and occlude the coronary orifice.

The porcine cylinder is versatile and can be used for aortic root replacement. Although it maximizes the valve orifice, root replacement may lead to difficult reoperations unless a generous button of recipient coronary is left to facilitate remobilization of the coronary buttons. Although the Freestyle™ valve benefits from both zero-pressure leaflet fixation and anti-mineralization treatment with α-amino oleic acid, the aortic wall will eventually calcify; this does not preclude reoperation as long as the coronary buttons can be remobilized. The Freestyle™ valve has also been used with the three sinuses scalloped, as with the Toronto SPV™ valve.

At present there is no information as to whether the porcine valve leaflet is happier within a human or a porcine valve sinus. Clearly, the leaflet within the porcine sinus is better than a leaflet mounted to an artificial stent. It has been proposed that the considerably better compliance of the native aortic wall is an environment superior to the poorly compliant glutaraldehyde-fixed porcine wall. However, the human pathological aorta also loses compliance, particularly when calcified. In addition, zero-pressure fixation and anti-mineralization for the porcine leaflet should ensure durability for patients over 70 years of age.

The most important aspect of stentless valve insertion is the assurance of commissural alignment and perfection in initial valve replacement. Malalignment of the valve through technical errors leads to increased stresses on the leaflets and aortic regurgitation. Of primary concern is a simple technique for implantation that can be performed with an ischaemic time allowing associated procedures, such as triple coronary bypass or double valve surgery, to be performed without prolonged cardiopulmonary bypass. The superiority of stentless devices in the aortic position will soon become apparent and it is important that the more demanding implant techniques do not preclude widespread adoption of these valves. Transvalvular gradients are significantly lower than for stented xenografts and resolution of left ventricular hypertrophy appears more rapidly in patients with stentless valves.

Chapter 22

Early results of the Edwards Prima™ stentless aortic bioprosthesis in 200 patients

H. Vanermen, K. Dossche, W. Daenen, R. Pillai and W. Konertz

Allografts have superior haemodynamics in aortic valve replacement, but the use of such grafts is usually restricted to young patients and cases of endocarditis because of their limited availability. The Edwards Prima™ stentless aortic bioprosthesis was designed to mirror the clinical performance characteristics of homografts while increasing availability and simplifying handling and implantation.

In June 1991 a clinical trial was initiated in four European centres for implantation of this newly designed stentless porcine valve. This retrospective study discusses the early clinical and haemodynamic results in 200 patients.

Materials

The stentless aortic bioprosthesis Model 2500 used in this study was manufactured by Baxter Healthcare Corporation, Edwards CVS Division, Irvine, California, USA. The valve is constructed from porcine aortic valves that have been preserved in 0.625% buffered glutaraldehyde, using low-pressure (2 mmHg) fixation techniques. It is essentially a porcine valve aortic root cylinder with resected coronary ostia. Dacron cloth is sewn around the inflow annulus to give additional support to the first suture line.

The design is intended to increase ease of implantation by helping to prevent distortion of the valve during the placement of the sutures, thus reducing the possibility of aortic incompetence (Figure 22.1).

Patients and methods

A multicentre clinical trial was initiated in June 1991 after approval of the ethics committee of each institution. Informed consent was obtained from every one of the 200 patients who have received this bioprosthesis. Inclusion and exclusion criteria are listed in Table 22.1.

The male/female ratio was 1.25/1. The mean age was 68.5 years (SD 8.1, range 24–83 years). Most of the patients were in NYHA class II or III.

The predominant aortic valve lesion was stenosis in 117 patients, mixed disease in 69, regurgitation in 13 and prosthetic dysfunction in one. Important pre-existing conditions were coronary disease in 33 patients, mitral insufficiency in 23, others in eight. All 200 patients underwent aortic valve replacement with a stentless porcine aortic valve. In addition, 33 patients underwent coronary artery

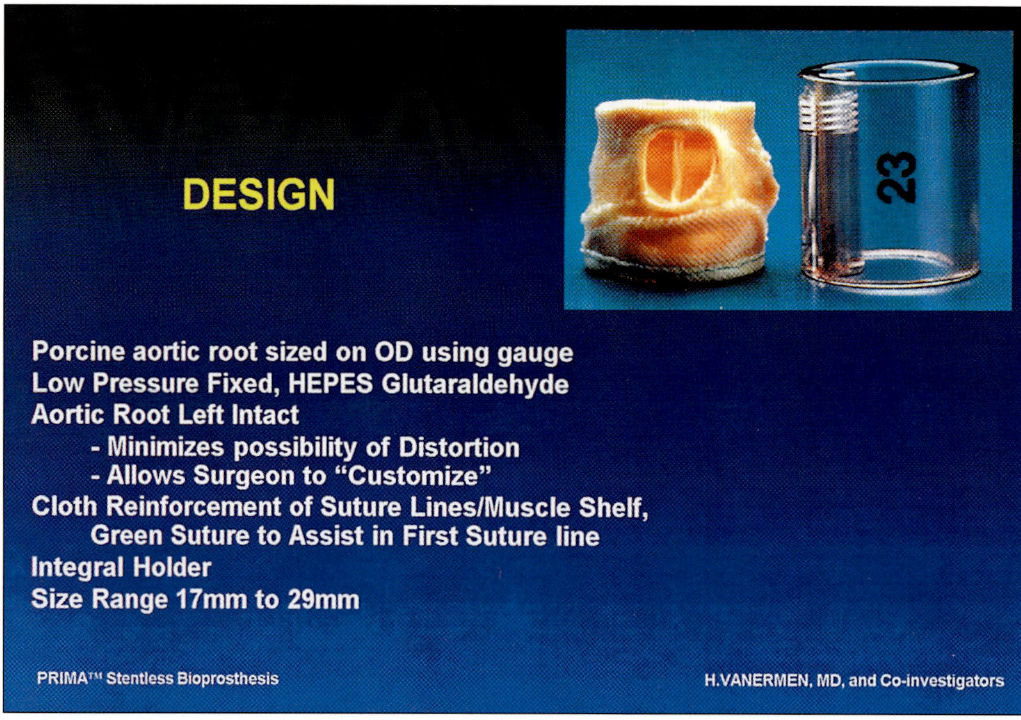

Figure 22.1. Design of the Edwards Prima™ stentless bioprosthesis Model 2500.

Table 22.1 Inclusion and exclusion criteria for implantation of the Edwards Prima™ stentless aortic bioprosthesis

Inclusion criteria

1. Patient is over 20 years of age
2. Patient requires an isolated aortic valve replacement

Exclusion criteria

1. Patient has a major, non-cardiac, life-threatening disease
2. Patient has active endocarditis
3. Patient has an abnormally dilated aortic root
4. Annular decalcification not possible
5. Patient's native coronary orifices are anatomically located in such a way that trimming into the cloth portion of the bioprosthesis is necessary
6. Patient is pregnant
7. Patient is an intravenous drug abuser or prison inmate

bypass grafting, one patient had mitral valve repair and one needed an intra-aortic balloon pump.

The aortic valve was exposed by either an oblique incision or a vertical 'hockey-stick' incision. After excision of the diseased aortic valve, meticulous decalcification and debulking of both the native annulus and the commissural areas was carried out, resulting in an almost complete removal of the native annular ridge (Figure 22.2).

The valve was secured in the subcoronary position in 161 patients by the same technique as that used for aortic valve allografts.[1] The inflow area of the valve was sutured with interrupted 3/0 or 4/0 polypropylene sutures. The second layer of sutures was done with a running 4/0 polypropylene suture after scalloping the left and right coronary sinuses. Before closure of the aortotomy, the non-coronary sinus of the bioprosthesis and the patient's aortic wall were approximated.[2]

In 39 patients, the inclusion technique was used. For this procedure, the lower suture line is the same as in the subcoronary position, but the coronary orifices

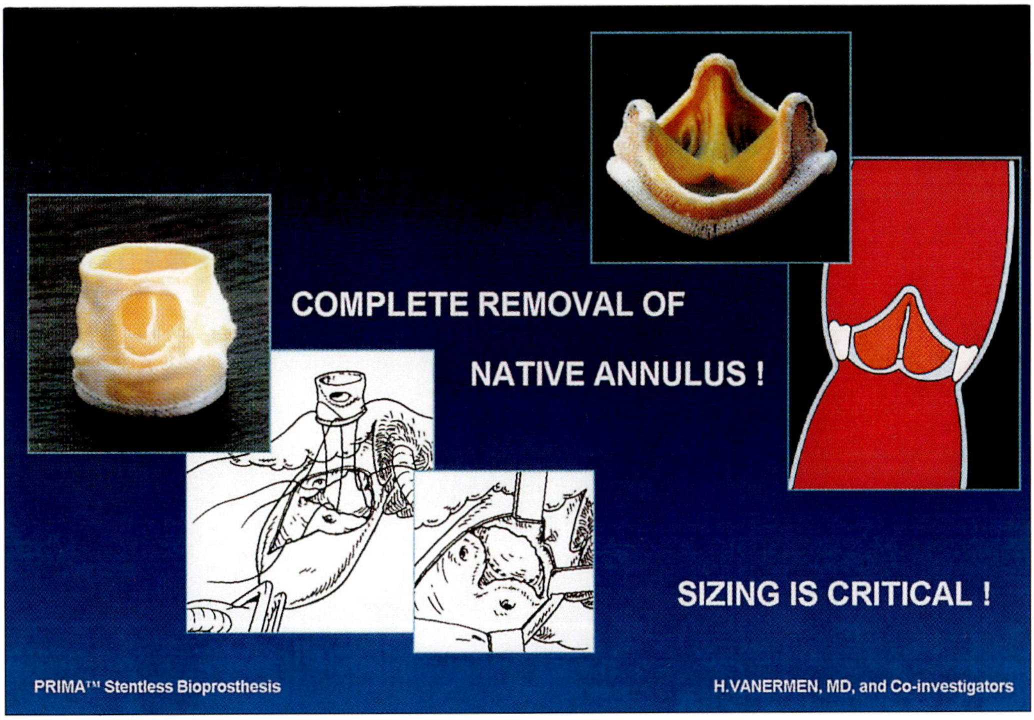

Figure 22.2. Complete decalcification of the native annulus is imperative.

are sewn to the precut holes in the valve cylinder with running 5/0 polypropylene as in aortic root replacement. The upper edge of the valve and the edges of the aortotomy are sewn together to attach the valve and close the aorta with one running 4/0 polypropylene suture[3] (Figure 22.3). An overview of the diameters of the bioprostheses implanted is shown in Figure 22.4.

As part of the study protocol, patients are interviewed and examined at discharge, and 6 and 12 months postoperatively. A Doppler echocardiographic

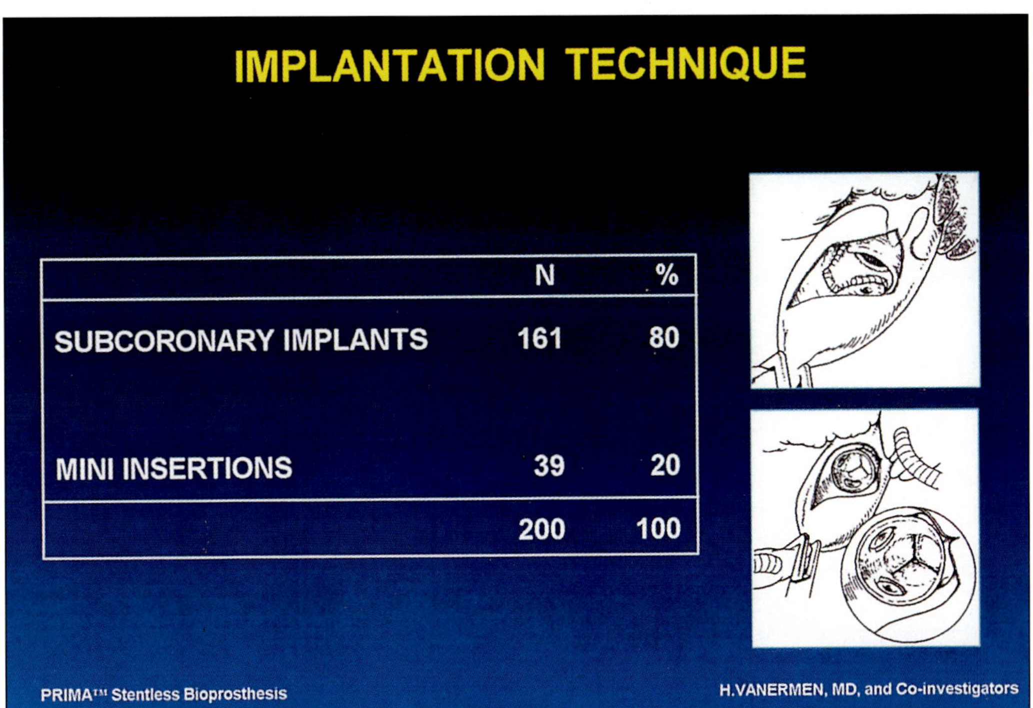

Figure 22.3. Modalities of implantation of the Edwards Prima™ stentless bioprosthesis Model 2500.

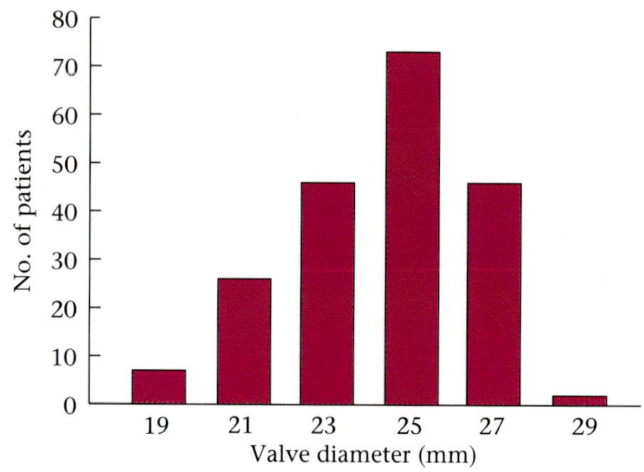

Figure 22.4. Diameters of prostheses implanted in 200 patients.

study is performed at the same time. The mean aortic valve gradient is obtained by planimetry and the peak systolic gradient across the bioprosthesis is calculated by the modified Bernoulli equation.[4] The aortic valve area is calculated by the continuity equation.[5]

Results

Tables 22.2–22.4 summarize the haemodynamic data (peak and mean gradient, effective orifice area) at discharge, 6 months and 12 months.

*Table 22.2 Peak transvalvular gradient at discharge, and at 6- and 12-month follow-up, of patients receiving various sizes of Edwards Prima™ stentless aortic bioprosthesis**

Prosthesis size (mm)	n	Peak transvalvular gradient (mmHg)		
		Discharge	6 months	12 months
19	7	30.9 ± 11.7	29.2 ± 14.5	35.2 ± 18.8
21	26	33.1 ± 18.2	25.3 ± 7.6	25.5 ± 8.1
23	46	24.1 ± 10.8	20.7 ± 8.8	23.1 ± 14.4
25	73	19.9 ± 11.2	17.4 ± 7.4	19.7 ± 12.5
27	46	15.9 ± 7.3	17.4 ± 6.9	11.7 ± 4.9
29	2	12.2	—	—

*Values are means ± SD

*Table 22.3 Mean transvalvular gradient at discharge, and at 6- and 12-month follow-up, of patients receiving various sizes of Edwards Prima™ stentless aortic bioprosthesis**

Prosthesis size (mm)	n	Mean transvalvular gradient (mmHg)		
		Discharge	6 months	12 months
19	7	15.4 ± 7.4	14.7 ± 7.7	17.4 ± 11.4
21	26	17.5 ± 12.0	13.1 ± 4.8	12.5 ± 4.4
23	46	11.6 ± 5.2	10.6 ± 4.6	10.9 ± 7.5
25	73	10.9 ± 10.4	8.4 ± 4.1	10.0 ± 6.4
27	46	7.1 ± 3.7	6.9 ± 3.4	5.6 ± 1.7
29	2	5.2	—	—

*Values are means ± SD

*Table 22.4 Effective orifice area at discharge, and at 6- and 12-month follow-up, of patients receiving various sizes of Edwards Prima™ stentless aortic bioprosthesis**

Prosthesis size (mm)	n	Effective orifice area (cm^2)		
		Discharge	6 months	12 months
19	7	1.0 ± 0.3	0.9 ± 0.2	0.8 ± 0.1
21	26	1.2 ± 0.3	1.0 ± 0.2	1.1 ± 0.5
23	46	1.4 ± 0.5	1.3 ± 0.4	1.5 ± 0.5
25	73	1.6 ± 0.6	1.7 ± 0.4	1.8 ± 0.7
27	46	2.0 ± 0.6	2.0 ± 0.6	2.3 ± 1.1
29	2	2.9	—	—

*Values are means ± SD

Complications were thromboembolism in six patients (transient in four cases), haemorrahge in three patients (two cases were valve related), endocarditis in two (of whom one needed explantation of the valve after 6 months), atrioventricular block in 14 (all needed endocavitary pacemaker implantation) and renal insufficiency in one.

Early postoperative mortality (less than 30 days after surgery) was six, late mortality was five (total after 1 year: 11/200 = 5.5%). Further details of causes of mortality are given in Table 22.5.

Table 22.5 Causes of early and late mortality at 12 months

Cause	No. of patients	Exitus (days after surgery)
Septic shock	3	2*, 23*, 59
Sudden death	2	65, 460
Intrapericardial bleeding	1	2*
Pulmonary artery embolism	1	10*
Suicide	1	421
Multiple organ failure	1	17*
Heart failure	1	460
Dissection for LV rupture	1⁻	1*

*Early mortality; **valve related

Aortic regurgitation at discharge was 24% for the overall group (first series of 100 implants 31%, second series of 100 implants 17%), none of which were clinically significant. At 1 year aortic regurgitation was stable at 27%, with no clinically significant cases apart from one patient with grade III+ aortic insufficiency (endocarditis).

Discussion

The indications for use of a stentless valve are, in principle, the same as those for stented bioprostheses. Because of the stentless valve's anticipated longer durability, younger patients might be considered to be candidates. Ineffective endocarditis remains a contraindication as long as Dacron fabric is used in the manufacturing process of these valves.

Peak and mean transvalvular gradients were higher than expected. Nevertheless, the absence of a mechanical stent is important. In stentless xenografts, as in homografts, the aortic root acts as a natural stent[6,7]. The anatomy and function of the aortic annulus and coronary sinuses are such that they may dampen the mechanical stress to which the leaflets of the aortic valve are subjected during the cardiac cycle.[8] In stented xenografts, stress is greatest along the leaflet attachments and it increases from the commissures to the base of the leaflets.[9,10] The absence of a stent should result in greater durability of the stentless valve.

A major advantage of the stentless valve was that, for a given aortic diameter, it was possible to implant a larger stentless valve (because one removes almost the entire native annulus). For this reason the resultant gradients need careful interpretation. In fact, for example, a stented 23 mm valve should be compared with a 25 mm stentless valve. To achieve optimal sizing of the annulus and correlating stentless valve, a tight fit of the sizer is required. Most surgeons are accustomed to leaving a slight remnant of the native valve annulus for the implantation of a stented xenograft, and a loose passage of the stented valve sizer is used for the best supra-annular valve placement. The Prima™ stentless bioprosthesis requires a snug fit to exploit its superior haemodynamic properties; a tight fit results in better coaptation of the leaflets.

As with allografts, a learning curve was observed in this study. Indeed, the implantation of a stentless valve is more demanding than that of a stented valve, but is easier than that of an allograft. The regurgitation rate was slightly higher in the first 100 patients than in the second group of 100 patients. None of the regurgitation reported was clinically significant at discharge. As this study was set up as a clinical trial for FDA approval, the Doppler echocardiographic criteria were very strict.

The rates of morbidity and mortality were acceptable in this series. Thromboembolic events were reported in six patients, but none of the causes was valve related. One patient with endocarditis underwent an aortic root replacement with a homograft 6 months after initial implantation of the stentless valve. In only one patient was the cause of death valve or implantation related.

Haemodynamic results and complications were not influenced by the implantation technique (subcoronary or mini-root).

Conclusions

The direct fixation of a stentless bioprosthesis to the aortic wall may improve durability. As far as gradients are concerned, the Edwards Prima™ stentless bioprosthesis performs better than its stented equivalents although these performances are not yet comparable to those of allografts.

Early results in the authors' series were encouraging, with few complications, although technically the implantation of a stentless valve is more demanding.

In the authors' opinion, the Edwards Prima™ stentless valve is a versatile tool. It can be used in a subcoronary position, as a mini-root or even in complete root replacement in more complex situations.[11]

To summarize, the following early conclusions may be drawn, after implantation of the Edwards Prima™ stentless aortic bioprosthesis in 200 patients:

1. Non-fatal postoperative complications are comparable to those with other bioprostheses;
2. Early and late mortality are acceptable;
3. Regurgitation is often detectable, but hardly ever (<1%) clinically significant;
4. The comparison between gradients of Baxter's stented and stentless bioprostheses is favourable;
5. The Edwards Prima™ stentless aortic bioprosthesis is a very versatile device and the surgical technique is easy and reproducible.

References

1. Ross D M 1964 Homotransplantation of aortic valve in subcoronary position. J Thorac Cardiovasc Surg 47: 713–719
2. Pillai R, Spriggings D, Amarsena N, O'Regan D, Parry A J, Westaby S 1993 Stentless aortic bioprosthesis? The way forward: early experience with the Edwards valve. Ann Thorac Surg 56: 88–91
3. Konertz W, Weynand M, Sidviopoulos A, Schwammenthal E, Breithardt G, Scheld H 1992 Technique of aortic valve replacement with the Edwards Stentless Aortic bioprosthesis 2500. Eur J Cardio-Thorac Surg 6: 274–277
4. Currie P J, Seward J B, Reeder G S, Vliestra R E 1985 Continuous-wave Doppler echocardiographic assessment of severity of calcific aortic stenosis: a simultaneous Doppler–catheter correlative study in 100 adult patients. Circulation 72: 810–818
5. Skjaerpe T, Hegrenaes L, Hatle L 1985 Non-invasive estimation of valve area in patients with aortic stenosis by Doppler ultrasound and two-dimensional echocardiography. Circulation 72: 810–818
6. Drury P J, Dobrin J, Bodnar E 1986 Distribution of flexibility in the porcine aortic root and in cardiac support frames. In: Bodnar E, Yacoub M (eds) Biologic and bioprosthetic valves. Yorke Medical, London, pp 580–587
7. Vesley D, Boughner D R 1986 A comparison of the bending behaviour of fresh and glutaraldehyde treated porcine aortic leaflets. Circulation 74: 339–351
8. David T E, Pollick C, Bos J 1990 Aortic valve replacement with stentless porcine aortic bioprosthesis. J Thorac Cardiovasc Surg 99: 113–118
9. Sabbah H N, Hamid M S, Steni P D 1986 Mechanical stress on closed cusp porcine bioprosthetic valves: correlation with sites of calcification. Ann Thorac Surg 42: 93–96
10. Thubrikar M, Deck J D, Aouad J, Nolan S P 1983 Role of mechanical stress in calcification of aortic bioprosthetic valves. J Thorac Cardiovasc Surg 86: 115–125
11. Sievers H, Podszus G, Lange P E, Bürsch J H, Bernhard 1985 Replacement of the aortic root by free implantation of a stentless aortic porcine bioprosthesis in a patient with aneurysm of the sinuses of Valsalva. J Thorac Cardiovasc Surg 88: 360–361

Chapter 23

Early valve haemodynamics following implantation of the Edwards Prima™ stentless aortic bioprosthesis

G. W. Parry, N. C. Batrick and F. P. Shabbo

Accumulated experience with the aortic homograft has indicated its advantages compared with the stent-mounted bioprostheses. Such advantages include good early haemodynamics and excellent long-term durability and performance.[1,2] The supply of homografts is, however, exhaustible compared with the population requiring aortic valve replacement; thus, the need for a viable stentless bioprosthesis is evident.

Stented xenografts, although having generally satisfactory haemodynamics, at least in the intermediate term, are associated with a number of potential problems, principally valve dysfunction, degeneration and valve failure.[3] Obstruction to forward (transvalvular) flow with stented xenografts may contribute to these problems.

This has led to the development of the stentless porcine xenograft, which may offer an adequate compromise, providing a source of bioprostheses with good to excellent haemodynamics without the wider problems inherent in the mechanical prostheses.

Patients and methods

A total of 25 patients (16 male, nine female) were studied prospectively following implantation of the Edwards Prima™ stentless aortic bioprosthesis (Baxter, Irvine, California, USA). Criteria for implantation of the Prima™ valve were identical to those for implantation of either a stented porcine xenograft or a homograft. The mean age of the study group was 64 years (range 34–80 years). The preoperative diagnoses were as shown in Table 23.1. Two of the 25 patients required redo procedures, having previously received stented porcine xenografts that had failed with time. Four of the 25 patients underwent simultaneous coronary artery bypass graft surgery.

Table 23.1 Preoperative diagnoses of 25 patients receiving the Edwards Prima™ stentless aortic bioprosthesis

Aortic regurgitaton	Aortic stenosis	Combined A R + A S	Infective endocarditis
7	10	6	2

Surgery

Cardiopulmonary bypass was established between right atrial drainage to ascending aortic return, with electrolyte prime and systemic cooling to 28°C. Myocardial protection was with St Thomas' crystalloid cardioplegic solution infused either directly into the aortic root or through the coronary ostia, depending on the underlying aortic valve disease. Cardioplegic solution was reinfused every 30 min while the aorta was crossclamped.

The native aortic valve was excised, and the annulus was prepared in the usual fashion and then sized for the Prima™ valve as appropriate.

Implantation technique

This was identical in all 25 patients. The stentless valve was implanted in a subcoronary position, with anastomosis of the host annulus to the xenograft annulus using interrupted horizontal mattress sutures of 3/0 Mersilene. Following completion of the annular anastomosis, the distal segment of the Prima™ valve was 'sculptured' around the orifices of the two coronary arteries and anatomosed with continuous 4/0 Prolene.

Valve sizes

The sizes of valve implanted are shown in Table 23.2.

Table 23.2 Sizes of Edwards Prima™ valves implanted

	Implanted valve size (mm)				
	21	23	25	27	29
No. of patients	2	6	7	7	3

Assessment techniques

All patients underwent echocardiographic assessment, with transthoracic echocardiography in all 25 patients and, in addition, transoesophageal echocardiography in 16 patients (Table 23.3). The presence of aortic regurgitation was specifically sought, using colour flow Doppler. Mean and peak aortic gradients were measured in all patients.

Table 23.3 Assessment techniques after implantation of the Edwards Prima™ valve

Echocardiography	Time after implantation (days)	Number of patients
Transthoracic	5 and 10	25
Transoesophageal	7	16

Anticoagulation

Patients had not received oral anticoagulants; following implantation of the Prima™ valve all patients were started on aspirin (75 mg daily).

Results

Mortality

The rates of mortality at 10 and 30 days were both zero. No deaths have occurred on long-term follow-up to date.

Morbidity

There was one cerebrovascular accident, which occurred on the 10th day after implantation. Cerebral computerized axial tomography (CAT) scan did not reveal any significant abnormality, and recovery was substantially complete.

One patient developed a wound infection 2 weeks after implantation and rapidly developed mediastinitis. Septicaemia and renal failure supervened despite aggressive medical and surgical therapy. This patient ultimately made a full and complete recovery without any stigmata of infective endocarditis being present, and with no impairment of valve function.

One patient developed large bowel obstruction on the sixth day after implantation. Investigations revealed a previously asymptomatic sigmoid carcinoma, and this patient underwent a sigmoid colectomy 1 week after implantation of the Prima™ valve, with no adverse effects.

Echocardiographic assessments

Doppler assessments of peak and mean aortic gradients 5 and 10 days after implantation revealed peaks (± SD) of 43 ± 5.36 mmHg at 5 days and 38.5 ± 5.7 mmHg at 10 days (Figure 23.1). Mean aortic gradients were 22 ± 3.9 mmHg at 5 days and 18.8 ± 4.1 mmHg at 10 days. The peak and mean aortic gradients were less than with comparable standard porcine xenografts (Figure 23.1). Transvalvular pressure gradient was unrelated to the size of prosthesis implanted.

Aortic regurgitation, either transvalvular or paravalvular, was not seen in any patient.

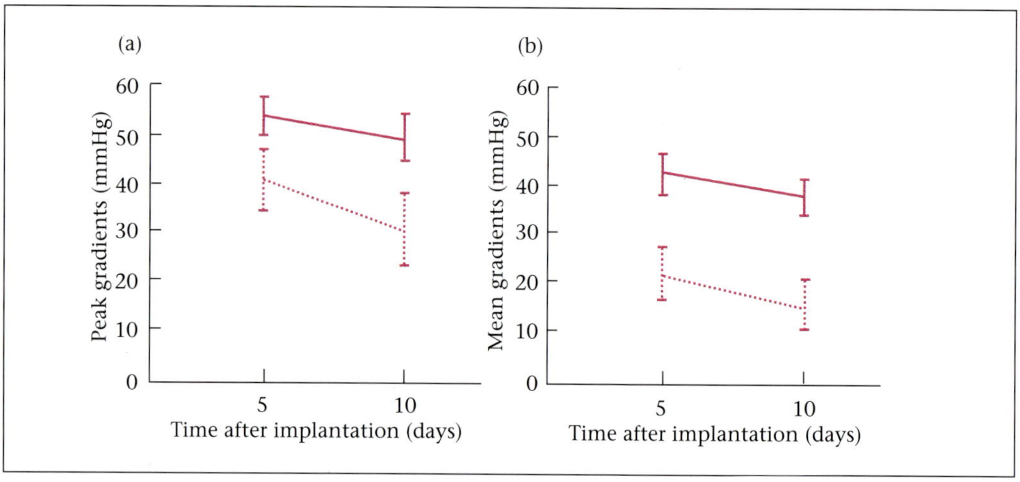

Figure 24.1. Peak (a) and mean (b) transvalvular pressure gradients across the Edwards Prima™ stentless aortic bioprosthesis (............) and a standard stented porcine xenograft (———). Vertical bars are SD.

Discussion

The results reported here confirm the impression gained from other studies with stentless xenografts,[4–8] but also reveal some other points worthy of emphasis.

First, peak and mean gradients are appreciably less with the Edwards Prima™ stentless bioprosthesis than with comparable stented bioprostheses (Figure 23.1). Both peak and mean gradients fall linearly with time following implantation, at least in the first 10 days. The exact reason for this is not clear, although some annular remodelling following implantation may account for this observation.

Secondly, aortic regurgitation was not detected in any of the patients in the authors' study. It remains to be seen whether aortic regurgitation, either clinically evident or echocardiographically detected, will develop later in the postimplantation period. Some other workers[8] have shown aortic regurgitation to be detectable on longer-term follow-up, but this may fluctuate. It does not, however, appear to represent a significant clinical problem.

Implantation technique in this series was straightforward and associated with a short learning curve. It appears to produce good early haemodynamics, although admittedly this may reflect the qualities of the valve itself as much as the insertion technique. Other methods of implantation technique are described[6] and appear to give equally satisfactory results.

The durability of bioprostheses involves many factors. The absence of a stent from the Prima™ (and other makes of stentless bioprostheses) may be the major reason for its excellent early haemodynamics, given the contribution of both annular and post stents to impaired haemodynamics with stented bioprostheses.[9] Equally, however, the method of preservation of the valve tissue itself may be of considerable significance, with zero-pressure stress-free fixation of importance.[10] This may allow for more effective retention of collagen orientation in the valve leaflets. Likewise, the absence of stents will, of course, eliminate movement between stent and prosthesis that may contribute to accelerated bioprosthetic degeneration.

The results of these studies demonstrate good early haemodynamics with the Edwards Prima™ stentless valve. Comparison with the closest conceptually and structurally equivalent bioprosthesis is with the aortic homograft, the long-term results with which are good to excellent.[2] It is to be hoped that similarly excellent results in the longer term will become evident with the stentless aortic bioprosthesis. If this is so, then the greater availability of this form of bioprosthetic valve will confer the benefits of homograft-style aortic valve replacement on a greater number of patients.

References

1. Ross D N 1962 Homograft replacement of the aortic valve. Lancet 2: 487
2. Matsuki O, Robles A, Gibbs S *et al* 1988 Long term performance of 555 aortic homografts in the aortic position. Ann Thorac Surg 46: 187–191
3. Thubrikar J, Deck J D, Aouad J, Nolan S P 1983 Role of mechanical stress in calcification of aortic bioprosthetic valves. J Thorac Cardiovasc Surg 86: 115–123
4. David T E, Pollick C, Bos J 1990 Aortic valve replacement with stentless porcine aortic bioprosthesis. J Thorac Cardiovasc Surg 99: 113–118
5. Hofig M, Ulrich N, Mehrdat M *et al* 1991 Performance of a stentless xenograft aortic bioprosthesis up to 4 years after implantation. J Thorac Cardiovasc Surg 103: 1068–1073
6. Konertz W, Weyand M, Sidiropoulos A *et al* 1992 Technique of aortic valve replacement with the Edwards stentless aortic bioprosthesis 2500. Eur J Cardio-Thorac Surg 6: 274–277
7. Casabona R, De Paulis R, Zaffera G, di Summa M, Bottone W, Stacchino C, Vrandecic M O, Morea M 1992 Stentless porcine and pericardial valve in the aortic position. Ann Thorac Surg 54: 681–685
8. Pillai R, Spriggings D, Amarasena N, O'Regan D J, Parry A S, Westaby S 1993 Stentless aortic bioprosthesis? The way forward. Early experience with the Edwards valve. Ann Thorac Surg 56: 88–91
9. Morris D C, Wickliffe C W, King S B 1976 Haemodynamic evaluation of the porcine heterograft aortic valve. Am J Cardiol 37: 157
10. Medtronic Heart Valve Commentary 1993 (March) 2–8

Chapter 24

Indications, implantation techniques and results of the various types of stentless bioprostheses: the surgeon's standpoint and the Sorin Pericarbon™ stentless valve

R. De Paulis, G. F. Zattera, F. Parisi, E. Leotta, R. Casabona, M. Galloni, G. Pugliese, M. di Summa and M. Morea

The use of a stentless (STL) valve has been a landmark in the history of surgery for heart valve disease. The first clinical experience in the early 1960s used stentless xenograft valves.[1,2] However, because of the development of better valve preservation methods, together with the advent of stent mounting, the stentless xenograft was virtually abandoned.

The years following showed that stented porcine or pericardial valves were not ideal valve substitutes because of their high rate of primary tissue failure in a period frame of about 10 years. For this reason, several methods of increasing the durability of a biological valve have been investigated. The knowledge that fresh or frozen aortic allografts, irrespective of the type of preservative, or for the same preservative, are more durable in the absence of a stent,[3] prompted renewed interest in stentless xenografts.[4,5] The known advantages of not having a rigid or even a flexible stent are in superior haemodynamic performance, which is especially significant in the small sizes of valve,[6,7] and probably a reduced incidence of perivalvular leakage, haemolysis and thromboembolism.

Furthermore, mounting of a valve in a stent causes increased stress on leaflets and commissures, which is responsible for non-calcific (typical of the first generation of pericardial valves) or calcific cusp tears. Leaflet abrasion against a cloth-covered stent is also a well-known complication.

As a consequence, stentless xenograft valves should, in theory, achieve increased durability through minimization of the mechanical stress on leaflets and commissures. Several authors have stated that the elastic aortic wall should dampen the mechanical stress on valve leaflets, contributing to its dissipation during diastole.[6] The potential improved durability of a stentless xenograft valve has prompted the development of several stentless valves, each with different and particular characteristics. The results of the clinical trials in progress will show which valve feature is most important for improved durability.

Construction of the Sorin Pericarbon™ stentless bioprosthesis

The Sorin Pericarbon™ stentless valve (Sorin Biomedica, Saluggia, Italy) is constructed from two sheets of glutaraldehyde-treated bovine pericardium. By a process of atraumatic tissue fixation without the use of mechanical moulds, the first sheet is shaped to form the three leaflets. It is then sutured to the second sheet of pericardium using a pyrolite carbon-coated suture. The suture line is especially designed to dissipate the mechanical stress at the level of the commissures. The final suturing and trimming of the two sheets sewn together will give the valve its cylindrical configuration. A small rim of tissue is present at the base of the valve in order to facilitate suturing to the aortic annulus.

Differences between the Sorin Pericarbon™ STL valve and other types of STL valves

To our knowledge the Sorin STL valve is the only commercially available stentless pericardial valve. Although stented pericardial valves have shown a higher incidence of structural deterioration in relation to the standard porcine valve, it is thought that a better design could improve durability. A new generation of pericardial stented valves is now undergoing clinical trials and the intermediate results seem encouraging.[8] Besides problems related to the tissue employed, for all types of STL valve the long-term outcome will probably depend not only on the intrinsic advantage of not having a stent but also on the valve design. Each available STL valve has particular features that may influence its choice in a particular clinical or anatomical situation.

Basically, porcine STL valves are available either as a complete porcine aortic root, or as a composite valve constructed by suturing together three non-coronary porcine leaflets, or as a porcine pulmonary valve. In the first case, the porcine aortic wall is very thick and a bulky muscle shelf is invariably present below the nadir of the coronary leaflets. This might be the cause of a transprosthetic gradient if implanted in a small aortic root. When this muscle shelf is thinned, a Dacron covering is usually required; the porcine aortic valve will therefore, in theory, have more difficulty in contouring the host aortic annulus, since it can not be crimped together by the sutures in order to accommodate small discrepancies in diameter with the host aorta. It can therefore be expected that this type of valve will not be the first choice in the presence of a small aortic root, and that the technique of total root replacement should be preferred to the mini-root or to the freehand insertion. On the other hand, this type of prosthesis seems most suitable in the case of a distorted aortic annulus, where maintaining perfect geometry is fundamental, yet may prove very demanding for the surgeon. The composite porcine STL valve offers the advantage of being less bulky, owing to the absence of muscular tissue, and the pulmonary porcine stentless valve has also the advantage of a thinner wall. The use of these two valve types would, therefore be indicated in small aortic roots. However, the best results would be obtained in patients with symmetric aortic roots and the preferred surgical technique should be the freehand insertion.[9]

The basic design characteristics of the Sorin STL pericardial valve are (a) a thin outer wall that ensures minimal obstruction inside the aortic root and makes it particularly suitable for the small aortic annulus and (b) the absence of any prosthetic material that might become relevant especially in the case of acute endocarditis. The tissue is soft and pliable, and permits different types of surgical techniques at the time of implantation. The ventricular side of the valve has a narrow rim of tissue that facilitates the lower suture and can easily be crimped in order to adapt better to the aortic annulus. Using this type of valve, the geometry of the aortic root also appears to be relevant. The presence of a distorted annulus or misjudgement in the valve sizing or positioning might result in valve incompetence. However, as a general rule it is important to stress that with all kinds of available STL bioprostheses, especially when implanted in a freehand fashion, perfect sizing is crucial. Undersizing of an allograft, that is capable of expanding in a pressurized aorta, has been shown to compensate in part for the discrepancies with the aortic annulus.[10] On the other hand, glutaraldehyde-fixed leaflets are less extensible, and valve incompetence would probably be present more often in cases of graft undersizing.

Surgical indications

At the University of Torino (Turin) there is a long history of experience with biological valves; these are usually the preferred type in patients over 60 years of age because of the low rate of thromboembolic and haemorrhagic complications. Besides those cases with relative contraindications to biological valves (renal insufficiency, need of anticoagulation for other medical reasons) or because of a specific request by the patient, age is the first criterion for using a bioprosthesis. Biological valves are also often used in cases of active endocarditis: mechanical valves are preferred in cases of a small aortic annulus. When considering STL valves and their theoretical advantages, the following are considered ideal candidates:

1. Elderly patients with a small aortic root;
2. Elderly patients with a heavily calcified aorta, where the aortic wall has become fragile after the process of decalcification;
3. Any patient with active endocarditis.

However, considering the need for a learning curve period, and with the hope that this valve could achieve better durability, patients with a large aortic annulus that would have otherwise received a stented bioprosthesis have received an STL pericardial valve.

Implantation technique

A transverse aortotomy is used to expose the aortic valve; in the first few cases the aortic incision was extended into the non-coronary sinus. However, the anatomical orientation was difficult and in several cases an aortic patch was required to close the aorta at the end of the procedure. On the other hand, a large simple transverse aortotomy 1 cm above the commissure gives a good exposure and maintains the symmetry of the aortic root. The valve is then excised and the aortic annulus decalcified. The valve size is chosen on the basis of the largest size that can be advanced smoothly through the valve annulus. In the presence of a dilated aortic root (i.e. when the distance between the commissure is much greater than the annulus diameter) a stented prosthesis is preferred. The valve is turned inside out and inverted, and three double armed 4/0 polypropylene sutures are passed through the nadir of each scallop of the host annulus, then through the corresponding point of the inferior rim of the valve. The valve is then lowered into the left ventricular outflow tract (LVOT) and each arm of the three sutures is used in a continuous over-and-over fashion. The suture is usually kept in a horizontal line, except at the level of the membranous septum where it is passed through the annulus to avoid any injury to the conduction system. The valve is then brought into normal orientation by pulling on the external pericardial cylinder. In cases of a small aortic annulus the lower suture is made without inverting the valve and by inserting single interrupted 4/0 polypropylene sutures; in this way the aortic annulus is more easily accessible. Even when the valve is inverted and the lower rim is doubled back on itself, the increase in tissue bulk in the LVOT is of no concern because the pericardial layers of the valve are extremely thin. The outer cylinder of the valve corresponding to the aortic wall is then scalloped for alignment of the coronary orifices. The three commissures are fixed to the aortic wall. The distal suture line is then made in a continuous over-and-over fashion, starting with three separate sutures below the two coronary ostia and halfway into the non-coronary sinus.

Clinical experience

Forty-three patients entered a clinical trial that started in February, 1990: four had acute endocarditis (age range from 20 to 55 years); the remainder had either a degenerative or a rheumatic disease (mean age 65.1 ± 9.2 years).

All patients underwent at transthoracic echocardiographic study every 12 months postoperatively. The maximum velocity across the valve was measured by continuous-wave Doppler echocardiography, the aortic valve area was calculated with the continuity equation, and colour Doppler echocardiography with pulse-wave mapping was used for detection of prosthetic valve regurgitation. As of June 1993, the mean follow-up period was 26.7 ± 10.3 months (range 5–41 months).

One patient died after postoperative myocardial infarction: autopsy showed total occlusion of the circumflex artery, which was probably caused by blood infiltration through the lower suture of the STL valve. Another patient died suddenly 6 months after surgery; a third patient, a drug addict, died of acute endocarditis 20 months after surgery and the autopsy showed multiple abscesses in the myocardium surrounding the aortic root; the valve leaflets were partially destroyed, but the aortic wall was intact. Another valve had to be removed shortly after its implantation because of excessive tension on the left main coronary artery. Finally, a patient underwent reoperation after 3 years because of acute aortic insufficiency; at surgery it was noted that one of the three leaflets had disappeared. Only a thin rim of mural leaflet was present on the aortic (pericardial) wall (Figure 24.1). The remaining patient had an uneventful postoperative course without thromboembolic complications or infective endocarditis. None of the patients received anticoagulants. The majority of the patients were in NYHA class I (36, 84%) and the remainder in class II.

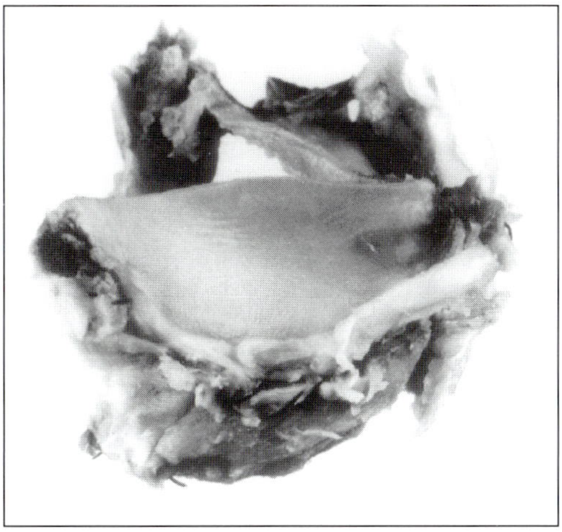

Figure 24.1. Sorin Pericarbon™ stentless valve explanted after 3 years because of acute aortic insufficiency. The site of the embolized leaflet and the normal appearance of the other two leaflets can be seen.

Doppler echocardiographic evaluation showed an average maximum velocity across the valve of 2.2 ± 0.5 m/s and a mean aortic valve area of

1.53 ± 0.4 cm^2. The values for each valve size are reported in Table 24.1. Trivial or mild aortic insufficiency, not clinically relevant, was present in 14 patients (33%).

Table 24.1 Haemodynamic data for the Sorin Pericarbon™ stentless valve

Valve size (mm)	No. of patients	Body surface area (cm^2)	V_{max} (m/s)	Aortic valve area (cm^2)
21	4	160.3 ± 21.5	2.22 ± 0.4	1.1 ± 0.1
23	13	159.5 ± 13.8	2.48 ± 0.4	1.3 ± 0.4
25	7	175.3 ± 22.3	2.38 ± 0.3	1.5 ± 0.4
27	11	180.5 ± 14.7	2.07 ± 0.5	1.8 ± 0.3
29	7	170.5 ± 14.3	1.92 ± 0.4	1.9 ± 0.1

Conclusions

In a previously published paper[7] the authors compared four groups of patients each receiving a tilting disc mechanical valve, a stented bioprosthesis, or two types of STL valves. After a mean follow-up of 17 ± 7 months, Doppler echocardiographic evaluation showed that the Sorin Pericarbon™ stentless (pericardial) valve had optimal haemodynamic performance. However, trivial or mild central aortic insufficiency was detected in a significantly higher number of patients (13/30 cases, 43%) with this type of valve. This finding was ascribed either to the surgeon's learning curve, or to the particular stiffness of the aortic leaflets requiring higher closing volumes and consequent higher regurgitation volume. However, in the study reported here there was a reduction in the incidence of regurgitation in more recent cases, stressing the importance of a surgeon confident with the valve type. On the other hand, the degree of valve insufficiency did not change with time, ruling out the possibility that stretching of the pericardial tissue with time could compensate for imperfect alignment of the leaflets. From these results, therefore, it can be concluded that glutaraldehyde-treated pericardial leaflets, like all glutaraldehyde-fixed porcine leaflets,[9] have a reduced ability to stretch in radial or circumferential directions.

As might be expected, good haemodynamic performances were maintained with an increased period of follow-up. At this early stage (follow-up period between 2 and 3 years) the pericardium does not show any tendency to thicken, fibrose or retract to the extent of causing suboptimal valve function. An interesting aspect that will find an answer with a longer period of follow-up is whether the presence of persistent aortic insufficiency would be responsible for earlier structural deterioration of the prosthesis. It can be assumed that valvular regurgitation is due to incorrect valve sizing, imperfect alignment of leaflets and commissures, malpositioning of suture lines, or dissymmetry of the aortic root. Each of these technical factors would lead to increased mechanical stress on some portion of the valve, potentially reducing its life expectancy.

Concern also remains about the fate of the non-mobile pericardium: if it eventually calcifies, the leaflets will be contained in a rigid tube, a condition very similar to that with pericardial stented valves; this would probably decrease valve durability.

Another potential surgical problem in patients with stentless valves is considered to be valve removal at reoperation. The presence of a cloth covering in some types of STL valves is thought to provide a good plane of dissection that facilitates the procedure. However, in the authors' only case that required reoperation after about 3 years, the pericardial stentless valve, which does not include any prosthetic material, was easily dissected free from the aortic wall. Nevertheless, as the outer pericardial cylinder is extremely thin, it could easily be left in place without significantly reducing the aortic inner diameter.

As with other types of STL valves, the authors' preliminary results with the STL pericardial valve are satisfactory. The next few years will probably answer some of the questions that are now crucial for surgeons dealing with different types of STL valves.

References

1. Binet J P, Duran C G, Carpentier A, Langloid J 1965 Heterologous valve transplantation. Lancet 2: 1275
2. O'Brien M F, Clarebrough J K 1966 Heterograft aortic valve transplantation for human valve disease. Aust Med J 2: 288
3. Angell W W, Oury J H, Lamberti J J, Koziol J 1989 Durability of the viable aortic allograft. J Thorac Cardiovasc Surg 98: 48
4. Sievers H H, Lange P E, Bernhard A 1980 Implantation of a xenogenic stentless aortic bioprosthesis. First experience. J Thorac Cardiovasc Surg 33: 225
5. David T E, Ropchan G C, Butany J W 1988 Aortic valve replacement with stentless porcine bioprostheses. J Card Surg 3: 501
6. David T E, Pollick C, Bos J 1990 Aortic valve replacement with stentless porcine aortic bioprostheses. J Thorac Cardiovasc Surg 99: 113
7. Casabona R, De Paulis R, Zatter G F, di Summa M, Bottone W, Stacchino C, Vrandecic M O P, Morea M 1992 Stentless porcine and pericardial valve in aortic position. Ann Thorac Surg 54: 681–685
8. Frater R W M, Salomon N W, Rainer W G, Cosgrove III D M, Weickham E 1992 The Carpentier–Edwards pericardial aortic valve: intermediate results. Ann Thorac Surg 53: 764
9. Barratt-Boyes B G, Christie G W, Raudkivi P J 1992 The stentless bioprosthesis: surgical challenges and implications for long-term durability. Eur J Cardio-Thorac Surg 6(Suppl 1): 39
10. Christie G W 1992 Computer modelling of bioprosthetic heart valves. Eur J Cardio-Thorac Surg 6(Suppl 1): 95

Chapter 25

O'Brien–Angell stentless valve: early results of 120 implants

U. Hvass

Over the last few years, increasing interest has developed in stentless porcine valves.[1–3] Compared with conventional xenografts, the stentless valves are less cumbersome and should demonstrate improved haemodynamics. They are also expected to offer mechanical features that enhance valve durability.[4] Despite the potential long-term advantages, many questions are still unanswered. Is the valve suitable for all cases? Is the procedure time consuming? Is the single running suture technique efficient and safe and what type of suture material should be recommended? What are the gradients and the incidence of regurgitation? This report on an initial 120 consecutive cases may enable some answers to be formulated.

Materials and methods

Valve

The O'Brien–Angell Cardiovascular Model 300 xenograft (distributed by Cryolife, Atlanta, USA) is a composite design, constructed with non-coronary leaflets obtained from three porcine valves. Leaflets are carefully excised from valves already fixed in glutaraldehyde under very low or near-zero pressure. Individual non-coronary leaflets are matched for size and symmetry to ensure synchronous opening and to promote maximum leaflet coaptation. The matched set of leaflets is sutured together along the free edges of the aortic wall at the leaflet commissures. The base of the valve is finished with a blanket stitch to ensure its integrity.

Patients (Table 25.1)

Since August 1991, the O'Brien-Angell stentless porcine valve has been used in all patients for whom an aortic bioprosthesis has been recommended. There is, therefore, no patient exclusion, and no patient was denied this type of valve. This systematic approach avoided uncertainty, reliance on the surgeon's 'feelings', and decision making in the operating theatre. All patients were operated on by the same surgeon. During the 2-year period reported, only one patient (80 years old) could not have the expected valve implanted owing to a calcified aorta; in this case the chest was closed and no surgery performed.

Table 25.2 shows the distribution of age groups. The largest group was of elderly patients and is correlated with the incidence of calcified aortic valve stenosis.

The diagnostic groups were as described below.

Table 25.1 Diagnostic groups

Diagnosis	No. of patients
Calcified aortic stenosis	54
Calcified aortic stenosis with associated lesions	(22)
Coronary	17
Mitral valve	2
Mitral valve and coronary	1
Subaortic septal obstruction	2
Aortic insufficiency	16
Aortic insufficiency with associated mitral lesions	13
Redo operations	(15)
Deterioration of aortic prosthesis	5
Aortic and mitral prosthesis	8
Aortic and coronary	1
Congenital	1

Table 25.2 Patient age groups

Age (years)	Percentage of patients
<40	15.0
40–60	16.6
60–70	24.1
>70	44.1

Isolated calcified aortic valve stenosis

A total of 54 patients presented a calcified aortic stenosis as the only cardiac complaint. Their ages ranged from 51 to 88 years; 25 were more than 75 years old. There were 29 male and 25 female patients.

Calcified aortic valve stenosis with associated cardiac lesions

A total of 22 patients had aortic valve stenosis with associated cardiac lesions. Coronary bypass with one to three grafts was necessary in 17 patients. The left internal mammary artery was always used on the left anterior descending artery. Two patients had mitral insufficiency and the valve was replaced with a bioprosthesis. One patient had mitral- and coronary-associated surgery. Two patients had an obstructive subaortic septal hypertrophy that was resected. Patients were 42–82 years old; 12 were more than 75 years old.

Isolated aortic insufficiency

There were 16 patients in this group, four of whom had endocarditis. The patients' ages ranged from 20 to 65 years.

Aortic insufficiency with associated cardiac lesions

In this group of 13 patients with mitral insufficiency, in two cases the mitral valves were repaired: in the other patients, the valves were replaced with a bioprosthesis. Three of the patients had aortic valve endocarditis.

Redo operations

In this group of 15 patients there were five cases of isolated redo aortic procedures, of which three were for deterioration of bioprostheses. One was for a thrombosed aortic valve and one was for endocarditis associated with a mechanical valve in a patient with mediastinitis. Eight patients underwent surgery for aortic and mitral degenerated bioprostheses. One patient had developed a calcified aortic stenosis 5 years after a coronary bypass procedure. The last patient in this group required reoperation for a congenital subaortic tunnel with aortic insufficiency.

Endocarditis

In the entire group of 120 patients, nine had endocarditis; in six of these cases the endocarditis was active with an aortic annular abscess (for which the patients had received antibiotics for less than 48 h preoperatively) and with positive blood cultures. The haemodynamic status of these patients was poor, justifying early operation. The infection was fungal in two patients, staphylococcal in one, streptococcal in one and rickettsial in one.

Operative technique

In all operations, the patient was placed on normothermic cardiopulmonary bypass maintaining the central temperature at 37°C. Myocardial preservation was achieved by antegrade cold blood cardioplegic infusion (700–1000 ml) through the coronary ostia. Repeat infusions of 300 ml were given every 30 min. Pericardial cooling was not used. Before the aortic clamp was released, a warm blood cardioplegic infusion (200 ml/min for 5 min) was delivered through the aortic wall; as a rule, spontaneous cardioversion occurred. The aortic incision was wide, extending to the non-coronary sinus. After excision of the valve and decalcifying the annulus, the aortic orifice was measured with a Hegar probe. The valve chosen was 1–2 mm larger than the orifice; it was rinsed in saline, like any other bioprosthesis, and then lowered to the aortic annulus and the commissures correctly positioned. Correspondence between the patient's annulus and the sewing area of the prosthesis was very clear. Starting with the right coronary sinus, a stitch was placed in the central portion and tied, thus allowing for moderate traction on the stentless valve to facilitate the supra-annular running suture. The same approach was made with the left, then the non-coronary sinus. Sutures were made with a 4/0 or 3/0 Prolene 18 mm semicircular needle.

Valve size (Table 25.3)

As described above, the aortic annulus was measured with a Hegar probe, which gives a very precise measurement, and a valve 1–2 mm larger was selected. This allowed for correct supra-annular positioning of the xenograft and avoided protrusion of any non-valvular tissue into the outflow tract. In the series reported here, the majority of valves were 23, 25 or 27 mm in size. Only four 21 mm valves were used, corresponding to an annulus of 18–19 mm.

Table 25.3 Sizes of valves used

Valve size (mm)	Percentage of patients
21	3.3
23	26.6
25	30.0
27	23.3
29	16.6

Special features (Table 25.4)

A bicuspid valve was found in nine cases; this was not considered a contraindication for a stentless valve. The mean age of patients with bicuspid valves was 76 years. An ascending aortic patch was used in 13 cases to enlarge the non-coronary sinus of Valsalva; the mean age of these patients was 37 years. Six patients presented an aortic annular abscess that had been diagnosed preoperatively by transoesophageal echography.

Table 25.4 Special pathological features

Feature	No. of patients
Bicuspid valve	9/120
Enlargement of ascending aorta	13/120
Aortic annular abscess	6/120

Is the procedure time consuming? (Figure 25.1)

There is concern that using a stentless porcine valve may prolong aortic crossclamp time and, consequently, duration of extracorporeal circulation, to an extent that may predispose to myocardial damage in multiple cardiac procedures. In cases of isolated aortic calcified valvular stenosis, using the single running suture technique, aortic crossclamp time ranges from 32 to 70 min with a mean of 39 min. This includes antegrade cardioplegia via the coronary ostia, excision of the valve and meticulous decalcification of the patient's annulus, insertion of the prosthesis, closure of the aorta, and 5 min warm blood reperfusion before unclamping. The time devoted exclusively to insertion of the valve varies from less than 20 to 30 min. The aortic crossclamp time in cases of calcified aortic stenosis (AS) with associated lesions varies from 50 (AS and subaortic assymmetric hypertrophy) to 100 min (AS with mitral valve replacement and two coronary bypasses), with a mean aortic crossclamp time of 65 min; in redo operations, the aortic crossclamp time ranges from 70 to 120 min with a mean of 80 minutes. In the authors' institution, therefore, the aortic crossclamping time has always been considered to be within a safe range.

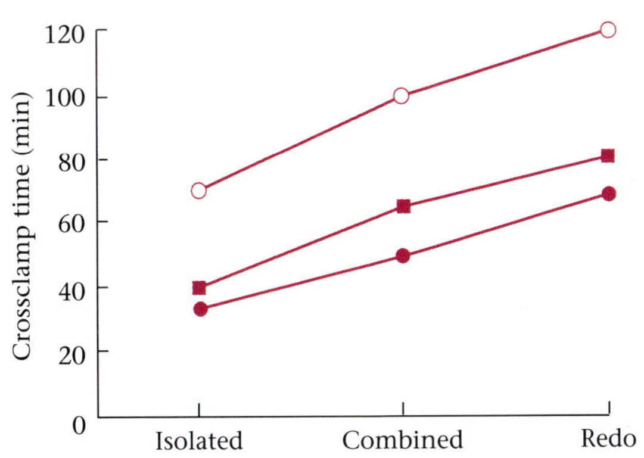

Figure 25.1. Aortic crossclamp time (●, minimum; ○, maximum; ■, mean) in patients receiving stentless porcine bioprostheses as isolated, combined or repeat ('redo') procedures.

Results

Operative mortality

There were no valve-related deaths. The 16 perioperative deaths were due in some cases to poor preoperative status, such as that encountered in valvular prosthetic endocarditis (two cases) and in emergency operations for critical aortic stenosis (two cases) or for thrombosis of a mechanical valve (one case). Other causes were intercurrent complications in the elderly, such as respiratory infections (four cases), renal insufficiency (three cases), multiple organ failure (two cases) and postoperative cholecystitis (one case). Of the six patients with active endocarditis and an associated aortic annular abscess, five survived, with no recurrence of infection; the sixth patient died at 3 weeks, after rupture of a splenic aneurysm (mycotic endocarditis).

Haemodynamics

In 100 patients, the transvalvular gradients and the presence of regurgitation were monitored during their hospital stay, or within 3 months of operation, by Doppler echocardiography. The extent of the leak into the left ventricle was recorded in the parasternal long axis, and the width of the leak compared with the diameter of the left ventricular outflow tract was assessed by M-mode colour Doppler, monitoring flow in the aortic arch and abdominal aorta. The regurgitation area was not evaluated in the parasternal short axis. For the 1-year follow-up, 50 of the 54 survivors living in France were available for assessment using the same parameters and the continuity equation to calculate the effective orifice area.

Early postoperative gradients (Table 25.5)

Early gradients not only are related to the valve but also are influenced by ventricular dynamics, subaortic muscular obstruction in hypertrophied hearts and by residual postoperative haemodilution and augmented cardiac output. Table 25.5 shows the distribution of early postoperative gradients, which were most frequently in the low range, being below 15 mmHg in 68% of cases. A small proportion of patients exhibited mean gradients about 20 mmHg, with no obvious correlation to valve size. On the other hand, lower gradients were more often encountered in the 27 and 29 mm valves.

Table 25.5 Early postoperative mean transvalvular gradients (100 patients)

Gradient (mmHg)	Percentage of patients
<10	35
10–15	33
15–20	24
>20	8

Mean transvalvular gradients after more than 1 year (Table 25.6)

In 50 of the 54 patients whose follow-up period exceeded 1 year, a greater percentage exhibited a mean gradient below 15 mmHg (82%, compared with 68% at the early postoperative assessment).

Table 25.6 Mean transvalvular gradients in 50 patients with follow-up period exceeding 12 months

Gradient (mmHg)	Percentage of patients
<10	44
10–15	38
15–20	12
>20	6

Incidence of early central regurgitation (Table 25.7)

There was no incidence of central regurgitation in 62 patients. A very localized central diastolic image was observed in 29 cases, in the form of a small cluster of coloured dots immediately below the closed leaflets; this was defined as microscopic regurgitation. Trivial to mild regurgitation was present in eight cases.

Patch enlargement of the ascending aorta did not have any adverse effect on the occurrence of trivial central regurgitation (1/13). The presence of a bicuspid valve gave rise to a slightly higher incidence of central regurgitation (one of nine cases). Of the six patients with an annular abscess, four had trivial central and three presented trivial peripheral regurgitation, arising from the anatomical abnormality and the fragility of the tissue in such cases. Patients receiving the larger-sized valves also showed a higher incidence of trivial central regurgitation (four of 20 cases). The only patient with a mild to moderate isolated central insufficiency had received a 27 mm valve to fit a 27 mm annulus; this underlines the importance of slightly oversizing the prosthesis when it is placed in the supra-annular position.

Table 25.7 Early postoperative (0–3 months) central regurgitation (100 patients)

Regurgitation	Percentage of patients
None	62
'Microscopic'	29
Trivial–mild	8
Mild–moderate	1

Incidence of early peripheral regurgitation (Table 25.8)
There was no evidence of any diastolic signal in 64 patients. A microscopic perivalvular leak was traced in 26 and a trivial to mild leak in seven. In three patients, mild to moderate perivalvular regurgitation was demonstrated in the first 3 months.

Table 25.8 Early (0–3 months) postoperative peripheral regurgitation (100 patients)

Regurgitation	Percentage of patients
None	64
'Microscopic'	26
Trivial–mild	7
Mild–moderate	3

Isolated central regurgitation at 12 months
The results of echographic studies performed 12 months after the patients' discharge did not differ from those of early postoperative studies.

Peripheral regurgitation at 12 months
The results were unchanged except for those in the patients who initially demonstrated mild to moderate peripheral regurgitation, two of whom developed moderately severe perivalvular regurgitation, accompanied by moderate to moderately severe central regurgitation.

Effective orifice area (EOA) (Table 25.9)
The EOA, measured by the continuity equation, is very dependent on precise measures of the subaortic diameter: wide variations in EOA will occur with only small differences in outflow tract estimation. Precise transthoracic measurements may not be possible in elderly patients with calcified costosternal cartilages. Transoesophageal echography is not always acceptable for routine monitoring. There is also a relation between the smaller valve sizes and old age, calcified aortic stenosis, calcified outflow tracts and mitral valves and female gender of the patients. The mean values found for each valve size are listed in Table 25.9.

Table 25.9 Effective orifice area (EOA; 50 patients)

Valve size (mm)	EOA (cm^3) Mean	Minimum	Maximum	No. of patients
21	1.25	1.1	1.36	3
23	1.34	1.1	1.50	12
25	1.29	0.8	1.45	12
27	1.65	1.4	2.10	13
29	1.84	1.5	2.40	10

Reoperations (Table 25.10)
Two patients required reoperation for peripheral regurgitation at 10 and 18 months, respectively. In one case this was due to rupture of a 4/0 Prolene suture in a patient who had initially a perfect result. The echocardiographic diagnosis in this case had been of massive central regurgitation. The error is understandable: a peripheral leak on a stented valve does not change the configuration of the leaflets and there will be no central leak, whereas when a stentless valve is not correctly held on the peripheral annulus the whole architecture of the valve and the central coaptation is unbalanced. The second reoperation was for a slack 4/0 Prolene running suture; the preoperative diagnosis was of peripheral and central regurgitation. In both cases the leaflets were normal and refixing the valves on the annulus would have been sufficient. Both patients received mechanical valves.

Table 25.10 Explanted xenografts

Valve size (mm)	Age of patient (years)	Diagnosis	Delay (months)	Cause
29	65	Aortic stenosis	10	Prolene 4/0 ruptured
27	43	Aortic insufficiency	18	Prolene 4/0 slack

Discussion

The majority of patients, and certainly of young patients, in the author's institution receive a mechanical prosthesis for aortic or mitral valve replacement. Aortic bioprostheses are mostly used in patients over 70 years of age or in younger patients for whom the referring cardiologist has contraindicated anticoagulants. In the series of cases described here, young patients needing a bioprosthesis received the porcine stentless valve because homografts were not available. The present policy, however, now that homografts are available, is to perform a pulmonary autograft in patients under 45 years of age, to use aortic homografts (root replacement technique) in patients aged between 45 and 65 years, and to use the porcine stentless valve in patients aged over 65 years. In the author's latest 30 cases of patients receiving porcine stentless valves (cases 101 to 130), 73.3% are over 70 years old.

The higher than normal mortality rate correlates closely with the patients' preoperative status. Elderly patients in a poor general state of health, who would have been denied surgery when this study was initiated, have become, for the cardiologists, an 'acceptable risk'. Mortality rates for valvular surgery in the author's institution during the study period (1991 and 1992) were 2.6 and 3.05% respectively.

Why use a stentless valve? The durability of bioprostheses is affected by biological and mechanical factors.[4] Mechanical stress is an important determinant of valve failure.[2] From being rigid, the frames of bioprostheses have become more flexible[5] in order to absorb the mechanical stress sustained by the valve leaflets. However, the aortic root itself is probably the best flexible stent for an aortic valve. Unstented homografts have superior results when compared with stented ones.[6] It is therefore hypothesized that a stentless porcine valve should last longer than a stented valve. Although there is no long-term evidence that the glutaraldehyde-fixed stentless porcine valve will give superior results, enhanced longevity of the valve is expected.

At present, these long-term considerations are weighed together with the immediate concern of the feasibility of everyday implantation. The series of patients described here has shown that the O'Brien–Angell stentless porcine valve can be used in a great variety of clinical settings, no patient having been denied this type of xenograft since the series started in August 1991. With the recommended technique involving one running suture, aortic crossclamp time does not differ from that with conventional valves. Transvalvular postoperative gradients are usually in the low range, and have fallen further when evaluated after 6 months. There has been no significant central regurgitation and no patient has needed reoperation for valvular deterioration. Central regurgitation is not a problem in these valves, owing to a high coaptation surface, whereas significant perivalvular regurgitation due to technical errors is of concern and will lead to leaflet dysfunction. Two patients required reoperation 10 and 18 months after implantation. In the author's opinion, the Prolene 4/0 suture is not sufficiently strong: several have snapped during the surgical procedure, necessitating repetition of the running suture in the sinus of Valsalva. As this suture is relatively fragile, it may not always be tightened correctly, leading to a slack suture. These reoperations may be considered as being associated with the learning curve. Only 3/0 Prolene sutures are used now, and no more cases of significant peripheral regurgitation have been encountered.

Careful evaluation of the O'Brien–Angell stentless porcine valve will continue in the author's institution.

References

1. David T E, Pollick C, Boss J 1990 Aortic valve replacement with stentless porcine aortic bioprosthesis. J Thorac Cardiovasc Surg 99: 113–118
2. Drury P J, Dobrin J, Bodnar E, Black M M 1986 Distribution of flexibility in porcine aortic root and in cardiac frames. In: Bodnar E, Yacoub M (eds) Biological prosthetic valves — Proceedings of the Third International Symposium. Yorke Medical Books, New York, pp 580–587
3. O'Brien M F, Clarebrough J K 1966 Heterograft aortic valve transplantation for human valve disease. Med J Aust 2: 228–230
4. Broom N D 1978 Fatigue induced damage in glutaraldehyde preserved heart valve tissues. J Thorac Cardiovasc Surg 76: 202–211
5. Reis R L, Hancock W D, Yarbrough J W, Glancy D L, Morrow A G 1971 The flexible stent: a new concept in the fabrication of tissue heart valve prostheses. J Thorac Cardiovasc Surg 62: 683–689
6. Angell W W, Oury J H, Lamberti J J, Koziol J 1989 Durability of the viable aortic allograft. J Thorac Cardiovasc Surg 98: 48–56

Chapter 26

Early experience with the Labcor stentless porcine aortic bioprosthesis

J. M. De Souza and S. A. Oliveira

The durability of a bioprosthesis can be affected by preservation and fixation methods. The durability and performance of the valve can also be affected by a rigid stent. Bioprostheses have been implanted in many patients, mainly because of the low incidence of acute complications associated with their use, and the low risk of thromboembolism, even without anticoagulants. This is a very important consideration in countries such as Brazil, where it is very difficult to monitor patients who live far from the cities. Long-term anticoagulant therapy impairs the quality of life and is a risk factor even in developed countries.[1] The main problems associated with bioprostheses are early deterioration in children and adolescents, and unacceptable transvalvular gradients when small valves are used;[2–4] in addition, the necessity for a stent reduces the effective prosthesis area.[5] Such modifications as a supra-annular valve[4] and flexible stent,[6] together with changes in fixation and preservation techniques, have resulted in some improvement.

The excellent long-term results with the aortic homografts introduced by Ross,[7] Barratt-Boyes[8] and Binet *et al.*,[9] and reported experience with stentless aortic valve heterografts,[3] encouraged the authors to use stentless porcine aortic valves as replacements for diseased human aortic valves.

Patients

Informed consent was obtained from all 26 patients before operation. The mean age was 51.8 years, with a range of 14–71 years. Sixteen patients were male. Six patients were in New York Heart Association functional class II, 17 were in class III and three were in class IV. The indications for aortic valve replacement are listed in Table 26.1. Associated operations were performed in 12 patients, as shown in Table 26.2.

Table 26.1 Preoperative aortic valve lesions in 26 patients receiving stentless porcine aortic bioprostheses

Indication	No. of patients	Percentage of patients
Aortic stenosis	11	42.3
Aortic insufficiency	7	26.9
Aortic insufficiency and stenosis	6	23.1
Bioprosthesis failure	2	7.7

Table 26.2 Associated operations in addition to implantation of stentless porcine aortic bioprostheses

Associated operations	No. of patients
Mitral valve replacement	5
Mitral valve commisurotomy	1
Myocardial revascularization	3
Left ventricle aneurysmectomy	1
AAA* resection	2

*AAA, aneurysm of ascending aorta

Preparation of bioprostheses

The stentless bioprostheses were manufactured by Laboratorio Labcor (Belo Horizonte, Brazil). The aortic valves were removed from pigs killed in a large slaughterhouse (8000 pigs killed daily), thus enabling valves in better condition to be selected. As soon as possible after removal, the valves were placed in a suitable medium for transportation to the laboratory. The 'tanning' process was started within 6 h, fixation being at zero pressure with 0.4% glutaraldehyde followed by 0.2% glutaraldehyde. The tanning process lasted at least 4 months, after which the bioprostheses were assembled using three similar-sized leaflets, and were then sterilized and preserved in 0.4% glutaraldehyde (Figure 26.1).

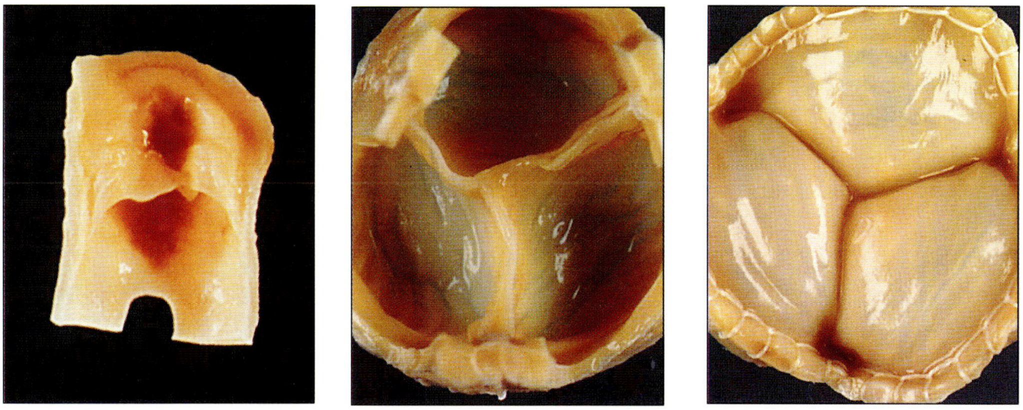

Figure 26.1. Labcor stentless porcine aortic bioprosthesis: (a) three component leaflets; (b) assembled valve showing leaflet coaptation; (c) assembled valve showing inflow area.

Operative procedure

In 24 patients the operations were performed using cardiopulmonary bypass and moderate hypothermia (nasopharyngeal temperature 30°C). In two patients undergoing associated resection of an aneurysm of the ascending aorta, hypothermia (25°C) was used. Buckberg cardioplegic solution was infused directly into the coronary ostia. The aortic valve was exposed by an oblique incision extending to the non-coronary sinus. After excision of the diseased valve, the stentless bioprosthesis was sewn to the aortic annulus using an interrupted 4/0 multifilament polyester suture. A second layer of continuous suture was done on the aortic wall, using 4/0 polypropylene (Figure 26.2). The aortotomy was closed directly in 22 patients; in two patients the aorta was enlarged with a pericardial patch, and a pericardial tube (Labcor) was used to replace the aorta after aneurysm resection in two other patients. The mean cardiopulmonary bypass time was 85 min and the mean crossclamp time was 61 min.

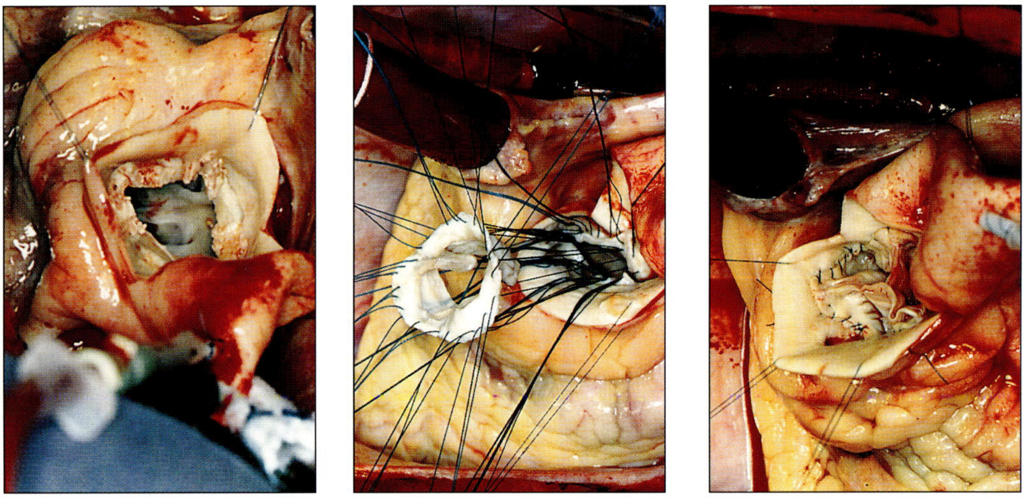

Figure 26.2. Technique of implantation of the Labcor stentless porcine aortic bioprostheses.

The diameters of the stentless bioprostheses are shown in Figure 26.3

Results

One patient developed severe and acute valve insufficiency on the third postoperative day. An echo-Doppler study revealed central valve insufficiency and the patient underwent emergency surgery. The suture on the aortic wall corresponding to the left coronary sinus was found to have ruptured because of the extreme fragility of the aorta. The 25 mm stentless bioprosthesis was resected and successfully replaced by a 23 mm (external diameter) stented bioprosthesis. Another patient developed atrial fibrillation followed by ventricular fibrillation on the day after operation. He was resuscitated but remained comatose until his death on the 20th postoperative day. An echocardiographic study of this patient after resuscitation revealed a transvalvular gradient of 20 mmHg.

One patient who underwent surgery in April 1993 (mitral valve replacement with a stented bioprosthesis in addition to implantation of an aortic stentless bioprosthesis) developed endocarditis of the mitral bioprosthesis and was reoperated on in August 1993. The aortic bioprosthesis was not infected. This

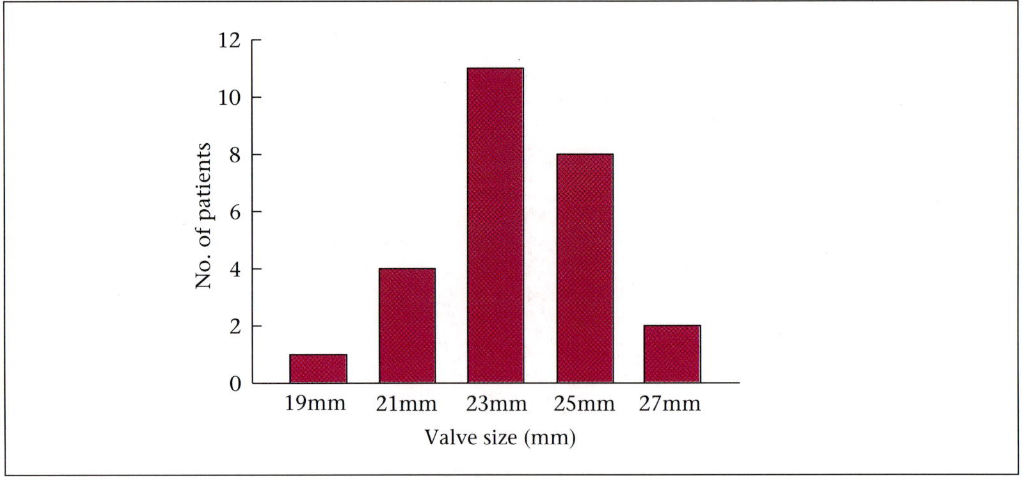

Figure 26.3. Diameters of bioprostheses implanted in 26 patients.

patient died from septicaemia on the 20th postoperative day. The remaining 23 patients had an uneventful early and intermediate (mean 4 months) postoperative course.

Echocardiographic analysis of all patients showed an average transvalvular gradient of 14.6 mmHg (range 4–20 mmHg). The early echocardiographic analysis showed the presence of trivial aortic insufficiency in five patients; this had no clinical effect and so far has not developed further. In another patient without insufficiency at the early postoperative assessment, examination by Doppler echocardiography 3 months after operation revealed newly developed but haemodynamically insignificant regurgitation. Oral anticoagulants were not administered and there were no thromboembolic episodes.

Discussion

Elimination of the stent not only increases the effective area of the valve but also permits insertion of a larger bioprosthesis. Only a 23 mm stented bioprosthesis could be implanted in the patient who required reoperation because of early insufficiency of a 25 mm (outside diameter) stentless valve. According to David *et al.*,[3] a 23 mm stented bioprosthesis corresponds to a 21 mm stentless bioprosthesis; thus there was a real decrease in valve diameter of 4 mm. Although implantation of a stentless valve is more difficult than that of a stented valve, and increases the cardiopulmonary and aortic crossclamp time, the surgical risk is not altered if adequate myocardial protection is provided. The durability of stentless aortic valve homografts is longer (average 12 years) than when they are mounted in a stent (average 8 years).[10] The 8-year average duration of stented porcine aortic valves has been established;[1,11] a longer duration is expected, therefore, for stentless porcine aortic valves.[3,10] The stress on the leaflets is reduced by attaching the bioprosthesis to the aortic wall, which means that there is less risk of rupture or calcification of the porcine aortic valve.[3,12,13] Reduction of blood turbulence by a larger effective orifice area of the stentless bioprosthesis can minimize the incidence of calcification.[4,14]

Carpentier has reported a patient survival of 17 years after implantation of a stentless bovine aortic valve bioprosthesis in 1968 (cited in reference 3). The implantation of a stentless aortic valve bioprosthesis may present some technical problems such as incorrect cusp orientation, or incorrect valve size because of disproportion between the annulus and the aortic root. A stentless valve must not be implanted in a patient with a friable aortic wall.

The early and medium-term results of implantation of this valve are satisfactory and it is expected that elimination of the stent will also improve the long-term outcome.

References

1. Cohn L H, Alfred B A, Di Sesa V J, Sawtelle K, Shenin R S, Collins J J 1984 Early and late risk of aortic valve replacement. J Thorac Cardiovasc Surg 88: 695–705
2. Pelletier C, Chaitman B R, Baillat R, Val P G, Bonan R, Dyeda I 1982 Clinical and haemodynamic results with Carpentier–Edwards porcine bioprosthesis. Ann Thorac Surg 34: 612–624
3. David T E, Pollick C, Bos J 1990 Aortic valve replacement with stentless porcine aortic bioprosthesis. J Thorac Cardiovasc Surg 99: 113–118
4. Carpentier A, Dubost C, Lane E, Nashef A, Carpentier S, Relland J, Deloche A, Fabiani J N, Chauvand S, Perier P, Maxwell S 1982 Continuing improvements in valvular bioprosthesis. J Thorac Cardiovasc Surg 83: 27–42

5. Morris D C, Wickliffe C W, King S B 1976 Haemodynamic evaluation of the porcine heterograft aortic valve. Am J Cardiol 37: 157–161
6. Reis R L, Hancock W D, Yarbrough J W, Glancy L, Morrow A G 1971 The flexible stent: a new concept in the fabrication of tissue heart valve prosthesis. J Thorac Cardiovasc Surg 52: 683–689
7. Russ D N 1962 Homograft replacement of the aortic valve. Lancet 2: 487
8. Barratt-Boyes B G 1964 Homograft aortic valve replacement in aortic incompetence and stenosis. Thorax 19: 131–150
9. Binet J P, Carpentier A, Lauglois J, Duran C, Calvez P 1965 Implantation de valves hétérologues dans le traitment de cardiopathies aortiques. C R Acad Sci (III) 261: 5733–5741
10. O'Brien M F, Clarebrough J K 1966 Heterograft aortic valve transplantation for human valve disease. Med J Aust 2: 228–230
11. Angell W W, Angell D D, Ohri J H, Lamberti J J, Grehl T M 1987 Long-term follow-up of viable frozen aortic homografts: a viable homograft bank. J Thorac Cardiovasc Surg 93: 815–822
12. Magiligan D J, Kemp S R, Stein P P, Peterson E 1987 Asynchronous primary valve failure in patients with porcine bioprosthesis aortic and mitral valves. Circulation 76(2): 141–145
13. Thubrikar M J, Deck D, Aouad J, Nolan S 1983 Role of mechanical stress in calcification of aortic bioprosthesis valves. J Thorac Cardiovasc Surg 86: 115–125
14. Pillai R, Spriggings D, Amarasena N, O'Regan D J, Parry A S, Westaby S 1983 Stentless aortic bioprosthesis? The way forward: early experience with the Edwards valve. Ann Thorac Surg 56: 88–91

Chapter 27

Porcine stentless heart valve substitutes, technical and clinical data: short- and mid-term follow-up

M. Vrandecic, B. F. Gontijo, F. A. Fantini, O. C. Oliveira, I. C. J. Martins, M. H. C. Oliveira, S. O. S. Avelar, E. A. Vrandecic and E. Vrandecic

The principle of using anatomically near-identical biological heart valve substitutes can be termed the 'stentless concept'. Currently, it has become an important option in heart valve replacement surgery, and represents a more 'natural' way of treating diseased heart valves.

The advantage of these new biological heart valve designs does not lie only in their superior valve function: the absence of a rigid stent may also enhance valve durability, as it is well known that a rigid support increases the mechanical stress on the tissue, particularly at the commissures.

As heart valves have a specific function and withstand pressure differences, their anatomical shape and form must be respected in order to obtain good function and longer durability.

The porcine aortic stentless valve is constructed from three selected aortic porcine leaflets, attached to a scalloped shaped pericardial patch which thus provides an upper and lower sewing ring. The porcine mitral stentless valve is tailored in a way that provides the annular valve and papillary continuity appropriate for the specific flow and left ventricular function.

This chapter reports the early and mid-term clinical follow-up of patients in whom Biocor (Biocor Industria e Pesquisas Ltda, Belo Horizonte, Brazil) porcine aortic and mitral stentless heart valves were implanted.

Patients and methods

Patients

After ethical committee approval, a prospective clinical trial of the aortic (Figure 27.1) and mitral (Figure 27.2) stentless valves was started in May 1990 and March 1992, respectively. Informed consent was obtained from all patients. A total of 120 patients underwent aortic valve replacement and 65 had their mitral valve replaced.

The aortic group included 83 male and 37 female patients, ranging in age from 11 to 76 years with a mean of 36 years. There were 85 patients under 40 years of age. The indications for aortic valve replacement were rheumatic heart disease sequelae in 64, congenital aortic bicuspid valve in 11, native aortic valve endocarditis in six, degenerative aortic valve disease in five and senile calcific aortic valve in four. There were 30 reoperations, of which 20 were necessitated primarily by prosthetic tissue failure and 10 by the presence of prosthetic valve

Figure 27.1. Porcine aortic stentless valve.

Figure 27.2. Porcine mitral stentless valve.

endocarditis. In 33 patients the aortic annulus was calcified, narrowed or infected.

The mitral group comprised 42 female and 23 male patients, ranging in age from 4 to 65 years with a mean of 26 years. Forty patients were less than 40 years of age. Indications for surgery were rheumatic heart disease in 50, native mitral valve endocarditis in six and myxomatous degeneration in four patients; five patients required replacement of a prosthetic heart valve owing to tissue deterioration.

The functional class (NYHA) for the aortic group included 61 in class III and 59 patients in class IV. The longest follow-up was 42 months with a mean of 26 months. In the mitral group, 35 patients were in class III and 30 in class IV. The longest follow-up was 21 months with a mean of 12 months.

Operative technique

The surgical technique for implantation of aortic and mitral stentless valves has been described elsewhere.[1–4] There were 77 associated procedures in the aortic group: these comprised aortic root enlargement in 47, mitral valve replacement in 15 second mitral valve replacement in four, mitral valvuloplasty in four, myocardial revascularization in five and mitral commissurotomy in two patients.

There were two associated procedures in the mitral group — myocardial revascularization in one patient and aortic commissurotomy in another. Twelve patients were operated on for the second time.

Patient follow-up

All patients were followed up at 3-month intervals. During the follow-up as well as the clinical examination of patients, valve function was assessed using colour flow Doppler echocardiography and routine laboratory blood tests.

Results

Of the 120 patients who underwent aortic valve replacement with the Biocor aortic stentless valve, six died in hospital. These deaths were not valve related: the hospital mortality was due to progressive organ failure related to poor preoperative status in two patients and to rhythm disturbance in three — two because of ventricular tachycardia and one while awaiting implantation of a permanent pacemaker. Mediastinitis was the cause of death in the sixth patient.

Complications occurred in 14 patients of the aortic group during their hospital stay: nine developed atrioventricular rhythm disturbance; of those, four returned to regular sinus rhythm and five required a permanent pacemaker. Revision of haemostasis was required in four patients and one patient suffered an acute cerebrovascular accident which probably was secondary to air emboli associated with extracorporeal circulation. All 14 patients recovered well and were discharged from the hospital.

In the mitral group, two of the 65 patients died in hospital: in one the cause of death was irreversible coma due to previous septic emboli; the second death was caused by intravascular coagulopathy due to congestive liver failure.

Hospital morbidity in the mitral group involved revision of haemostasis in one patient and replacement of a mitral valve prosthesis in another patient 2 weeks after the original implantation, because the wrong size of mitral stentless valve had been used.

With regard to late mortality in the aortic group, four of the 114 remaining patients died: one death was from pancreatic carcinoma 3 months after surgery and one was due to recurrent fungal (*Candida albicans*) endocarditis 2 months after surgery; one patient sustained a hard blow on the pacemaker site while playing football 6 months after surgery and the fourth died from acute fulminating pulmonary emboli, 6 months after operation.

In the mitral group there were two late deaths: the first was due to liver failure and pneumonia secondary to pancreatic carcinoma 11 months after surgery; the second patient, who had undergone concomitant myocardial revascularization, died after an acute myocardial infarction, 8 months after surgery. All patients who died at the late follow-up stage underwent post-mortem examination.

Reoperations

There were four reoperations in the aortic group: two were secondary to valve endocarditis, in patients who had first received surgery during the active disease; the other two reoperations were necessary because of partial dehiscence of the annular suture, probably through faulty technique.

In the mitral group there were three late reoperations, two due to recurrent endocarditis and the third because of partial annular dehiscence attributable to the technique used.

In all patients undergoing reoperation the valve was replaced with a Biocor standard No-React™ valve; all patients made a good recovery from this second valve replacement.

Most of the aortic patients were in functional class I — 99 patients or 90% — nine patients or 8% in class II and two patients or 2% in class III.

Currently there are 58 mitral patients being followed: (49) patients in functional class I (84.5%) and nine patients in class II (15.5%).

Long-term follow-up

Currently, 106 patients in the aortic group and 58 in the mitral group are being followed up.

Clinical and laboratory examination has revealed a satisfactory degree of recovery, in all patients.

Colour Doppler echocardiography has provided evidence of a competent aortic valve in 96% of patients; in 4% a trivial to mild degree of regurgitation was detected with no haemodynamic disturbance. Of the valves implanted, 64% were either 23 or 25 mm in diameter, with 8, 26 and 2% measuring 21, 27 and 29 mm, respectively. The peak transvalvular gradient ranged from 10 to 24 mmHg and the mean gradient was 10.5 mmHg. Five patients underwent cardiac catheterization, revealing competent aortic valves, good ventricular function and zero gradients (Figure 27.3).

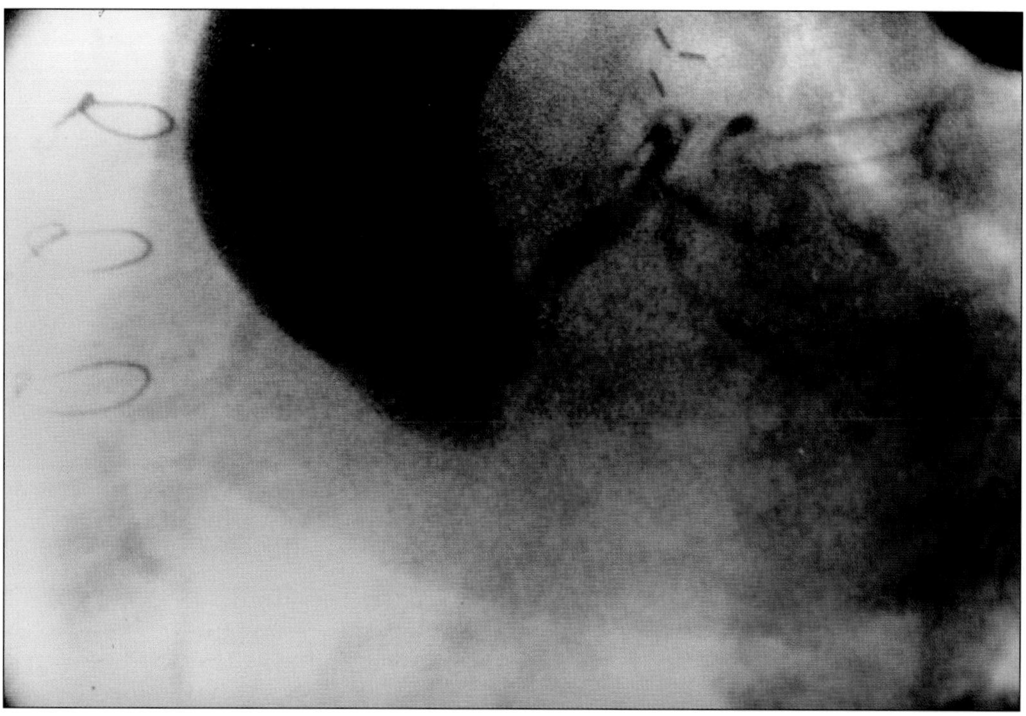

Figure 27.3. Postoperative cardiac catheterization shows competent aortic stentless valve.

In the mitral group, colour Doppler echocardiography revealed competent valves in 86%; in only 10% of patients was there a minor degree of regurgitation seen by echo and in 4% there was a mild insufficiency, also seen by echo. There was a progressive improvement of the left atrial size and left ventricular contractility. The valve sizes used most frequently were 27 and 29 mm. The valve effective orifice area measured by echo-Doppler ranged from 2.3 to 4.0 cm^2 with a mean of 2.6 cm^2. Five patients who underwent postoperative cardiac catheterization displayed excellent left ventricular contractility, zero transvalvular gradients and good performance of the stentless mitral valve (Figure 27.4).

No patient in either the aortic or the mitral group has subsequently shown any increase in the degree of insufficiency detected by echocardiography: minor regurgitation has either disappeared or remained stable.

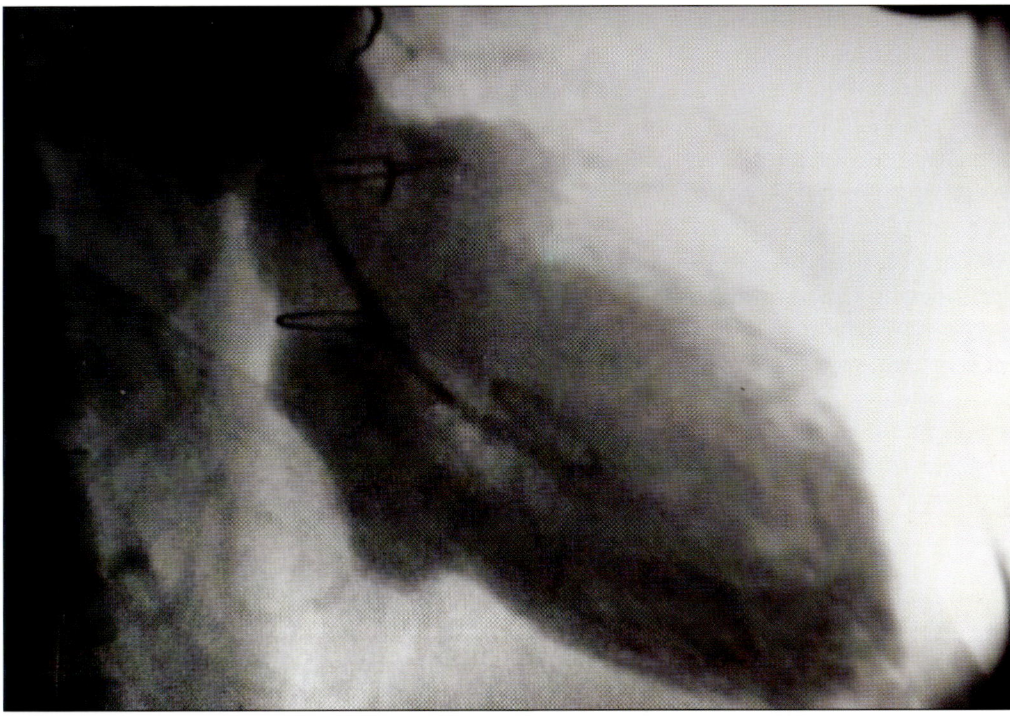

Figure 27.4. Postoperative cardiac catheterization shows competent mitral stentless valve.

Comments

The principle of using a tissue valve that is anatomically similar to a native valve in design aims to improve the performance of the prosthesis by obviating the well-known drawbacks of the heart valve substitutes currently available. A recent paper by Hammermeister[5] that found similar rates of survival and complication rate for both mechanical and bioprostheses shows that neither of the current models can be considered ideal: the search for new, better-performing and more durable heart valve substitutes must therefore continue.

Results of long-term follow-up of patients who have been given the homograft valve as an aortic heart valve substitute are encouraging; although this valve is not always available in all sizes and has shown a relatively significant degree of late valve insufficiency, it is the model that naturally fits the aortic root better. With this in mind, the porcine aortic stentless valve was developed to address the problems of availability and late insufficiency, and to overcome the technical difficulties encountered with use of the homograft in general.

With regard to the mitral stentless valve, Carabello[6] stated: 'The ejection fraction is always maintained or improved whenever annulo-ventricular integrity is preserved or restored'. In this sense the surgical technique using the mitral stentless valve provides the means for better durability and longevity.

The present results with the aortic stentless valve have proved the validity of this new design and the reproducibility of the technique, with excellent follow-up results over 3 years. The rates of mortality and morbidity are acceptable, taking into consideration the advanced stage of the disease in many patients. The atrioventricular disturbances encountered also reflect preoperative annular involvement; in the last 40 patients treated there was no incidence of atrioventricular conduction disturbance. The four reoperations necessary (two because of recurrent endocarditis) underlined the importance of selecting the appropriate suture technique for each individual annulus.

Current clinical results in the mitral stentless patients are excellent and, again, have confirmed the validity of the concept, the reproducibility of the surgical technique and the excellent performance of the valve, in reducing atrial chamber size and improving left ventricular performance. The low rates of hospital mortality and morbidity, in spite of the advanced stage of disease in many patients, reflects favourably on the concept, design and surgical technique. The early reoperation emphasized the importance of choosing the correct valve size. The late reoperations in two patients were secondary to recurrent endocarditis; in the fourth case, in which partial annular dehiscence necessitated reoperation, the use of interrupted sutures may prevent this complication, in future.

The results obtained on echocardiography and heart catheterization have borne witness to the quality of the valve design and surgical procedure described in this chapter.

Both aortic and mitral stentless valves have provided excellent clinical results and have been most rewarding for the authors and their patients. The long-term validity of this technique awaits the test of time.

References

1. Vrandecic M P, Gontijo B F, Fantini F A, Gutierrez C, Paula e Silva J, Barbosa J T and Andrade C A O 1992 Anatomically complete heterograft mitral valve substitute: surgical technique and immediate results. J Heart Valve Dis 1: 254–259
2. Vrandecic M P, Morea M Radegran K, Gontijo B F, Paula E Silva J A, Fantini F A and Barbosa J T 1992 Clinical use of a new stentless porcine aortic bioprosthesis: a multicenter study. In: Grossi A, Donatelli F, Corno A, Brodman R (eds) Cardiology and cardiac surgery: current topics. Futura, Mount Kisco, New York, pp 223–234
3. Vrandecic M P, Gontijo B F, Fantini F A, Paula E Silva J A, Barbosa J T, Gutierrez C, Barbosa M and Morea M 1993 Stentless porcine aortic valve: mid-term clinical follow-up. In: D'Alessandro L C (ed) Heart surgery. Casa Editrice Scientifica Internazionale, Rome, vol 7: 43–56
4. Casabona R, Paulis R, Zaterra G, Summa M, Vrandecic M and Morea M 1992 Stentless porcine and pericardial valve in aortic position. Ann Thorac Surg 54: 681–685
5. Hammermeister K E, Sethi G K, Henderson W G, Oprian C, Kim T, Rahimtoola S A 1993 Comparison of outcomes in men 11 years after heart-valve replacement with a mechanical valve or bioprosthesis. N Engl J Med 328: 1289–1296
6. Carabello B A 1992 What exactly is 2+ to 3+ mitral regurgitation? J Am Coll Cardiol 2: 339–340

Chapter 28

Echocardiography and the aortic stentless valves

A. Nguyen Van-Cao

Stentless aortic bioprostheses are new devices recently offered to surgeons. Technically, they lie between the cryopreserved aortic homografts and the glutaraldehyde-fixed stented xenografts. Absence of the stent means that excessive mechanical stress is not generated at the annular suture line; however, glutaraldehyde fixation stiffens the biological tissues, unlike the homograft. Experimental studies have shown that stentless bioprostheses have better haemodynamics and greater longevity than stented bioprostheses.[1,2]

Clinical evaluation of stentless bioprostheses is under way; echocardiography seems to be the best tool, for several reasons: (a) there is no ultrasonic attenuation by the stent; (b) ultrasound technology has progressed rapidly in recent years, with better image quality; (c) the follow-up of stentless valves will probably be very long-term, with repetitive examinations.

Echocardiography will help to determine the best prosthetic design such as the composite valve, the complete porcine aorta or the pericardial stentless valve. In addition, it will determine the precise indications for the various surgical procedures (total aortic replacement, intra-aortic cylinder or subcoronary replacement).

Progress in ultrasound technology

In recent years, transducer technology and the colour Doppler flow imaging process have rapidly improved, together with the extensive use of digital electronics. Transthoracic transducers have a better focus and a higher number of crystals, yet are still small in size. Thus, all the echocardiographic and Doppler modes are greatly enhanced.

The most important advance in transoesophageal echocardiography is the multiplane probe with a rotating exploration plane perpendicular to the long axis of the fibroscope.[3] During open-heart surgery, epicardial echocardiography can be used to assess the surgical procedure.[4]

Colour Doppler flow imaging is very sensitive and can be used for diagnostic and quantitative purposes. Small valvular regurgitations are now easily detected and localized, particularly with transoesophageal echocardiography.

Digital technology has facilitated storage of the video images of echocardiography and their further comparison. The sharing of data and the transmission of images are also easier.

Echo-Doppler examination

The purpose of the aortic echo examination in this context is to obtain details of the anatomy and haemodynamics of the stentless xenograft. The observer must be aware of the design of the prosthesis and of the surgical procedure in order to conduct a complete examination, as some complications are related to these factors.

Whatever the stentless valve and the implantation technique used, a complete examination is usually possible with the combination of transthoracic and transoesophageal echocardiography. The transthoracic approach explores the anterior part of the prosthesis (Figure 28.1) whereas the transoesophageal approach visualizes its posterior side (Figure 28.2).

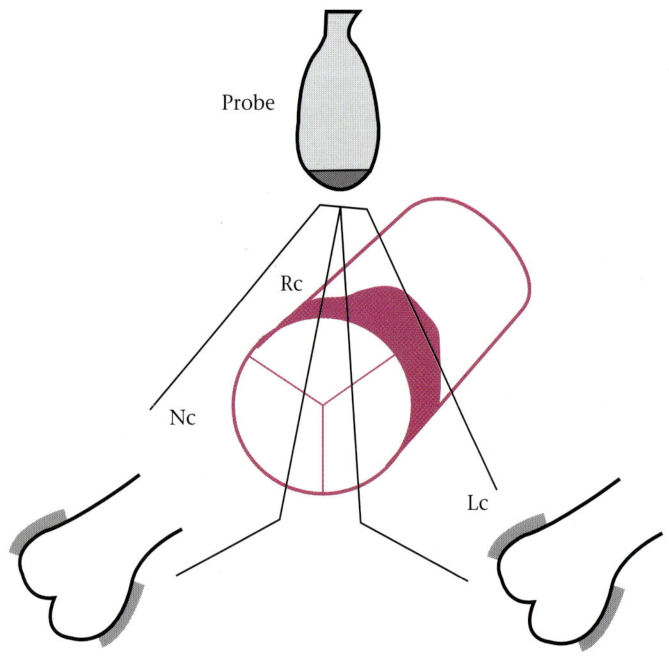

Figure 28.1. Transthoracic and epicardial views: Nc, non-coronary cusp; Rc, right coronary cusp; Lc, left coronary cusp.

A cross-section of the annulus and a series of perpendicular views cross-secting the annulus can be obtained, whatever the method. With multiplane transoesophageal echocardiography all these views are obtained by rotating both the cross-section plane of the transducer and the probe on itself. Thus the annulus, the commissures, the free edges and the body of the cusps are detailed anatomically. Not only their thickness but also their dynamics must be noted, whether normal or altered by aortic cusp prolapse.

Multiplane transoesophageal echocardiography is the best way to explore the aortic root. Several diameters can be measured at different levels — the aortic annulus, the sinus level, the sinotubular junction and the distal anastomosis.

An aortic stentless bioprosthesis has an echocardiographic aspect that is very similar to that of a homograft or a native valve. The annulus and the aortic walls are thickened after an intra-aortic cylinder procedure. The cusps may be slightly asymmetric or distorted; they are wrinkled or prolapsed when there is a size mismatch between the native annulus and the bioprosthesis.

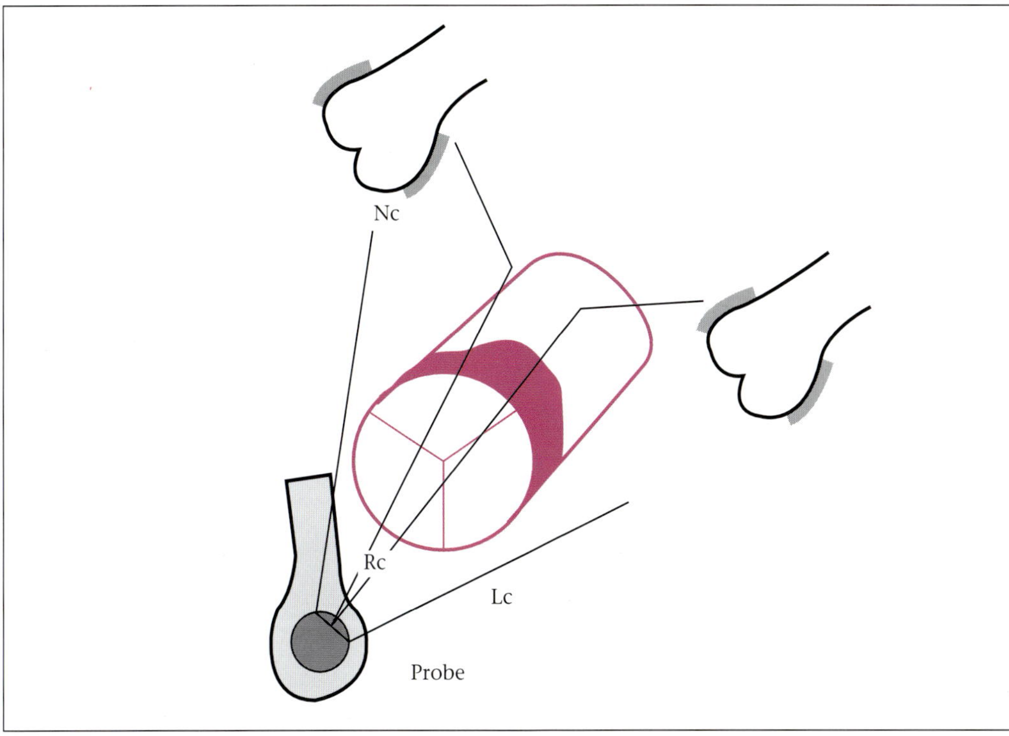

Figure 28.2. Transoesophageal views: abbreviations as in Figure 28.1.

Haemodynamic examination

The haemodynamic examination relies on colour Doppler flow imaging and on pulsed and continuous-wave Doppler for the quantitative measurements. The purpose is to follow the transaortic pressure gradients and to detect any aortic regurgitation due to structural failure.

Colour Doppler is used for the diagnosis and localization of aortic regurgitation, which is the most common abnormal finding. The leak may be at a commissure near the annulus or more central. Semiquantitative assessment of regurgitation is necessary, for comparison with subsequent findings.[5]

The peak and mean pressure gradients are the most easily recorded and reproducible Doppler measurements; they compare well with invasive pressure measurements.[6,7] For the aortic orifice the widely used simplified Bernoulli equation neglects the proximal velocity that is the mean left ventricular outflow velocity and thus overestimates the mean transprosthetic pressure gradient. This is particularly the case when there are high left ventricular outflow velocities (above 1 m/s). The true mean transprosthetic pressure gradient is the difference between the mean aortic pressure gradient and the mean left ventricular outflow pressure gradient (Table 28.1). These gradients must be interpreted after correction for the cardiac output, as pressure gradients increase with increased cardiac output, for the same prosthesis. The principle of effective orifice area (EOA) is used for this purpose; it permits follow-up of the same prosthesis and the comparison of different stentless valves.

The EOA is calculated using a continuity equation that states an equality between the cardiac output through the left ventricular outflow tract and the cardiac output measured at the aortic orifice.[8] The EOA formula is as follows:

$$EOA = LV_{outflow}\ VTI*LV_{outflow}\ Area/Ao\ VTI$$

Table 28.1 Equations for calculating transvalvular pressure gradients

Bernoulli equation:

$$\text{Pressure} = 4^* (V_{distal}^2 - V_{proximal}^2)$$

V_{distal} = Blood velocity distal to the orifice (aortic orifice)
$V_{proximal}$ = Blood velocity proximal to the orifice (left ventricular outflow tract)

Simplified Bernoulli equation:

$$\text{Pressure (mmHg)} = 4^* \text{Velocity (m/s)}^2$$

In the simplified Bernoulli equation the $V_{proximal}$ factor is neglected so that the pressure gradient is dependent on the distal velocity that is the aortic velocity.

$$\text{Mean pressure gradient} = \text{Ao Mean pressure} - \text{LV Mean pressure}$$

Ao Mean pressure = Mean aortic pressure gradient derived from the aortic velocities
LV Mean pressure = Mean left ventricular outflow pressure derived from the left ventricular outflow velocities

where $LV_{outflow}VTI$ and Ao VTI are, respectively, the integral of the left ventricular outflow velocity and the integral of the aortic velocity.

For each measurement there are several technical requirements: the aortic pressure gradient is the highest measurement from the apical, subcostal, left and right parasternal, right subclavian and suprasternal windows; each mean pressure gradient or velocity is the average of at least three cardiac cycles; the left ventricular outflow diameter is measured just beneath the cusps of the aortic prosthesis.

Haemodynamic data

Short-term and mid-term haemodynamic data are still scarce but appear to verify the experimental assumptions.[1,2] The transprosthetic pressure gradients of the stentless valves are lower than those for stent-mounted bioprostheses and their EOA values are higher.[9,10] Casabona and co-workers[10] compared the haemodynamics of a porcine and a pericardial stentless bioprosthesis with those of a stented xenograft and a mechanical prosthesis. The EOA values of the stentless valves, 1.83 ± 0.3 and 1.72 ± 0.2 cm^2, respectively, were significantly higher than those of the stented and mechanical valves (1.56 ± 0.1 and 1.38 ± 0.1 cm^2, respectively). This difference was more significant for the smaller valve sizes. The published mean transprosthetic gradients are generally under 15 mmHg.[11,12]

Complications

The most frequent abnormal finding is aortic regurgitation. The reported incidence is wide ranging: Casabona *et al.*,[10] and Hvass *et al.*[11] reported a 30% incidence of mild or 'microscopic' regurgitation with stentless composite valves; on the other hand Konertz *et al.*[13] reported no aortic regurgitation with their first 40 patients whereas a 26% incidence has been reported with the same model.[12] Regurgitation is located centrally and is generally mild or microscopic; it is attributable to floppy leaflets caused by a size mismatch between the prosthesis and the native annulus. The incidence of regurgitation diminishes with the learning curve of the surgeon and proper sizing of the stentless prostheses;[14,15] it is very low with the total aortic replacement procedure.

Some complications are related to the surgical procedure, such as haematomas between the native aorta and the wall of the prosthesis when an intra-aortic root cylinder technique has been used. They can lead to central or paravalvular aortic leaks. Pseudoaneurysms at the distal part of a total aortic replacement are possible, as in aortic homografts. After a total aortic valvular replacement, major calcification of the grafted aorta is possible, in theory.

The incidence of endocarditis is expected to be reduced with the stentless aortic valves. Transoesophageal echocardiography is the best imaging method for the diagnosis of vegetations largely superior to the transthoracic approach.[16]

Follow-up

A complete baseline echo-Doppler examination is necessary before, or soon after, discharge of the patient. Transthoracic and transoesophageal examinations are used to obtain the most detailed description of the anatomy and haemodynamics of the prosthesis. When this baseline echocardiogram is normal a transthoracic examination is necessary every year; more frequent transthoracic examinations are needed when moderate aortic regurgitation is present. Larger degrees of regurgitation call for transoesophageal echocardiography.

The mid-term follow-up data are scarce up to 2 years, but all the reported early follow-up data are favourable. No premature structural dysfunction has been reported and the haemodynamic data are stable. During the first months a decrease in the mean transvalvular gradient is sometimes observed.[11] Mild aortic regurgitation is also stable and increased only rarely.[10–12]

Digital storage of the echocardiographic video images is a major improvement. They can be indexed for easier retrieval and comparison. These digital files can be transmitted through ordinary telephone lines or high-speed communication networks.

Conclusions

Echocardiography is the main clinical tool for the clinical follow-up of stentless aortic prostheses, the anatomy and haemodynamics of which are precisely detailed with the combination of transthoracic and multiplane transoesophageal echocardiography. Further progress in digital ultrasound will probably enhance the image quality and the follow-up of such stentless bioprostheses.

References

1. Hazekamp M G, Goffin Y A, Huysmans H A 1993 The value of the stentless biovalve prosthesis. An experimental study. Eur J Cardio-Thorac Surg 7 (10): 514–519
2. Sung H W, Kingsbury C J, Quintero L J, Myers K E, Le T N, Yoganathan A P, Quijano R C 1991 In vitro evaluation of a stentless porcine aortic bioprosthesis. A preliminary study. ASAIO Trans 37(3): M172–M174
3. Seward J B, Khandheria B K, Freeman W K, On J K, Enriquez S M, Miller F A, Edwards W D, Tajik A J 1993 Multiplane transesophageal echocardiography: image orientation, examination technique, anatomic correlations, and clinical applications. Mayo Clin Proc 68(6): 523–551
4. Lazar H L, Plehn J 1990 Intraoperative echocardiography. Ann Thorac Surg 50(6): 1010–1018
5. Perry G J, Helmcke F, Nanda N C, Byard C, Soto B, 1987 Evaluation of aortic insufficiency by Doppler color flow mapping. J Am Coll Cardiol 9(4): 952–959
6. Currie P J, Seward J B, Reeder G S 1985 Continuous-wave Doppler echocardiographic assessment of severity of calcific aortic stenosis: a simultaneous Doppler–catheter correlative study in 100 adult patients. Circulation 71: 1162

7. Hatle L, Angelsen B 1984 Doppler ultrasound in cardiology: physical principles and clinical applications, 2nd edn. Lea and Febiger, Philadelphia
8. Richards K L, Cannon S R, Miller J F, Crawford M H 1986 Calculation of aortic valve area by Doppler echocardiography: a direct application of the continuity equation. Circulation 73: 964
9. Konertz W, Hamann P, Schwammenthal E, Breithardt G, Scheld H H 1992 Aortic valve replacement with stentless xenografts. J Heart Valve Dis 1(2): 249–252
10. Casabona R, De Paulis R, Zattera G F, di Summa M, Borrone W, Stacchino C, Vrandecic M O, Morea M 1992 Stentless porcine and pericardial valve in aortic position. Ann Thorac Surg 54(4): 681–684
11. Hvass U, Chatel D, Ouroudji M, Pansard Y, Laperche T, Depoix J P, Lenormand C 1994 The O'Brien–Angell stentless valve. Early results of 100 implants. Eur J Cardio-Thorac Surg 8: 384–387
12. Pillai R, Spriggings D, Amarasena N, O'Regan D J, Parry A J, Westaby S 1993 Stentless aortic bioprosthesis? The way forward: early experience with the Edwards valve. Ann Thorac Surg 56(1): 88–90
13. Konertz W, Weyand M, Sidiropolous A, Schwammenthal E, Breithardt G, Scheld H H 1992 Technique of aortic valve replacement with the Edwards stentless aortic bioprosthesis 2500. Eur J Cardio-Thorac Surg 6(5): 274–277
14. David T E, Bos J, Rakowski H 1992 Aortic valve replacement with the Toronto SPV bioprosthesis. J Heart Valve Dis 1(2): 244–248
15. Barratt-Boyes B, Christie G W, Raudkivi P J 1992 The stentless bioprosthesis: surgical challenges and implications for long-term durability. Eur J Cardio-Thorac Surg 6(Suppl 1): S39–S43
16. Mugge A, Daniel W G, Frank G, Lichten P R 1989 Echocardiography in infective endocarditis: reassessment of prognostic complications of vegetation size determined by the transthoracic and the transesophageal approach. J Am Coll Cardiol 14: 631

Index

ABO blood group compatibility, 128
Accelerated wear testing
    Medtronic Freestyle™ valve, 50, 53–4
    No-React™ processed tissues, 96, 97
    porcine pulmonary valves, 81–92
Acylazide treatment, 118–22
    processing techniques, 119–20
    tissue properties after, 120–2
Age
    choice of prosthesis and, 14, 188
    durability of stented bioprostheses and, 25–31, 33, 34
α-Amino oleic acid (AOA), 48, 100
    biocompatibility, 101, 102
    see also Medtronic Freestyle™ valve, α-amino oleic acid (AOA)-treated
Animal models
    biocompatibility tests, 101
    see also Calf implant model; Rat subcutaneous model; Sheep implant model
Anisotropy, fatigue-tested porcine pulmonary valves, 87, 89, 91
Annulo-aortic ectasia, 131
Anticoagulants, 155, 173, 190
Antimineralization agents, 13, 112, 115
    biocompatibility tests, 100–4
    new, 93–8
    see also α-Amino oleic acid
AOA *see* α-Amino oleic acid
Aorta, simulated, 49
Aortic annulus
    decalcification, 166, 167
    diameter
        Medtronic Freestyle™ valve sizing, 159
        pulmonary autografts and, 131
        Toronto SPV™ valve sizing, 142–3, 148
Aortic crossclamp times
    Medtronic Freestyle™ valve insertion, 149, 161–3
    O'Brien–Angell valve insertion, 185
    Toronto SPV™ insertion, 140, 147
Aortic homografts, 3–5
    durability, 8, 9, 64, 137
    frame-mounted, 5, 17–18
    hand-sewn, 9
    indications, 128
    mitral valve replacement, 4–5, 18
    new types, 17–18
    patient selection, 14
    postoperative left ventricular function, 149–51
    stent-mounted, 5, 9
Aortic regurgitation/insufficiency, 204
    Biocor porcine stentless valves, 198
    colour Doppler assessment, 203
    Edwards Prima™ valve, 169, 170, 174–5
    Labcor stentless porcine valve, 192, 193
    Medtronic Freestyle™ valve, 155–6
    O'Brien–Angell valve, 186–7, 189
    pulmonary autografts, 136
    Sorin Pericarbon™ valve, 179, 180
    Toronto SPV™ valve, 46, 142–3, 147, 148
Aortic root
    glutaraldehyde treatment effects, 10
    inclusion method *see* Mini-root technique
Aortic root replacement, 12, 19, 20
    Medtronic Freestyle™ valve, 163–4
        clinical studies, 152–6
        hydrodynamic effects, 54, 55, 56
    pulmonary and aortic xenografts, 22, 23
    pulmonary autografts, 19, 20, 132–4
Aortic valve
    anatomy, 93, 94
    porcine stentless, 21–2, 23, 94
Aortic valve replacement (AVR)
    aortic homografts, 3–5, 149–51
    Biocor porcine stentless valves, 195–200
    Edwards Prima™ valve, 165–70, 172–5
    Labcor stentless porcine valve, 190–3
    Medtronic Freestyle™ valve, 152–6, 157–64
    O'Brien–Angell Model 300 valve, 182–9
    porcine pulmonary valves, 21–2, 23, 81
    pulmonary autografts *see* Pulmonary autografts
    Sorin Pericarbon™ Stentless Valve, 67, 176–81
    stented porcine bioprostheses, 25–8, 29 31–4, 149–51
    stentless valves
        echocardiography, 201–5
        postoperative LV function, 149–51

Toronto SPV™ stentless porcine valve, 139–45, 146–8, 149–51
Autogenics™ system, 37

Bernoulli equation, 203, 204
Medtronic Freestyle™ valve and, 50, 51–3
Biocompatibility tests, 101
acylazide (DPPA)-treated pericardium, 120–1, 122
α-amino oleic acid (AOA), 101, 102
AOA-treated Medtronic Freestyle™ valves, 100–4
No-React™ processed tissues, 96, 97
Biocor porcine stentless valves, 195–200
clinical outcome, 197–8
haemodynamics, 198
patients and methods, 195–6
Biocor standard No-React™ valve, 197
Bioflow™ valve, 36
Bioprostheses, 17, 58
see also Stented porcine bioprostheses; Stentless bioprostheses
Borate buffer, 106–8

Calcification
acylazide (DPPA)-treated pericardium, 121–2
diepoxide-fixed porcine valve tissues, 112–14, 115
Edwards Prima™ valve, 61
glutaraldehyde and, 9, 114, 118–19
homografts, 137
mechanisms, 9
Medtronic Freestyle™ valve, 154
No-React™ processed tissues, 96
pericardial valves, 39
porcine pulmonary valves, 90
pulmonary autografts, 136–7
stented porcine bioprostheses, 31–3
stentless bioprostheses, 64
see also Antimineralization agents
Calf implant model
No-React™-treated bioprostheses, 96
Toronto SPV™, 44–5
Carpentier–Edwards pericardial bioprosthesis, 36, 38, 39, 58
Carpentier–Edwards standard porcine bioprosthesis (CE-S), 7, 24, 25
durability, 13, 25, 27–31, 33, 34
haemodynamics, 11
postoperative complications, 25, 26–7, 31, 32–3
Carpentier–Edwards supra-annular valve (CE-SAV), 24, 25, 33
durability, 13, 25, 27–31, 34
postoperative complications, 25, 26–7, 31, 32–3
Children, 14
homografts, 127–9
pulmonary autografts, 19, 130, 137–8
Colladerm, 122
Collagen, 118
pericardial, 38
thermal stability *see* Thermal stability
Collagenase digestion
acylazide (DPPA)-treated pericardium, 120, 121
diepoxide-fixed porcine valve tissues, 109–10, 111
Colour Doppler flow mapping (CDFM), 201
aortic regurgitation, 203
Medtronic Freestyle™ valve, *in vitro*, 50–1, 53
Complications, postoperative
Biocor porcine stentless valves, 197
echocardiographic assessment, 204–5
Edwards Prima™ valve, 169, 170, 174
homografts, 128–9
Medtronic Freestyle™ valve, 153–5
pulmonary autografts, 136–7
stented bioprostheses, 28, 31, 33
Toronto SPV™ valve, 141, 147
Congenital heart disease, homografts, 127–9
Cryopreservation, homografts, 127
Cylinder within a cylinder method, 12–13
Medtronic Freestyle™ valve, 157–64
Cytotoxicity/cytocompatibility tests
acylazide (DPPA)-treated pericardium, 120–1, 122
α-amino oleic acid (AOA)-treated Medtronic Freestyle™ valves, 101–3
No-React™ treated tissues, 96, 97

Dacron cloth, 9, 20–1, 22
Medtronic Freestyle™ valve, 48
Toronto SPV™ valve, 44
Denacol EX-810 *see* Ethylene glycol diglycidyl ether
Diepoxide *see* Ethylene glycol diglycidyl ether

Diphenylophosphorylazide (DPPA)
properties of treated pericardium, 120–2
tissue processing technique, 119–20
Doppler echocardiography, 201
examination methods, 202
haemodynamic examination, 203–4
in vitro, Medtronic Freestyle™ valve, 50–4
patient follow-up, 205
postoperative complications, 204–5
DPPA *see* Diphenylophosphorylazide
Durability
aortic homografts, 8, 9, 64, 137
pericardial bioprostheses, 38–9
pulmonary autografts, 8, 136–8
pulmonary homografts, 137
stented porcine bioprostheses, 8–9, 25–8, 29–30, 31–4
stentless bioprostheses, 64, 175, 176, 188–9, 193
Toronto SPV™ valve, 143
Dura mater valves, 18

Echocardiography, 201–5
examination techniques, 202
patient follow-up, 205
postoperative complications, 204–5
progress in technology, 201
transoesophageal *see* Transoesophageal echocardiography
see also Doppler echocardiography
Edwards Prima™ stentless porcine valve, 10, 58–62
clinical studies, 165–70, 172–5
outcome, 168–70, 173–4
patients and methods, 165–8, 172–3
design, 59–60, 61, 165, 166
haemodynamics, 168–9, 174
implantation techniques, 12, 166–7, 173, 175
preclinical testing, 61, 62
Effective orifice area (EOA)
Biocor porcine stentless valves, 198
Doppler measurement, 203–4
Edwards Prima™ valve, 169
Medtronic Freestyle™ valve
clinical studies, 155, 162
in vitro studies, 51, 52, 55, 56
O'Brien–Angell valve, 79, 187–8
stentless bioprostheses, 204
Toronto SPV™ valve, 45, 46, 142
Elastic modulus, 83
diepoxide-fixed porcine valve tissues, 110, 111, 113
fatigue-tested porcine pulmonary valves, 87–8, 90–1
Elderly patients, 14, 20
Sorin Pericarbon™ valve, 178
stented porcine bioprostheses, 28, 29, 31, 33, 34
Endocarditis
O'Brien–Angell valve insertion, 184, 185–6
postoperative, 205
Biocor porcine stentless valves, 197
Edwards Prima™ valve, 169, 170
Medtronic Freestyle™ valve, 154
pericardial bioprostheses, 38
stented porcine bioprostheses, 25, 27
Toronto SPV™ valve, 141, 147
pulmonary autograft (Ross) operation, 138
Sorin Pericarbon™ valve insertion, 178
stentless bioprostheses and, 64
Enzymatic digestion
acylazide (DPPA)-treated pericardium, 120, 121
diepoxide-fixed porcine valve tissues, 109–10, 111
Epoxide *see* Ethylene glycol diglycidyl ether
Ethylene glycol diglycidyl ether (EDGE), 100, 104, 115
chemistry, 104–5, 106
fixation kinetics, 105, 107
-fixed porcine valve tissues, 108–14
biaxial mechanical properties, 110–12
preparation, 108
resistance to enzymatic digestion, 109–10, 111
resistance to mineralization, 112–14, 115
thermal stability, 108–9
increasing rate of fixation, 106–8
Extensibility, 83
diepoxide-fixed porcine valve tissues, 111–12
fatigue-tested porcine pulmonary valves, 88, 91

Fascia lata valves, 5, 6, 18
Fatigue testing *see* Accelerated wear testing
Fibroblast cultures, 96, 97, 101
Fixation, 6–7, 77

acylazide, 118–22
ethylene glycol diglycidyl ether (EDGE), 100, 105–8
formalin (formaldehyde), 6, 77
glutaraldehyde *see* Glutaraldehyde, fixation
Medtronic Freestyle™ valve, 48–9, 100
No-React™ process, 95–8
'root pressure', 48, 100, 108
zero-pressure leaflet, 13
Formalin (formaldehyde) fixation, 6, 77
Fracture tests
acylazide (DPPA)-treated pericardium, 120
fatigue-tested porcine pulmonary valves, 89, 91
Freestyle™ valve *see* Medtronic Freestyle™ valve

Georgia Tech. left heart pulse duplicator system, 49, 50
Glutaraldehyde
calcification and, 9, 114, 118–19
fixation, 7, 95
disadvantages, 10, 118–19
Labcor stentless valve, 191
O'Brien–Angell Model 300 valve, 77
porcine pulmonary valves, 81
Sorin Pericarbon™ Stentless Valve, 65
stented porcine bioprostheses, 33–4
Toronto SPV™ valve, 44
versus acylazide, 120–2
residual, biocompatibility, 103–4
Growth, pulmonary autografts, 137–8

Haemodynamics
Biocor porcine stentless valves, 198
Edwards Prima™ valve, 168–9, 174
Labcor stentless porcine valves, 73, 193
measurement methods, 203–4
Medtronic Freestyle™ valve, 11, 155, 162, 163
O'Brien–Angell valve, 186–8
pericardial bioprostheses, 37–8
pulmonary autografts, 136
Sorin Pericarbon™ valve, 179–80
stentless valves, 10–11, 64, 204
Toronto SPV™ valve, 45, 142, 143, 147
Haemorrhagic complications
Edwards Prima™ valve, 169
stented porcine bioprostheses, 25, 26
Toronto SPV™ valve, 141
Hancock II valve, 13, 33–4
Hancock MO valve, 11, 13
Hancock pericardial bioprosthesis, 36, 38
Hancock standard valve (H-S), 24, 25
durability, 25, 27–30, 33
hydrodynamic performance, 49–51, 52
postoperative complications, 25, 26–7, 31–2
HLA typing, 128
Homografts, 17, 58, 127–9
congenital heart disease, 127–9
clinical outcome, 128–9
indications, 128
improvements, 58
new types, 17–18
preparation and storage, 127–8
pulmonary autograft (Ross) operation, 130, 134–5
versus stentless xenografts, 20, 21
see also Aortic homografts
Hydrodynamics
Labcor stentless aortic valve, 69, 70
Medtronic Freestyle™ valve, 49–56
O'Brien–Angell Model 300 valve, 79
see also In vitro testing

Implantation techniques *see* Surgical techniques
In vitro testing
Labcor stentless aortic valve, 69, 70, 71
Medtronic Freestyle™ valve, 49–56
O'Brien–Angell Model 300 valve, 79
Toronto SPV™ valve, 44–5
Ionescu-Shiley Low Profile valve, 36
Irritancy assays, AOA-treated Medtronic Freestyle™ valves, 102, 103–4
Ischaemic times *see* Aortic crossclamp times
Isopropyl alcohol, 106, 107

Labcor stentless aortic valve, 69–75
clinical studies, 190–3
outcome, 192–3
patient data, 190
construction, 70–1, 73–5
haemodynamics, 73, 193
implantation technique, 191–2
preclinical testing, 69, 70, 71
presentation, 71–2, 75

processing, 69–70, 191
Leaflet thickness, fatigue-tested porcine pulmonary valves, 86–7, 90
Leaks, paravalvular
Labcor stentless aortic valve, 69, 73
Medtronic Freestyle™ valve, 153, 154
Left ventricular (LV) function, postoperative, 149–51
Left ventricular outflow reconstruction
homografts, 128
see also Aortic root replacement
Leonardo Da Vinci, 3, 4, 93

Marfan's syndrome, 14, 131
Mechanical prostheses, 5, 14
Mechanical studies
acylazide (DPPA)-treated pericardium, 120
diepoxide-fixed porcine valve tissues, 110–12
fatigue-tested porcine pulmonary valves, 82–3, 86–9, 90–1
Medox-Gebbay unicusp pericardial bioprosthesis, 36
Medtronic Freestyle™ valve, 10, 48–57
accelerated wear testing, 53–4
α-amino oleic acid (AOA)-treated
biocompatibility, 100–4
hydrodynamics, 50, 53–4
clinical studies, 152–6, 157–64
outcome, 153–5
patients, 152
design, 48
fixation techniques, 48–9, 100
haemodynamics, 11, 155, 162, 163
hydrodynamic performance, 49–56, 57
implantation techniques, 12, 13, 54–5, 152–3
cylinder within a cylinder method, 157–64
hydrodynamic effects, 54–6, 57
sizing
hydrodynamic effects, 55, 56, 57
method, 158–9
Medtronic Intact™ valve, 24, 25, 33–4
durability, 13, 25, 27–31, 34
postoperative complications, 25, 26–7, 31, 32–3
Mercurochrome, 6
Mineralization *see* Calcification
Mini-root technique (root inclusion method), 12
Edwards Prima™ valve, 166–7
Medtronic Freestyle™ valve, 54, 55, 56, 152–6
pulmonary autografts, 134
Mitral regurgitation, Biocor porcine stentless valves, 198
Mitral valve
anatomy, 93–4
porcine stentless, 94–5
Mitral valve replacement (MVR), 24
aortic homografts, 4–5, 18
Biocor porcine stentless valves, 195–200
stented porcine bioprostheses, 25, 28, 29–31, 32, 33, 34
Mitroflow™ valve, 36, 38, 39
Morbidity, postoperative *see* Complications, postoperative
Morphological studies, fatigue-tested porcine pulmonary valves, 82, 84–5, 86, 90
Mortality, late postoperative
Biocor porcine stentless valves, 197
Edwards Prima™ valve, 169
homografts, 129
Medtronic Freestyle™ valve, 154
Toronto SPV™ valve, 141
Mortality, perioperative
Biocor porcine stentless valves, 197
Edwards Prima™ valve, 169, 173
Labcor stentless porcine valve, 192–3
Medtronic Freestyle™ valve, 153
O'Brien–Angell valve, 185–6, 188
Sorin Pericarbon™ valve, 179
Toronto SPV™ valve, 141, 147

New York Heart Association (NYHA) functional class
Biocor porcine stentless valve recipients, 197
Medtronic Freestyle™ valve recipients, 155, 163
Sorin Pericarbon™ valve recipients, 179
Toronto SPV™ valve recipients, 141, 147
No-React™ tissue-tanning process, 95–98
accelerated fatigue testing, 96, 97
animal studies, 95–7
cytocompatibility testing, 96, 97
in vitro tensile testing, 95, 96, 97

O'Brien–Angell Model 300 valve, 77–80
 assembly, 78
 clinical study, 182–9
  outcome, 185–8
  patients and methods, 182–5
 design, 78, 182
 haemodynamics, 186–8
 implantation technique, 79–80, 184
 in vitro testing, 79
 tissue recovery and processing, 77
Orifice area, effective *see* Effective orifice area

Paroguide, 122
Perforations, 9
 fatigue-tested porcine pulmonary valves, 84, 90
 Labcor stentless aortic valve, 69, 72
 stented porcine bioprostheses, 31–3
Pericarbon™ Stentless Valve *see* Sorin Pericarbon™ Stentless Valve
Pericardial bioprostheses (PB), 18, 36–9, 58, 177
 advantages, 37–8
 Autogenics™ system, 37
 characteristics, 36–7
 clinical follow-up, 38–9
 current situation, 39
 history, 5, 6, 36
 see also Sorin Pericarbon™ Stentless Valve
Pericardium, 38
 acylazide processing, 119–20
 properties of acylazide (DPPA)-treated, 120–2
 wraps, pulmonary autografts, 134
pH, diepoxide fixation and, 106–8
Polyester/plastic cloth, 20–1, 59–60
Preclinical testing *see* Calf implant model; *In vitro* testing; Sheep implant model
Prima™ valve *see* Edwards Prima™ stentless porcine valve
Protease digestion, diepoxide-fixed porcine valve tissues, 109, 110
Pulmonary autografts, 3, 5, 18–19, 130–8
 aortic root replacement, 19, 20, 132–4
 clinical outcome, 136–8
 contraindications, 131
 durability, 8, 136–8
 growth, 137–8
 haemodynamics, 136
 implantation technique (Ross operation), 131–6
Pulmonary homografts
 durability, 137
 pulmonary autograft (Ross) operation, 130, 134–5
Pulmonary tract reconstruction, 128
Pulmonary valve replacement, homografts, 128
Pulmonary xenografts (porcine)
 advantages, 21–2, 23
 fatigue-tested, 81–92
  mechanical studies, 82–3, 86–9, 90–1
  morphological study, 82, 84–5, 86, 90

Rat subcutaneous model
 acylazide (DPPA)-treated pericardium, 121–2
 diepoxide-fixed porcine valve tissues, 112–14
 No-React™ processed tissues, 95–6
Regurgitation
 Medtronic Freestyle™ valve, *in vitro* studies, 51, 52, 53–4, 56, 57
 see also Aortic regurgitation/insufficiency; Mitral regurgitation
Reoperations
 Biocor porcine stentless valves, 197
 homografts, 128–9
 Labcor stentless porcine valve, 192–3
 Medtronic Freestyle™ valve, 153, 154
 O'Brien–Angell valve, 188, 189
 pulmonary autografts, 136–7
 Sorin Pericarbon™ valve, 179, 180–1
 stented porcine bioprostheses, 31, 32
 Toronto SPV™ valve, 141
Right ventricular outflow reconstruction
 aortic xenografts, 22, 23
 homografts, 128
 pulmonary autograft (Ross) operation, 130, 134–5
 pulmonary xenografts, 22, 23
'Root pressure fixation', 48, 100, 108
Ross operation *see* Pulmonary autografts

Sheep implant model
 Edwards Prima™ valve, 61, 62
 Labcor stentless aortic valve, 70, 71–3
 No-React™-treated bioprostheses, 96
 Toronto SPV™, 44
Shiley pericardial bioprostheses, 36, 38
Shrink temperature testing *see* Thermal stability

Sinotubular junction diameter, 139, 143, 148
Sizes, valve
  Edwards Prima™ valve, 167, 168, 173
  Labcor stentless porcine valve, 192
  Toronto SPV™ valve, 46, 47
Sizing
  Edwards Prima™ valve, 170
  Medtronic Freestyle™ valve *see* Medtronic Freestyle™ valve, sizing
  O'Brien–Angell valve, 184
  pulmonary autografts, 131, 136
  Sorin Pericarbon™ stentless valve, 178
  stentless xenografts, 20, 21, 22
  Toronto SPV™ valve, 139, 142–3, 148
Sorin Pericarbon™ Stentless Valve, 36, 37, 39, 64–7, 176–81
  advantages, 67
  clinical experience, 66–7, 179–80
  construction, 65–6, 177
  design, 65, 177–8
  haemodynamics, 179–80
  implantation technique, 178–9
  surgical indications, 178
  versus other stentless valves, 177–8
Stented porcine bioprostheses, 7–9, 17, 24–34, 58
  clinical results, 25–31
  durability, 8–9, 25–8, 29–30, 31–4
  patient data, 24–5
  postoperative LV function, 149–51
Stentless bioprostheses, 3–14
  advantages, 19–21, 64, 176
  anatomy and function, 93–5
  disadvantages, 14, 19–21
  durability, 64, 175, 176, 188–9, 193
  echocardiography, 201–5
  haemodynamics, 10–11, 64, 204
  implantation techniques, 11–13, 164
  patient selection, 13–14
  pulmonary versus aortic, 21–2, 23
  resurrection, 9–11
  versus aortic homografts, 20, 21
  versus Sorin Pericarbon™ valve, 177
  see also specific types
Stress relaxation curves, slope, 83, 87, 88–9
Stress relaxation tests, 83–4
  fatigue-tested porcine pulmonary valves, 87–9, 91
Surgical techniques
  Biocor porcine stentless valves, 196
  cylinder within a cylinder method *see* Cylinder within a cylinder method
  Edwards Prima™ valve, 12, 166–7, 173, 175
  full root replacement *see* Aortic root replacement
  homografts, 128, 129
  Labcor stentless porcine valve, 191–2
  Medtronic Freestyle™ valve *see* Medtronic Freestyle™ valve, implantation techniques
  mini-root method *see* Mini-root technique
  O'Brien–Angell Model 300 valve, 79–80, 184
  pulmonary autograft (Ross) procedure, 131–6
  Sorin Pericarbon™ valve, 178–9
  stentless valves, 11–13, 164
  subcoronary method, 12
  Toronto SPV™ valve, 12, 139, 147–8
Suture dehiscence
  Labcor stentless aortic valve, 70, 71
  O'Brien–Angell valve, 188, 189

Tanning, 95, 118–19
  see also Fixation
Tears, 9
  fatigue-tested porcine pulmonary valves, 84, 85
  Labcor stentless aortic valve, 69, 72
  pericardial valves, 38
  stented porcine bioprostheses, 31–3
Tensile stress/strain tests, 83–4
  acylazide (DPPA)-treated pericardium, 120
  diepoxide-fixed porcine valve tissues, 110–12, 113
  fatigue-tested porcine pulmonary valves, 87–9
  No-React™ processed tissues, 95, 96, 97
Thermal stability
  acylazide (DPPA)-treated pericardium, 120
  diepoxide-fixed porcine valve tissues, 108–9
Thromboembolism (TE)
  Edwards Prima™ valve, 169, 170
  Medtronic Freestyle™ valve, 153, 154

pericardial bioprostheses, 38
pulmonary autografts, 137
stented porcine bioprostheses and, 25, 26
Toronto SPV™ valve, 141, 147
Toronto SPV™ valve, 9–10, 11, 43–7
clinical trials, 45–7, 139–45, 146–8
outcome, 141–2, 146–7
patients and methods, 139–40, 146
postoperative LV function, 149–51
design, 43–4
durability, 143
haemodynamics, 45, 142, 143, 147
implantation technique, 12, 139, 147–8
preclinical testing, 44–5
sizes implanted, 46, 47
sizing, 139, 142–3, 148
Transoesophageal echocardiography, 201, 205
examination methods, 202, 203
patient follow-up, 205
Transvalvular pressure gradients
Biocor porcine stentless valves, 198
Doppler measurement, 203, 204
Edwards Prima™ valve, 168, 174
Labcor stentless valves, 69, 193
Medtronic Freestyle™ valve
clinical studies, 155, 162, 163
in vitro studies, 51–3, 55, 56
O'Brien–Angell valve, 79, 186
stentless bioprostheses, 10–11, 204
stentless versus stented bioprostheses, 150, 151
Toronto SPV™ valve, 45, 142, 147
Tricuspid valve replacement, homografts, 128

Ultrasound *see* Echocardiography

Viscosity
fatigue-tested porcine pulmonary valves, 88–9
measurement, 83

Wear testing *see* Accelerated wear testing

Xenobioprostheses, 36